T0406607

Springer Theses

Recognizing Outstanding Ph.D. Research

For further volumes:
http://www.springer.com/series/8790

Aims and Scope

The series "Springer Theses" brings together a selection of the very best Ph.D. theses from around the world and across the physical sciences. Nominated and endorsed by two recognized specialists, each published volume has been selected for its scientific excellence and the high impact of its contents for the pertinent field of research. For greater accessibility to non-specialists, the published versions include an extended introduction, as well as a foreword by the student's supervisor explaining the special relevance of the work for the field. As a whole, the series will provide a valuable resource both for newcomers to the research fields described, and for other scientists seeking detailed background information on special questions. Finally, it provides an accredited documentation of the valuable contributions made by today's younger generation of scientists.

Theses are accepted into the series by invited nomination only and must fulfill all of the following criteria

- They must be written in good English.
- The topic should fall within the confines of Chemistry, Physics, Earth Sciences, Engineering and related interdisciplinary fields such as Materials, Nanoscience, Chemical Engineering, Complex Systems and Biophysics.
- The work reported in the thesis must represent a significant scientific advance.
- If the thesis includes previously published material, permission to reproduce this must be gained from the respective copyright holder.
- They must have been examined and passed during the 12 months prior to nomination.
- Each thesis should include a foreword by the supervisor outlining the significance of its content.
- The theses should have a clearly defined structure including an introduction accessible to scientists not expert in that particular field.

Jonathan Pearson

Generalized Perturbations in Modified Gravity and Dark Energy

Doctoral Thesis accepted by the University of Manchester, UK

Author
Dr. Jonathan Pearson
Department of Mathematical Sciences
Durham University
Durham
UK

Supervisor
Prof. Richard Battye
School of Physics and Astronomy
University of Manchester
Manchester
UK

ISSN 2190-5053 ISSN 2190-5061 (electronic)
ISBN 978-3-319-01209-4 ISBN 978-3-319-01210-0 (eBook)
DOI 10.1007/978-3-319-01210-0
Springer Cham Heidelberg New York Dordrecht London

Library of Congress Control Number: 2013942999

Printed on acid-free paper

Springer is part of Springer Science+Business Media (www.springer.com)

Publications

The work in this thesis is partly based upon the following publications:

- *Massive gravity, the elasticity of space-time, and perturbations in the dark sector* [259], with Richard Battye
- *Effective action approach to cosmological perturbations in dark energy and modified gravity* [187], published in JCAP with Richard Battye.
- *Effective field theory for perturbations in dark energy and modified gravity* [216], conference proceedings.
- *Equations of state for dark sector perturbations for general high derivative scalar-tensor theories*: in preparation with Richard Battye.
- *Parameterizing dark sector perturbations via equations of state*: in preparation with Richard Battye
- *Observational signatures and constraints on generalized cosmological perturbations*: in preparation with Richard Battye and Adam Moss.

That there exist rules to be checked at all is some sort of miracle

—Richard Feynman

Book-keeping, continued to all eternity, could not produce one farthing

—C. S. Lewis

The creator does not appear in the scientific account, but it is clearly ridiculous to deduce from this that the creator did not exist

—John Lennox

In the beginning God created the heavens and the earth. Now the earth was formless and empty, darkness was over the surface of the deep, and the Spirit of God was hovering over the waters

—Genesis 1

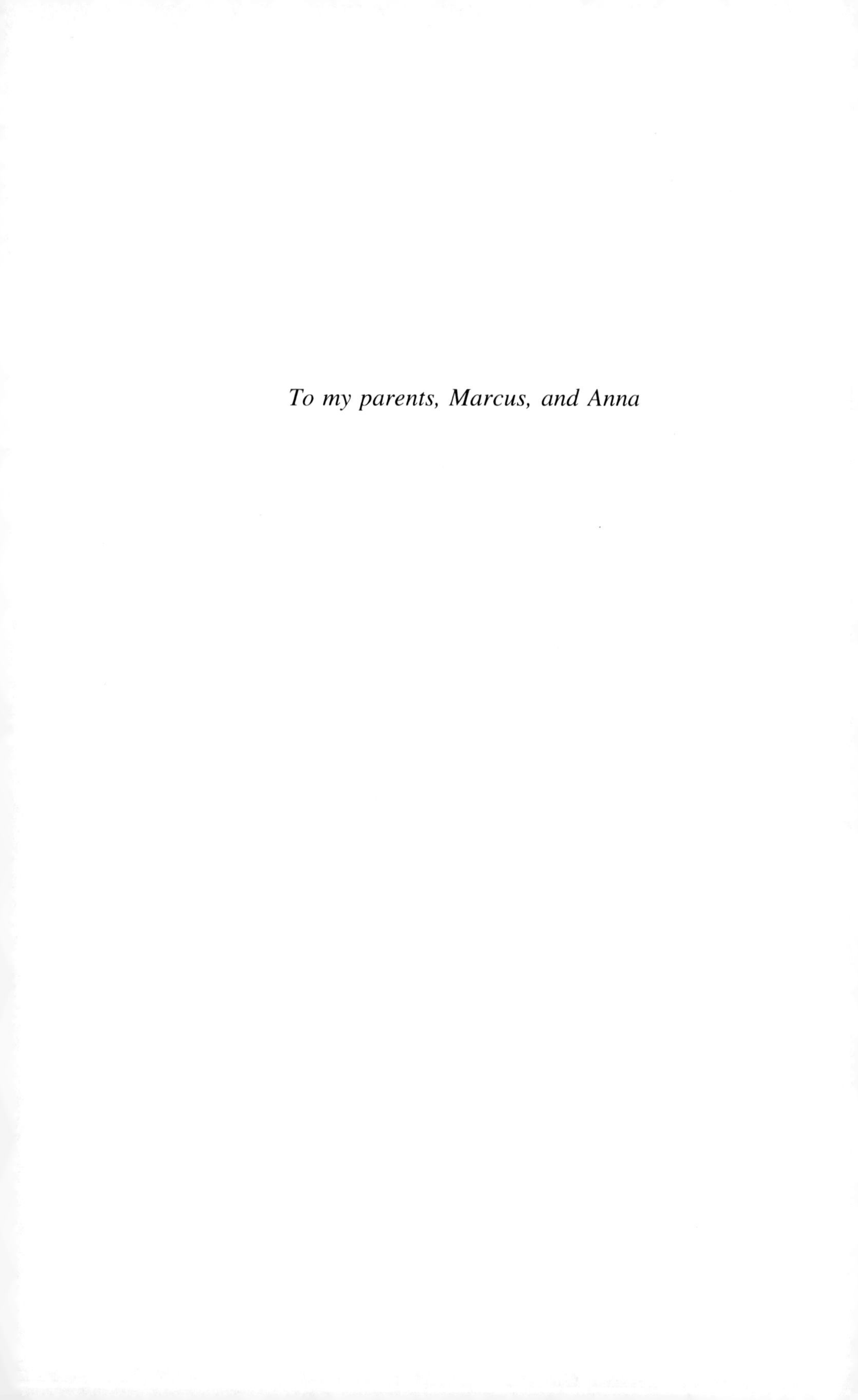

To my parents, Marcus, and Anna

Supervisor's Foreword

The Nobel Prize for Physics in 2011 was awarded jointly to Saul Perlmutter, Adam Reiss, and Brian Schmidt for the discovery of cosmic acceleration. However, its origin is still one of the greatest mysteries in modern physics. A cosmological constant is, perhaps, the simplest explanation but it is difficult to reconcile with our understanding of modern particle physics theories that predicts a vacuum energy 120 orders of magnitude larger than observed. It has been suggested that a dynamical component of the Universe, known as dark energy, or modifications to Einstein's theory of General Relativity in the infrared could be responsible.

There are now many theories that have been proposed to describe dark energy and modified gravity. Observations are likely to play a key role in distinguishing the different theories. There are two classes of observations: (i) those that are only sensitive to the background evolution; (ii) those that are, in addition, sensitive to the evolution of perturbations. In particular, weak gravitational lensing is affected by the distribution of the dark energy, or the evolution of perturbations in modified gravity. The observed distortions of galaxy shapes, or those observed in the Cosmic Microwave Background, can be very different to those predicted for models involving a cosmological constant.

The goal of this thesis was to develop a general formalism for modeling the evolution of perturbations in dark energy and modified gravity. The basic idea was to try to make as few theoretical assumptions as possible in deriving the observational predictions. The hope would then be that those analyzing observations would measure the free parameters in the model and allow us to hone in the correct theory.

Starting from the field content of a model and some basic arguments of symmetry, Jonathan was able to derive expressions for the effective fluid variables in terms of the active degrees of freedom of whatever is causing the cosmic acceleration. In a number of cases, it was possible to eliminate the internal degrees of freedom and deduce an equation of state for the perturbations that is parameterized by a number of free functions of time. He then went on to perform a survey of the possible observational consequences. In work done since the submission of the thesis, we have extended the basic idea to a wider class of models and, at present, we are investigating the constraints imposed from the recently released data from the Planck satellite.

Manchester, June 2013 Prof. Richard Battye

Abstract

When recent observational data and the GR + FRW + CDM model are combined we obtain the result that the Universe is accelerating, where the acceleration is due to some not-yet-understood "dark sector". There has been a considerable number of theoretical models constructed in an attempt to provide a description of the dark sector: Dark energy and modified gravity theories. The proliferation of modified gravity and dark energy models has brought to light the need to construct a "generic" way to parameterize perturbations in the dark sector. In this thesis we discuss our new way of approaching this problem. We write down an effective action for linearized perturbations to the gravitational field equations for a given field content and use it to compute generalized gravitational field equations for linearized perturbations. Our approach is inspired by that taken in particle physics, where the most general modifications to the standard model are written down for a given field content that is compatible with some assumed symmetry (which we take to be isotropy of the background spatial sections). After applying the formalism we obtain equations of state for dark sector perturbations, where the number of free parameters for wide classes of theories is identified.

Acknowledgments

First and foremost, I want to say thank you to my parents, brother, and sister for loving, encouraging and being patient with me over the years. They never tired of answering my annoying questions when I was little! I feel incredibly blessed and privileged to have such a family. Also, thank you to Dr. Paul (doc) Dale for inspiring and teaching me from such a young age. It has been an honour and a privilege to have been part of Lee Abbey, Ivy Manchester, Churchinaclub, and Vine Life during my time in Manchester; youth work has kept my feet firmly planted on the ground! I have learnt so much and have made so many wonderful friends from these incredible places. Special thanks to Alan Taylor and Emma Gaze for encouraging, teaching, and guiding me over the years.

I am indebted to my friends and various housemates for making my 7 years in Manchester as incredible as they have been. I have had many conversations, lunches, beers, and fun with loads of people. To mention but a few: Nathan Betteridge, Sam Martin, Simon (wozza) Worrell, Rick Newton, Simon Pike, Chris Welshman, James Agnew, Jen Gupta, Mareike Haberichter, Adam Webber, Sotirios Sanidas, and Peter Cuttell. Of course, saving the best to last: The lovely Iona. Manchester would not have been the same without any of you.

I would like to thank my supervisor Prof. Richard Battye for his patience, ideas, criticism, guidance, rants, and conversations on ski lifts. I have learnt a lot from Richard about science, scientists, who invented what first, skiing, hammers and nails, and policies for a better Britain.

I have enjoyed and benefitted from collaboration, conversations, and comments with leaders in the field: Stephen Appleby, Tessa Baker, Rachel Bean, Ed Copeland, Pedro Ferreira, Michael Kopp, Anthony Lewis, Adam Moss, Levon Pogosian, Alkistis Pourtisdou, Costas Skordis, Paul Sutcliffe, and Jochen Weller.

Contents

Chapter 1
Introduction to Gravitational Theories and Cosmology

Gravity is one of the most familiar of the fundamental forces of nature. It is the presence of gravity which enables the earth to orbit the Sun and keeps objects firmly planted on the surface of the earth. Whilst gravity is the weakest of the fundamental forces, it is gravity which dictates the structure and shape of the universe on a vast range of scales.

For a long time, the vast expanses of the natural world was imagined as a spacetime arena, upon which objects with mass were placed and moved; spacetime did not respond to the presence of the massive objects. The motion of a particle in spacetime was affected by its vicinity to massive objects which exerted gravitational forces upon the particle. On the whole, this paradigm was very successful and conceptually very simple, but was found to be flawed in situations where the gravitational forces were very strong. This picture was supplanted by Einstein's theory of gravity, where the important conceptual difference is that spacetime *responds* to the presence of massive objects, causing objects to travel along curves rather than straight lines.

In this chapter we will introduce the notation and necessary mathematics used to describe, understand and obtain an intuitive idea of the modern view of gravity: tensor calculus, curvature and General Relativity. We then apply these tools to cosmology, motivating why the dark sector has been introduced and finally review modified gravity theories. No claim of uniqueness is laid to the content in this chapter. There are many classic texts and reviews on gravity and cosmology [1–11]; we have selected a few to review to enable a relatively self-contained understanding of gravity to be obtained by reading of this chapter.

1.1 Tensor Calculus, the Metric and Curvature

Modern gravitational theories rely on a mathematical framework known as differential geometry; our aim is to provide an illustration of the mathematical construction which is used to model our universe and not of the technical details of differential

J. Pearson, *Generalized Perturbations in Modified Gravity and Dark Energy*,
Springer Theses, DOI: 10.1007/978-3-319-01210-0_1,

geometry (for that, see e.g. [4, 12]). A manifold can be thought of as a smooth collection of points, where each point describes an "event" in the universe (an event is a point which has a unique "time" and "space" coordinate). These events may be connected by smooth curves. Objects in the universe travel along these curves. Some geometrical structure in the universe may provide special sets of curves which are unique to a particular universe—if two universes contain different material substances then the "special sets of curves" will be different in these two universes. If, by experimental observation of the universe we inhabit, one can obtain information about these paths then one can determine the underlying geometrical structure of our universe.

1.1.1 Notation, Coordinates and Tensors

An event on a four dimensional manifold requires four numbers to be given to uniquely identify that event; these four numbers are collected and arranged into the components of the contravariant position vector:

$$x^{\mu} = (x^0, x^1, x^2, x^2). \tag{1.1}$$

In the Cartesian basis the components of this vector are the time coordinate, $x^0 = t$, and the spatial coordinates, $x^i = (x^1, x^2, x^3) = (x, y, z)$. Differentiation of a function Z with respect to the spacetime coordinate x^{μ} is written as

$$\partial_{\mu} Z \equiv \frac{\partial Z}{\partial x^{\mu}}. \tag{1.2}$$

Performing a coordinate transformation finds the values of the coordinates at that event, from a different reference frame; this different frame could correspond to an observer that is accelerating relative to the first observer, or where the second observer is at a different point in a gravitational potential. The coordinate transformation is

$$x^{\mu} \longrightarrow x'^{\mu} = x'^{\mu}(x^{\nu}). \tag{1.3}$$

The Jacobian $J^{\mu}{}_{\nu}$, and its inverse, $J_{\nu}{}^{\mu}$, is associated with a given coordinate transformation and is constructed via

$$J^{\mu}{}_{\nu} = \frac{\partial x'^{\mu}}{\partial x^{\nu}}, \qquad J_{\nu}{}^{\mu} \equiv \frac{\partial x^{\mu}}{\partial x'^{\nu}}, \qquad J^{\mu}{}_{\alpha} J_{\nu}{}^{\alpha} \equiv \delta^{\mu}{}_{\nu}. \tag{1.4}$$

Tensors are defined according to the way in which they transform under the coordinate transformation. The rank of a tensor is given by the number of indices in a particular location and is written as $\binom{n}{m}$. A scalar should be thought of as a tensor of rank-0 and a vector as a tensor of rank-1. A tensor of rank $\binom{n}{m}$ has n "upper" contravariant indices and m "lower" covariant indices, and transforms as

$$A'^{\mu_1\mu_2\ldots\mu_n}{}_{\nu_1\nu_2\ldots\nu_m} = \left(\prod_{i=1}^{n}\prod_{j=1}^{m} J^{\mu_i}{}_{\alpha_i} J_{\nu_j}{}^{\beta_j}\right) A^{\alpha_1\alpha_2\ldots\alpha_n}{}_{\beta_1\beta_2\ldots\beta_m}. \tag{1.5}$$

For example, a rank-2 contravariant tensor would look like $A^{\mu\nu}$, and a rank-2 covariant tensor like $A_{\mu\nu}$, and transform as

$$A'^{\mu\nu} = J^{\mu}{}_{\alpha} J^{\nu}{}_{\beta} A^{\alpha\beta}, \qquad A'_{\mu\nu} = J_{\mu}{}^{\alpha} J_{\nu}{}^{\beta} A_{\alpha\beta}. \tag{1.6}$$

$$\breve{A}' \equiv A'^{\mu\nu} A'_{\mu\nu} = J^{\mu}{}_{\alpha} J^{\nu}{}_{\beta} J_{\mu}{}^{\rho} J_{\nu}{}^{\sigma} A^{\alpha\beta} A_{\rho\sigma} = A^{\alpha\beta} A_{\alpha\beta} \equiv \breve{A}, \tag{1.7}$$

which is invariant under coordinate transformation. Their contraction forms a scalar,

Let us consider the partial derivative operator acting upon the components of a vector, $\partial_\nu A^\mu$. Performing a coordinate transformation yields

$$\begin{aligned}\partial'_\nu A'^\mu &= J_\nu{}^\alpha \partial_\alpha (J^\mu{}_\beta A^\beta)\\ &= J_\nu{}^\alpha J^\mu{}_\beta \partial_\alpha A^\beta + A^\beta J_\nu{}^\alpha \partial_\alpha J^\mu{}_\beta.\end{aligned} \tag{1.8}$$

The final term on the right hand side reveals that $\partial_\nu A^\mu$ does not transform as a $\binom{1}{1}$ tensor, as one may have thought. The "physical" reason for this is that the derivative operation takes the value of the components of the vector two different locations. Tensorial quantities are only defined at a given location on a manifold. The way of remedying this is to introduce the connection, which we will do in the next section.

The components of a tensor may not be independent. For example, if $A_{\mu\nu} = A_{\nu\mu}$ then we say that $A_{\mu\nu}$ is a *symmetric* tensor, and if $B_{\mu\nu} = -B_{\nu\mu}$ then $B_{\mu\nu}$ is *anti-symmetric*. Symmetric and anti-symmetric tensors can be formed from general tensors:

$$S_{\mu\nu} = T_{(\mu\nu)} \equiv \frac{1}{2}\left(T_{\mu\nu} + T_{\nu\mu}\right) = S_{\nu\mu}, \quad A_{\mu\nu} = T_{[\mu\nu]} \equiv \frac{1}{2}\left(T_{\mu\nu} - T_{\nu\mu}\right) = -A_{\nu\mu} \tag{1.9}$$

1.1.2 The Metric

One of the simplest questions one can ask is: "*what is the distance between two points?*" Pythagoras' theorem enables distances to be computed on a Euclidean manifold. For two points on an N-dimensional flat space, $\mathbb{R}^N$ whose coordinates differ by the infinitesimal amounts $\mathrm{d}x^1, \mathrm{d}x^2, \ldots, \mathrm{d}x^N$, the square of the distance between the two points is given by the sum of the squares of the incremental variations in each of the coordinates:

$$d\ell^2 = (dx^1)^2 + (dx^2)^2 + \cdots + (dx^N)^2 = \sum_{i,j=1}^{N} \delta_{ij} dx^i dx^j = \delta_{ij} dx^i dx^j. \quad (1.10)$$

The first equality is just the usual statement of Pythagoras' theorem; to be able to write the second equality the Kronecker-delta, δ_{ij}, is introduced, which enables Pythagoras' theorem to be more compactly written down; the final equality follows by using the "Einstein summation convention" where repeated indices are summed over. The Kronecker-delta is an identity matrix, whose components are given by

$$\delta_{ij} = \begin{cases} 1 & \text{if} \quad i = j, \\ 0 & \text{if} \quad i \neq j. \end{cases} \quad (1.11)$$

The quantity $d\ell^2$ is interpreted as the elemental distance between two infinitesimally separated points, and by introducing the quantity δ_{ij} one becomes able to compactly write this distance down. In this example of a Euclidean space, the quantity δ_{ij} plays the role of the metric, and it is constant throughout the space; this property need not be true in general. An important feature of the Kronecker-delta is that its determinant is positive definite; a manifold endowed with a metric whose determinant is positive-definite is called a Riemannian manifold. The metric defines the geometry of a space.

The Minkowski metric is the simplest example of a metric which can be used to incorporate the time coordinate into the distance measure. The components of the Minkowski metric are given by

$$\eta_{\mu\nu} = \text{diag}(-1, 1, 1, 1), \quad (1.12)$$

and the distance measure is

$$ds^2 = \eta_{\mu\nu} dx^\mu dx^\nu = -dt^2 + dx^2 + dy^2 + dz^2. \quad (1.13)$$

The Minkowksi metric $\eta_{\mu\nu}$ is, like the Kronecker-delta, constant throughout spacetime, but its determinant is negative-definite; a manifold endowed with a metric whose determinant is negative-definite is called a pseudo-Riemannian manifold.

To define distances on a general (possibly curved) manifold, the manifold is endowed with a symmetric rank-2 tensor, the metric. The components of the metric are written as $g_{\mu\nu}$, and their values may depend upon the spacetime location at which they are evaluated: $g_{\mu\nu} = g_{\mu\nu}(x^\alpha)$. The metric is used to write down the infinitesimal interval between two spacetime events,

$$ds^2 = g_{\mu\nu} dx^\mu dx^\nu. \quad (1.14)$$

The line element (1.14) is invariant under coordinate transformations (1.3). The norm of a 4-vector is defined

$$\mathsf{A} \cdot \mathsf{A} \equiv g_{\mu\nu} A^{\mu} A^{\nu} = A^{\mu} A_{\mu}. \tag{1.15}$$

If $A^{\mu} A_{\mu} < 0$ then the vector whose whose components are A^{μ} is called *time-like*, if $A^{\mu} A_{\mu} = 0$ then A^{μ} is *null* and if $A^{\mu} A_{\mu} > 0$ then A^{μ} is *space-like*. The components of the inverse metric are $g^{\mu\nu}$, and is defined so that $g^{\mu\nu} g_{\nu\alpha} = \delta^{\mu}{}_{\alpha}$. The metric is used to lower contravariant indices into covariant indices and the inverse metric is used to raise covariant indices into contravariant indices. For example,

$$g_{\mu\nu} A^{\mu\alpha\beta} = A_{\nu}{}^{\alpha\beta}, \qquad g^{\mu\nu} B_{\mu\alpha\beta} = B^{\nu}{}_{\alpha\beta}. \tag{1.16}$$

The ordering of the indices is preserved by index raising and lowering.

When the components of the metric depend upon the spacetime coordinates, the derivative of the metric with respect to the spacetime coordinates does not vanish:

$$g_{\mu\nu} = g_{\mu\nu}(x^{\alpha}) \qquad \Rightarrow \qquad \partial_{\alpha} g_{\mu\nu} \neq 0. \tag{1.17}$$

When this is the case, the operations of partial differentiation and index raising/lowering do not commute. For example,

$$\partial_{\mu} A^{\alpha} = \partial_{\mu}(g^{\alpha\beta} A_{\beta}) = g^{\alpha\beta} \partial_{\mu} A_{\beta} + A_{\beta} \partial_{\mu} g^{\alpha\beta}. \tag{1.18}$$

Obviously, the final term vanishes in a Minkowski spacetime, since there $\partial_{\alpha} \eta_{\mu\nu} = 0$. In the general case where the components of the metric vary throughout the spacetime manifold, $\partial_{\alpha} g_{\mu\nu} \neq 0$, a new type of differential operator which preserves the metric is defined. That is, in addition to using the partial derivative ∂, we introduce the *covariant derivative* ∇, such that

$$\nabla_{\alpha} g_{\mu\nu} = \partial_{\alpha} g_{\mu\nu} - \Gamma^{\lambda}{}_{\alpha\mu} g_{\lambda\beta} - \Gamma^{\lambda}{}_{\alpha\nu} g_{\mu\lambda} = 0. \tag{1.19}$$

This is called the metricity condition and enables the operation of covariant differentiation and index raising/lowering to commute: $\nabla_{\mu} A^{\alpha} = \nabla_{\mu}(g^{\alpha\beta} A_{\beta}) = g^{\alpha\beta} \nabla_{\mu} A_{\beta}$. The covariant derivative of a tensor transforms as a tensor under coordinate transformations (whereas the partial derivative of a tensor does not transform as a tensor). Using this information, one is able to construct the covariant derivative of contravariant and covariant vectors,

$$\nabla_{\mu} A^{\nu} = \partial_{\mu} A^{\nu} + \Gamma^{\nu}{}_{\mu\alpha} A^{\alpha}, \qquad \nabla_{\mu} A_{\nu} = \partial_{\mu} A_{\nu} - \Gamma^{\alpha}{}_{\mu\nu} A_{\alpha}, \tag{1.20a}$$

$$\nabla_{\alpha} B_{\mu\nu} = \partial_{\alpha} B_{\mu\nu} - \Gamma^{\beta}{}_{\alpha\mu} B_{\beta\nu} - \Gamma^{\beta}{}_{\alpha\nu} B_{\mu\beta}. \tag{1.20b}$$

This can be extended to the covariant derivative of mixed rank tensors,

$$\nabla_{\mu} T^{\alpha}{}_{\beta} = \partial_{\mu} T^{\alpha}{}_{\beta} + \Gamma^{\alpha}{}_{\mu\nu} T^{\nu}{}_{\beta} - \Gamma^{\nu}{}_{\mu\beta} T^{\alpha}{}_{\nu}. \tag{1.20c}$$

These covariant derivatives transform as tensors by construction.

We now ask: "*what is the shortest distance between two points?*". We use the notion of distance, as defined using the metric, to compute the "shortest" distance between two points on a spacetime manifold of arbitrary geometry. We join the two points by a trajectory, where along that trajectory the spacetime coordinates have the values X^μ. By integrating the infinitesimal path length ds (1.14) along the length of the trajectory the total length of the trajectory is obtained, through a spacetime having a given metric:

$$S = \int \mathrm{d}s = \int \sqrt{g_{\mu\nu}\mathrm{d}X^\mu \mathrm{d}X^\nu}. \tag{1.21}$$

We like to know what the values of the X^μ are along the trajectory. Of course, along an arbitrary trajectory the coordinates X^μ are arbitrary. We can however pick a particular trajectory from the set of all possible trajectories, where this particular trajectory is the shortest possible line joining any two points. What this means is that the path length is extremized with respect to variations in the coordinates that lie on this particular trajectory,

$$\frac{\delta S}{\delta X^\mu} = 0. \tag{1.22}$$

To calculate the values of the coordinates along this trajectory we introduce an affine parameter λ along the curve, so that $X^\mu = X^\mu(\lambda)$. This allows the path length (1.21) to be written as

$$S = \int \mathrm{d}\lambda \sqrt{g_{\mu\nu}(X^\alpha)\dot{X}^\mu \dot{X}^\nu}, \tag{1.23}$$

where an overdot denotes derivative with respect to the affine parameter, $\dot{X}^\mu \equiv \mathrm{d}X^\mu/\mathrm{d}\lambda$. Treating the integrand in (1.23) as a Lagrangian, the principle of least action can be used to obtain the Euler-Lagrange equation whose solution (i.e. the $X^\mu(\lambda)$) extremizes the path length (1.21). The Euler-Lagrange equation one obtains is

$$\ddot{X}^\mu + \Gamma^\mu{}_{\alpha\beta}\dot{X}^\alpha \dot{X}^\beta = 0, \tag{1.24}$$

where $\Gamma^\mu{}_{\alpha\beta}$ are the components of the Christoffel symbol and are given by

$$\Gamma^\mu{}_{\alpha\beta} \equiv \frac{1}{2} g^{\mu\nu}\left(\partial_\alpha g_{\beta\nu} + \partial_\beta g_{\alpha\nu} - \partial_\nu g_{\alpha\beta}\right) = \Gamma^\mu{}_{\beta\alpha}. \tag{1.25}$$

Equation (1.24) is called the *equation of an affinely parameterized geodesic*. It is important to realise that the $\Gamma^\mu{}_{\alpha\beta}$ do not form the components of a tensor. Notice that for flat space, $\Gamma^\alpha{}_{\mu\nu} = 0$, so that the geodesic (1.24) becomes $\ddot{X}^\mu = 0$. What this means is that geodesics in flat space are straight lines.

The geodesic Eq. (1.24) can only be solved once the components of the metric tensor, $g_{\mu\nu}$, are known. To know what the components $g_{\mu\nu}$ are for a particular spacetime, gravitational field equations are required which we will discuss shortly.

1.1.3 Curvature

The manifolds discussed are, in general, curved. The curvature of a manifold is characterized by the second derivative of the fundamental object defining the manifold, which is the metric. A manifold whose metric has non-zero second derivatives is curved. One usually imagines curvature of a surface that is embedded in a higher-dimensional space. This is *extrinsic* curvature. It is also possible to develop the notion of *intrinsic* curvature without needing to refer to some higher-dimensional space. We will begin our discussion by constructing the intrinsic curvature tensors and then we construct extrinsic curvature tensors.

To quantify intrinsic curvature, tensors are constructed from the second derivatives of the metric; the so-called curvature tensors. The first tensor is known as the Riemann tensor, and arises by considering the commutator of covariant derivatives. One is able to obtain the Ricci identity

$$\left(\nabla_\mu \nabla_\nu - \nabla_\nu \nabla_\mu\right) A^\alpha = R^\alpha{}_{\beta\mu\nu} A^\beta, \tag{1.26}$$

where the Riemann tensor, $R^\alpha{}_{\beta\mu\nu}$, is given by

$$R^\alpha{}_{\beta\mu\nu} \equiv 2\partial_{[\mu} \Gamma^\alpha{}_{\nu]\beta} + 2\Gamma^\alpha{}_{\rho[\mu} \Gamma^\rho{}_{\nu]\beta}. \tag{1.27}$$

In flat space the Christoffel symbols vanish so that covariant derivatives reduce to partial derivatives, which commute and so $R^\alpha{}_{\beta\mu\nu} = 0$ for flat space. The Ricci tensor is formed by setting the first and third indices of the Riemann tensor to be equal,

$$R_{\mu\nu} \equiv R^\alpha{}_{\mu\alpha\nu}, \tag{1.28}$$

The contracted Ricci identity is

$$(\nabla_\mu \nabla_\nu - \nabla_\nu \nabla_\mu) A^\mu = R_{\alpha\nu} A^\alpha. \tag{1.29}$$

Finally, by contracting the Ricci tensor one obtains the Ricci scalar,

$$R \equiv R^\mu{}_\mu. \tag{1.30}$$

The Ricci tensor and scalar are combined to produce the Einstein tensor,

$$G_{\mu\nu} \equiv R_{\mu\nu} - \frac{1}{2} g_{\mu\nu} R, \tag{1.31}$$

whose indices are symmetric, $G_{\mu\nu} = G_{(\mu\nu)}$. One can use the above definitions to show that the Einstein tensor satisfies a conservation equation,

$$\nabla_\mu G^{\mu\nu} = 0, \tag{1.32}$$

which is called the Bianchi identity.

The components of a tensor change under a coordinate transformation, but the value of a scalar does not. This means that, for example, the values of the components of the Ricci tensor $R_{\mu\nu}$ change, but the value of the Ricci scalar does not. With this in mind, we write down three curvature invariants,

$$R^{\mu\nu\alpha\beta} R_{\mu\nu\alpha\beta}, \qquad R^{\mu\nu} R_{\mu\nu}, \qquad R, \tag{1.33}$$

whose values remain the same regardless of coordinate system. Usefully, any singularities present in these curvature invariants are intrinsic to the manifold, and are unable to be removed by coordinate transformation.

Whilst the theories of gravity we will consider are covariant (i.e. space and time are on an "equal footing"), it is useful to write down expressions which give 3D "space" and "time" some meaning in a covariant way. We imagine 3D sheets as being embeded in the 4D universe, where to travel along a given sheet means to travel through space, and to different sheets means to travel through time. This gives rise to the notion of the extrinsic curvature of the 3D sheets, due to their embedding in a higher dimensional space. We foliate the 4D universe by 3D sheets, where there is a metric $\gamma_{\mu\nu}$ that exists on these sheets and a time-like unit vector u^μ is everywhere orthogonal to these sheets. The metric and curvature tensors are rewritten by imposing the following structure

$$g_{\mu\nu} = \gamma_{\mu\nu} - u_\mu u_\nu, \qquad u^\mu u_\mu = -1, \qquad u^\mu \gamma_{\mu\nu} = 0. \tag{1.34}$$

The vector u^μ is a time-like geodesic, and so

$$u^\mu \nabla_\mu u_\nu = 0. \tag{1.35}$$

We then define the extrinsic curvature tensor

$$K_{\mu\nu} = \nabla_\mu u_\nu, \tag{1.36}$$

and, by using (1.34, 1.35), can be shown to satisfy

$$u^\mu K_{\mu\nu} = 0 \qquad \Longrightarrow \qquad K_{\mu\nu} = \gamma^\alpha{}_\mu \nabla_\alpha u_\nu. \tag{1.37}$$

The trace of the extrinsic curvature, $K \equiv K^\mu{}_\mu = \gamma^{\mu\nu} K_{\mu\nu}$ will correspond to the curvature scalar of the 3D sheets due to their embedding in a higher dimensional space. Differentiation along the time-like direction and differentiation confined to the 3D sheets are defined respectively as

$$\dot{A}^{\mu} \equiv u^{\alpha}\nabla_{\alpha}A^{\mu}, \qquad \bar{\nabla}_{\mu}A^{\nu} \equiv \gamma^{\alpha}{}_{\mu}\gamma^{\nu}{}_{\beta}\nabla_{\alpha}A^{\beta}. \tag{1.38}$$

With this definition we find that $\bar{\nabla}_{\mu}$ is the covariant derivative which preserves the 3D metric,

$$\bar{\nabla}_{\mu}\gamma_{\alpha\beta} = 0. \tag{1.39}$$

By combining the various definitions given above, we compute the anti-symmetric sheet-confined derivative of a vector which it itself confined to the 3D sheet (i.e. $u_{\mu}V^{\mu} = 0$),

$$2\bar{\nabla}_{[\alpha}\bar{\nabla}_{\beta]}V_{\nu} = -\gamma^{\pi}{}_{\alpha}\gamma^{\rho}{}_{\beta}\gamma^{\sigma}{}_{\nu}R^{\epsilon}{}_{\sigma\pi\rho}V_{\epsilon} + (K_{\beta\nu}K^{\epsilon}{}_{\alpha} - K_{\alpha\nu}K^{\epsilon}{}_{\beta})V_{\epsilon}. \tag{1.40}$$

The left-hand-side of this expression is defined to be the Riemann tensor on the 3D sheet, $2\bar{\nabla}_{[\alpha}\bar{\nabla}_{\beta]}V_{\nu} \equiv -{}^{(3)}R^{\epsilon}{}_{\nu\alpha\beta}V_{\epsilon}$. Hence, a relationship is obtained between the Riemann tensor on the 3D sheet, in the 4D manifold in which the sheet is embedded and the extrinsic curvature tensors:

$${}^{(3)}R^{\epsilon}{}_{\nu\alpha\beta} = \gamma^{\pi}{}_{\alpha}\gamma^{\rho}{}_{\beta}\gamma^{\sigma}{}_{\nu}R^{\epsilon}{}_{\sigma\pi\rho} + K_{\alpha\nu}K^{\epsilon}{}_{\beta} - K_{\beta\nu}K^{\epsilon}{}_{\alpha}. \tag{1.41}$$

By contraction, we obtain the 3D Ricci tensor,

$${}^{(3)}R_{\nu\beta} = \gamma^{\rho}{}_{\beta}\gamma^{\sigma}{}_{\nu}R_{\sigma\rho} + \gamma^{\rho}{}_{\beta}\gamma^{\sigma}{}_{\nu}u^{\pi}u_{\epsilon}R^{\epsilon}{}_{\sigma\pi\rho} + K_{\epsilon\nu}K^{\epsilon}{}_{\beta} - K_{\beta\nu}K, \tag{1.42}$$

and computing ${}^{(3)}R \equiv g^{\mu\nu}{}^{(3)}R_{\mu\nu}$ we obtain the 3D Ricci scalar,

$${}^{(3)}R + K^2 - K^{\mu\nu}K_{\mu\nu} = R + 2u^{\mu}u^{\nu}R_{\mu\nu} = 2u^{\mu}u^{\nu}G_{\mu\nu}. \tag{1.43a}$$

Similarly, one can compute

$$\bar{\nabla}_{\mu}K^{\mu}{}_{\alpha} - \bar{\nabla}_{\alpha}K = \gamma^{\mu}{}_{\alpha}u^{\nu}R_{\mu\nu} \tag{1.43b}$$

The Eq. (1.43) are known as the Gauss-Codacci relations, and enable the geometry of 3D sheets to be determined by their embedding in a 4D manifold.

1.2 General Relativity

General Relativity (GR) is arguably one of the greatest advances in theoretical physics from the past century. Although the mathematical formalism existed well before Albert Einstein laid out what we now call GR, the key insight which Einstein had was to understand that the geometry of the universe is determined by, and responds to, the gravitating content of the universe. Einstein constructed a set of dynamical rules for spacetime with arbitrary geometry; these rules are now what we call *Einstein's field equations of General Relativity*. Gravity is interpreted as the manifestation of the location-dependance of the metric. The field equations determine the values of the components of the metric of a spacetime for some known content. Once the metric is known one can begin to compute geodesics in the spacetime: these are the paths that bundles of light rays travel along or the orbits that planets trace out. This enables GR, as a gravitational theory, to predict directly observable quantities. This also applies to the entire universe: Einstein's field equations allow the metric for the entire universe to be computed once the content of the universe is known.

Einstein's field equations of GR provide a specific way in which the metric is determined from the content of the spacetime. The information about the content is contained within the energy-momentum tensor $T_{\mu\nu}$. Einstein's field equations of GR relate $G_{\mu\nu}$ and $T_{\mu\nu}$ linearly,

$$G_{\mu\nu} = 8\pi G T_{\mu\nu}, \tag{1.44}$$

where the constant of proportionality, $8\pi G$, is determined by comparing the Newtonian limit of the GR field equations with the Newtonian field equations. The field Eq. (1.44) are second order evolution equations for the metric, and they are sourced by the energy-momentum tensor. Because the Einstein tensor satisifies a Bianchi identity, (1.32), so too must the energy-momentum tensor,

$$\nabla_\mu T^{\mu\nu} = 0. \tag{1.45}$$

This can be interpreted as either a constraint equation or an evolution equation for the matter content. In the absence of gravitational fields, the conservation equation reads $\partial_\mu T^{\mu\nu} = 0$, but in the presence of gravitational fields,

$$\partial_\mu T^{\mu\nu} = -\Gamma^\mu{}_{\mu\alpha} T^{\alpha\nu} - \Gamma^\nu{}_{\mu\alpha} T^{\mu\alpha}, \tag{1.46}$$

so that gradients of the metric contribute to the dynamics of the fluid. Without knowledge of this sourcing, it would appear that the fluid does not satisfy a conservation equation.

1.2.1 Exact Analytic Solutions to Einstein's Field Equations

No general solution to Einstein's field Eq. (1.44) is known to exist. However, solutions are known for rather specific configurations of the content. The simplest solution is for a spacetime which is completely empty (of gravitating matter) is called a vacuum, and whose metric is given by the Minkowski metric,

$$\mathrm{d}s^2 = -\mathrm{d}t^2 + \mathrm{d}x^2 + \mathrm{d}y^2 + \mathrm{d}z^2. \tag{1.47}$$

The next simplest metric is that for a spacetime containing a homogeneous and isotropic fluid, whose energy-momentum tensor is of the form

$$T_{\mu\nu} = \rho u_\mu u_\nu + P\gamma_{\mu\nu}, \tag{1.48}$$

where $\rho = \rho(t)$, $P = P(t)$, and is given by Friedmann-Robertson-Walker's solution [13],

$$\mathrm{d}s^2 = -\mathrm{d}t^2 + a^2(t)\left(\mathrm{d}x^2 + \mathrm{d}y^2 + \mathrm{d}z^2\right). \tag{1.49}$$

Once distributions of matter which are confined to particular locations are included the solutions become much less attractive. The energy-momentum tensor for a single stationary mass can be written as $T^\mu{}_\nu = M\delta^{(3)}(\mathbf{r})u^\mu u_\nu$. The metric of a spacetime which asymptotes to vacuum at infinity which contains a single stationary black hole of mass M at the origin, is given by Schwarzschild's solution [14],

$$\mathrm{d}s^2 = -\left(1 - \frac{2M}{r}\right)\mathrm{d}t^2 + \left(1 - \frac{2M}{r}\right)^{-1}\mathrm{d}r^2 + r^2\mathrm{d}\Omega^2, \tag{1.50}$$

where $\mathrm{d}\Omega^2 \equiv (\mathrm{d}\theta^2 + \sin^2\mathrm{d}\phi^2)$ is the solid angle element. The metric for a spacetime containing a stationary black hole of mass M in a universe containing a cosmological constant Λ is given by Kottler's solution [15],

$$\mathrm{d}s^2 = -\left(1 - \frac{2M}{r} - \frac{\Lambda}{3}r^2\right)\mathrm{d}t^2 + \left(1 - \frac{2M}{r} - \frac{\Lambda}{3}r^2\right)^{-1}\mathrm{d}r^2 + r^2\mathrm{d}\Omega^2, \tag{1.51}$$

also called a Schwarzschild de-Sitter metric. A single black hole of mass M immersed in a homogeneous isotropic fluid has a metric which is given by McVittie's solution [16, 17],

$$\mathrm{d}s^2 = -\left(\frac{2a(t)r - M}{2a(t)r + M}\right)^2\mathrm{d}t^2 + a^2(t)\left(1 + \frac{M}{2a(t)r}\right)^4\left(\mathrm{d}r^2 + r^2\mathrm{d}\Omega^2\right). \tag{1.52}$$

Kerr's solution [18] provides the metric for a spacetime containing a single black hole of mass M rotating with angular momentum $J = M\omega$,

$$\begin{aligned}\mathrm{d}s^2 &= -\frac{r^2 - 2Mr + \omega^2 - \omega^2\sin^2\theta}{r^2 + \omega^2\cos^2\theta}\mathrm{d}t^2 - \frac{4M\omega r\sin^2\theta}{r^2 + \omega^2\cos^2\theta}\mathrm{d}\phi\mathrm{d}t \\ &+ \frac{\sin^2\theta}{r^2 + \omega^2\cos^2\theta}\left[\left(r^2 + \omega^2\right)^2 - \omega^2\sin^2\theta\left(r^2 - 2Mr + \omega^2\right)\right]\mathrm{d}\phi^2 \\ &+ \left(r^2 + \omega^2\cos^2\theta\right)\left[\frac{\mathrm{d}r^2}{r^2 - 2Mr + \omega^2} + \mathrm{d}\theta^2\right].\end{aligned} \tag{1.53}$$

1.2.2 Perturbative Gravity

As remarked above, it is not known how to solve the gravitational field equations in general. This is largely because they are highly complicated non-linear differential equations. The problem can be simplified by using perturbation theory so that the gravitational field equations become linear in the metric. Suppose, for example, that the metric $\bar{g}_{\mu\nu}$ is known for some simple physical system. If the physical system of interest has a metric given by $g_{\mu\nu}$, and can be considered as a small deviation about that simple system, $g_{\mu\nu} = \bar{g}_{\mu\nu} + \delta g_{\mu\nu}$, then perturbation theory can be used to obtain the linearized version of the gravitational field equations. We will now provide formulae for the linearized gravitational field equations.

The background value of the metric is written with an overline, $\bar{g}_{\mu\nu}$, and the perturbation to that background is denoted by $h_{\mu\nu}$. The metric thus written as

$$g_{\mu\nu} = \bar{g}_{\mu\nu} + h_{\mu\nu}, \qquad g^{\mu\nu} = \bar{g}^{\mu\nu} + h^{\mu\nu}. \tag{1.54}$$

All quantities calculated from the metric will also be perturbed. For example, inserting (1.54) into the Christoffel symbol (1.25) we find

$$\Gamma^{\mu}{}_{\alpha\beta} = \bar{\Gamma}^{\mu}{}_{\alpha\beta} + \delta\Gamma^{\mu}{}_{\alpha\beta}, \tag{1.55}$$

where the background and perturbed contributions to the overall Christoffel symbol are given by

$$\bar{\Gamma}^{\mu}{}_{\alpha\beta} = \frac{1}{2}\bar{g}^{\mu\nu}\left(\partial_\alpha \bar{g}_{\beta\nu} + \partial_\beta \bar{g}_{\alpha\nu} - \partial_\nu \bar{g}_{\alpha\beta}\right), \tag{1.56a}$$

$$\delta\Gamma^{\mu}{}_{\alpha\beta} = \frac{1}{2}\bar{g}^{\mu\nu}\left(\partial_\alpha h_{\beta\nu} + \partial_\beta h_{\alpha\nu} - \partial_\nu h_{\alpha\beta}\right) + \bar{\Gamma}^{\rho}{}_{\alpha\beta} h^{\mu}{}_{\rho}. \tag{1.56b}$$

The formula for the perturbed Christoffel symbol (1.56b) can be shown to be equivalent to

$$\delta\Gamma^{\mu}{}_{\alpha\beta} = \frac{1}{2}\bar{g}^{\mu\nu}\left(\nabla_\alpha h_{\beta\nu} + \nabla_\beta h_{\alpha\nu} - \nabla_\nu h_{\alpha\beta}\right), \tag{1.57}$$

where ∇_α is the covariant derivative with respect to the background metric $\bar{g}_{\mu\nu}$. This way of writing the perturbed Christoffel symbol is more convenient than (1.56b). The perturbed Christoffel symbol can also be written as

$$\delta\Gamma^{\mu}{}_{\alpha\beta} = \left(\bar{g}^{\mu\pi}\delta^{\rho}{}_{(\alpha}\delta^{\sigma}{}_{\beta)} - \tfrac{1}{2}\bar{g}^{\mu\rho}\delta^{\sigma}{}_{\alpha}\delta^{\pi}{}_{\beta}\right)\nabla_\rho h_{\sigma\pi}. \tag{1.58}$$

After similar manipulations one obtains the perturbation to the Riemann tensor,

$$\delta R^{\alpha}{}_{\mu\beta\nu} = \left(g^{\alpha\rho}\delta^{\sigma}{}_{\mu}\delta^{[\pi}{}_{\beta}\delta^{\xi]}{}_{\nu} + g^{\alpha\pi}\delta^{\sigma}{}_{\mu}\delta^{[\rho}{}_{\nu}\delta^{\xi]}{}_{\beta} + g^{\alpha\pi}\delta^{\rho}{}_{\mu}\delta^{[\sigma}{}_{\nu}\delta^{\xi]}{}_{\beta}\right)\nabla_{\xi}\nabla_{\rho}h_{\sigma\pi} \tag{1.59}$$

In summary, the perturbations to the Christoffel symbol, Ricci tensor, Ricci scalar and Einstein tensor are given by

$$\delta\Gamma^{\alpha}{}_{\mu\nu} = \frac{1}{2}\bar{g}^{\alpha\beta}\left(\nabla_{\mu}h_{\nu\beta} + \nabla_{\nu}h_{\mu\beta} - \nabla_{\beta}h_{\mu\nu}\right), \tag{1.60a}$$

$$\delta R_{\mu\nu} = \nabla^{\alpha}\nabla_{(\mu}h_{\nu)\alpha} - \frac{1}{2}\left(\Box h_{\mu\nu} + \nabla_{\mu}\nabla_{\nu}h\right), \tag{1.60b}$$

$$\delta R = \nabla^{\mu}\nabla^{\nu}h_{\mu\nu} - \Box h - R^{\mu\nu}h_{\mu\nu}, \tag{1.60c}$$

$$\begin{aligned} 2\delta G_{\mu\nu} = {} & \nabla^{\alpha}\nabla_{\mu}h_{\nu\alpha} + \nabla^{\alpha}\nabla_{\nu}h_{\mu\alpha} - \Box h_{\mu\nu} + \bar{g}_{\mu\nu}\Box h - \nabla_{\mu}\nabla_{\nu}h \\ & -\bar{g}_{\mu\nu}\nabla^{\alpha}\nabla^{\beta}h_{\alpha\beta} - Rh_{\mu\nu} + \bar{g}_{\mu\nu}R^{\alpha\beta}h_{\alpha\beta}, \end{aligned} \tag{1.60d}$$

where we wrote $h = h^{\mu}{}_{\mu} = \bar{g}^{\mu\nu}h_{\mu\nu}$. The covariant derivative, ∇_{μ}, Ricci tensor and scalar $R_{\mu\nu}$, R are those of the background spacetime. For perturbations about Minkowski spacetime, $\bar{g}_{\mu\nu} = \eta_{\mu\nu}$, the perturbed field equations of General Relativity yield

$$\begin{aligned} 2\delta G_{\mu\nu} &= \partial^{\alpha}\partial_{\mu}h_{\nu\alpha} + \partial^{\alpha}\partial_{\nu}h_{\mu\alpha} - \Box h_{\mu\nu} + \eta_{\mu\nu}\Box h - \partial_{\mu}\partial_{\nu}h - \eta_{\mu\nu}\partial^{\alpha}\partial^{\beta}h_{\alpha\beta} \\ &= 16\pi G\delta T_{\mu\nu}. \end{aligned} \tag{1.61}$$

1.2.3 Field Equations from an Action

The formulation of the gravitational field equations we have presented has been geometrical: the Einstein tensor was constructed from various combinations of derivatives of the metric, and was equated to the energy-momentum tensor. The field equations can be derived in another way, using the principle of least action. This strategy is algorithmic and it becomes obvious as to how we can construct "modified gravity" theories. We start from the Einstein-Hilbert action

$$S = \int \mathrm{d}^4x\sqrt{-g}\left[R - 16\pi G\mathcal{L}_{\mathrm{m}}\right], \tag{1.62}$$

where g is the determinant of the metric tensor, R is the Ricci scalar and $\mathcal{L}_{\mathrm{m}}$ is the Lagrangian density of all sources of energy and momentum. Applying the principle of least action to (1.62) yields the field equations of General Relativity. The field equations (of the metric) are found by extremizing the variation of the action with respect to variations in the metric,

$$\frac{\delta S}{\delta g^{\mu\nu}} = 0. \tag{1.63}$$

Varying the action (1.62) yields

$$\delta S = \int \mathrm{d}^4 x \sqrt{-g} \left[\frac{R - 16\pi G \mathcal{L}_\mathrm{m}}{\sqrt{-g}} \delta(\sqrt{-g}) + \delta R - 16\pi G \delta \mathcal{L}_\mathrm{m} \right]. \tag{1.64}$$

To go further, we need the following results:

$$\frac{1}{\sqrt{-g}} \delta\sqrt{-g} = -\frac{1}{2} g_{\mu\nu} \delta g^{\mu\nu}, \qquad \delta R = \Big(R_{\mu\nu} + g_{\mu\nu} \Box - \nabla_\mu \nabla_\nu \Big) \delta g^{\mu\nu}. \tag{1.65}$$

Using (1.65) in (1.64) provides a formula for the variation of the Einstein-Hilbert action (1.62),

$$\delta S = \int \mathrm{d}^4 x \sqrt{-g} \left[G_{\mu\nu} \delta g^{\mu\nu} + (g_{\mu\nu} \Box - \nabla_\mu \nabla_\nu) \delta g^{\mu\nu} - 16\pi G \frac{1}{\sqrt{-g}} \delta\big(\sqrt{-g} \mathcal{L}_\mathrm{m}\big) \right], \tag{1.66}$$

where the Einstein tensor, $G_{\mu\nu} \equiv R_{\mu\nu} - \frac{1}{2} g_{\mu\nu} R$ is identified. The second term in the integrand is a total derivative, which can be rewritten as a surface integral and neglected (in the following section we will show how to deal with this surface integral). In theories where the Ricci scalar does not appear linearly these terms do not correspond to total derivatives, cannot be removed and will therefore represent a modification to the field equations. By setting the variation of the action with respect to the variation of the metric to zero, $\delta S / \delta g^{\mu\nu} = 0$, and assuming that the variation of the metric vanishes on the boundary, we obtain from (1.66) the field equations for the metric

$$G_{\mu\nu} = 8\pi G T_{\mu\nu}, \tag{1.67}$$

where the energy-momentum tensor is defined as

$$T_{\mu\nu} \equiv \frac{2}{\sqrt{-g}} \frac{\delta}{\delta g^{\mu\nu}} \big(\sqrt{-g} \mathcal{L}_\mathrm{m}\big). \tag{1.68}$$

Hence, we observe that by applying the variational principle (1.63) to the Einstein-Hilbert action (1.62) we obtained the Einstein field Eqs. (1.67). By obtaining the gravitational field equations from the principle of least action of a Lagrangian, we gain some intuition that the values of the metric which solve the field Eqs. (1.67) extremize the value of some global measure.

1.2.4 Review of the Gibbons–Hawking Term

In the discussion preceding (1.67) it was mentioned that the surface term requires special attention: here we show how to deal with the surface term by reviewing the Gibbons-Hawking term. We begin with a discussion from classical mechanics of actions containing a total derivative [19] which will build up some intuition of total derivatives before moving on to discuss the total derivative in GR.

The action of a Lagrangian containing at most first derivatives of the generalized coordinate is

$$S = \int_{t_1}^{t_2} \mathrm{d}t\, L(q, \dot{q}), \tag{1.69}$$

and can be varied to yield

$$\begin{aligned}\delta S = \int_{t_1}^{t_2} \mathrm{d}t \left(\delta q \frac{\partial L}{\partial q} + \delta \dot{q} \frac{\partial L}{\partial \dot{q}} \right) &= \int_{t_1}^{t_2} \mathrm{d}t \left\{ \left(\frac{\partial L}{\partial q} - \frac{\mathrm{d}}{\mathrm{d}t} \frac{\partial L}{\partial \dot{q}} \right) \delta q + \frac{\mathrm{d}}{\mathrm{d}t} \left(\delta q \frac{\partial L}{\partial \dot{q}} \right) \right\} \\ &= \int_{t_1}^{t_2} \mathrm{d}t \left\{ \left(\frac{\partial L}{\partial q} - \frac{\mathrm{d}}{\mathrm{d}t} \frac{\partial L}{\partial \dot{q}} \right) \delta q \right\} \\ &\quad + \delta q \frac{\partial L}{\partial \dot{q}} \bigg|_{t_1}^{t_2} . \end{aligned} \tag{1.70}$$

To be able to use the variational principle to obtain the equations of motion *boundary data*, prescribing the values of the variations at the endpoints of the trajectory, $\delta q(t_1)$, $\delta q(t_2)$, must be specified. These variations are usually taken to vanish. Let us now construct a Lagrangian by adding a total derivative, L_2, to a pre-existing theory, L_1:

$$L = L_1 + L_2 = -\frac{1}{2} q \ddot{q}, \qquad L_1 \equiv \frac{1}{2} \dot{q}^2, \quad L_2 \equiv -\frac{\mathrm{d}}{\mathrm{d}t} \left(\frac{1}{2} q \dot{q} \right). \tag{1.71}$$

The variation of the Lagrangian L yields

$$\delta L = \ddot{q} \delta q + \frac{1}{2} \frac{\mathrm{d}}{\mathrm{d}t} (\dot{q} \delta q - q \delta \dot{q}). \tag{1.72}$$

Notice that the equation of motion of L and of L_1 are identical ($\ddot{q} = 0$): the addition of L_2 does not affect the dynamical equations of motion because L_2 is a total derivative and only contributes on the boundary. To obtain the equations of motion for L four pieces of boundary data are required, (i.e. the values of $\delta q(t_1)$, $\delta q(t_2)$ and $\delta \dot{q}(t_1)$, $\delta \dot{q}(t_2)$), whilst for the theory with L_1 only two pieces of boundary data are required (the values of the variation at the endpoints, $\delta q(t_1)$, $\delta q(t_2)$.

The point is: by adding on a total derivative (which does not change the equations of motion) the amount of data that is required to be specified on the boundary is changed, and this can be thought of as introducing an "inconsistency". To bring the theory back into "consistency" we modify the theory by adding on a boundary term to the Lagrangian which will kill off the offending term. So, the theory L must be modified to

$$L \to \check{L} = L + \frac{\mathrm{d}}{\mathrm{d}t}\left(\frac{1}{2} q \dot{q}\right). \tag{1.73}$$

The Lagrangian L can be thought of as being analogous to the Einstein-Hilbert action, and the boundary term $\frac{\mathrm{d}}{\mathrm{d}t}\left(\frac{1}{2} q \dot{q}\right)$ can be thought of as being analogous to the Gibbons-Hawking term, as we will now show.

Our discussion now takes inspiration from a number of sources [7, 20–23]. We will start by showing what the problem is with only specifying the Einstein-Hilbert action

$$S = \int \mathrm{d}^4 x \sqrt{-g}\, R = \int_M R, \tag{1.74}$$

and then go on to show the popular way to resolve the problem. Without removing any terms, the variation of the Einstein-Hilbert action yields

$$\delta S = \int_M \left[G_{\mu\nu} \delta g^{\mu\nu} + g^{\mu\nu} \delta R_{\mu\nu} \right], \tag{1.75}$$

where one can obtain

$$g^{\mu\nu} \delta R_{\mu\nu} = \nabla_\alpha v^\alpha, \qquad v^\alpha \equiv g_{\mu\nu}\left(\nabla^\alpha \delta g^{\mu\nu} - \nabla^\nu \delta g^{\alpha\mu}\right). \tag{1.76}$$

Hence, the variation of the Einstein-Hilbert action (1.75) using (1.76) can be written as

$$\delta S = \int_M \left[G_{\mu\nu} \delta g^{\mu\nu} + \nabla_\alpha v^\alpha \right]. \tag{1.77}$$

By applying the divergence theorem to the second term, this becomes

$$\delta S = \int_M G_{\mu\nu} \delta g^{\mu\nu} + \int_{\partial M} n_\alpha v^\alpha, \tag{1.78}$$

where n_α is a unit normal to the boundary. Within the surface integral we have

$$n_\alpha v^\alpha = n_\alpha g_{\mu\nu}\big(\nabla^\alpha \delta g^{\mu\nu} - \nabla^\nu \delta g^{\alpha\mu}\big). \tag{1.79}$$

Therefore, after varying the metric in the Einstein-Hilbert action, there is a "surface" contribution. The metric on ∂M is given by $\gamma_{\mu\nu} = g_{\mu\nu} \pm n_\mu n_\nu$, where n^μ is a normal unit vector to the boundary ∂M. Equation (1.79) can be rewritten as

$$n_\alpha v^\alpha = n_\alpha \gamma_{\mu\nu}(\nabla^\alpha \delta g^{\mu\nu} - \nabla^{(\mu} \delta g^{\nu)\alpha}). \tag{1.80}$$

The point of this was to show that the metric $g_{\mu\nu}$ in (1.79) can be replaced with the induced (boundary) metric $\gamma_{\mu\nu}$ in $n_\alpha v^\alpha$. After lowering the indices appropriately (1.80) is used to obtain the following formula for the variation of the Einstein-Hilbert action (1.78):

$$\delta S = \delta \int_M R = \int_M G_{\mu\nu}\delta g^{\mu\nu} + \int_{\partial M} \Big[n^\alpha \gamma^{\mu\nu}\big(\nabla_{(\mu}\delta g_{\nu)\alpha} - \nabla_\alpha \delta g_{\mu\nu}\big)\Big]. \tag{1.81}$$

As it stands, the behaviour of the derivative of $\delta g_{\mu\nu}$ must be specified normal to the surface, $n^\alpha \nabla_\alpha \delta g_{\mu\nu}$, as well as the the transverse derivative $\gamma^{\mu\nu}\nabla_{(\mu}\delta g_{\nu)\alpha}$; note that the transverse derivative of the perturbed metric tells us how the metric varies on the boundary. The requirement of vanishing normal derivative can be removed by introducing an extra term into the action, where the extra term only resides on the boundary and encodes information about the geometry of the boundary.

The Gibbons-Hawking term [20] (sometimes called the Gibbons-Hawking-York term [24]) is an explicit example of a boundary term which can be added to the Einstein-Hilbert action to provide the field equations of General Relativity after employing the variational principle. Without adding the term more stringent conditions must be imposed upon the behaviour of the varied metric at the boundary. The Gibbons-Hawking term is the trace of the extrinsic curvature of the boundary, $K = K^\mu{}_\mu$, where $K_{\mu\nu} \equiv \nabla_\mu n_\nu$. The Einstein-Hilbert action with the Gibbons-Hawking term is given by

$$S = \frac{1}{2}\int \mathrm{d}^4x\sqrt{-g}\,R + \oint \mathrm{d}^3x\sqrt{-\gamma}\,K = \frac{1}{2}\int_M R + \int_{\partial M} K. \tag{1.82}$$

The variation of (1.82) yields

$$\delta S = \frac{1}{2}\int_M G_{\mu\nu}\delta g^{\mu\nu} + \int_{\partial M}\left(\frac{1}{2}n_\alpha v^\alpha + \delta K + \frac{1}{2}K\gamma^{\mu\nu}\delta\gamma_{\mu\nu}\right), \tag{1.83}$$

where $v_\alpha n^\alpha$ is the surface contribution from the variation of the Ricci scalar, defined in (1.80). With the definitions of the extrinsic curvature tensor and $n^\mu n_\mu = 1$, one can deduce the following identities:

$$\nabla_\alpha \gamma_{\mu\nu} = -2K_{\alpha(\mu}n_{\nu)}, \qquad \delta n_\mu = \frac{1}{2}n_\mu n^\alpha n^\beta \delta g_{\alpha\beta}, \tag{1.84a}$$

$$\delta\gamma_{\mu\nu} = \delta g_{\mu\nu} \pm n_\mu \delta n_\nu \pm n_\nu \delta n_\mu. \tag{1.84b}$$

The variation of the trace $K = \gamma^{\mu\nu} K_{\mu\nu}$ yields

$$\begin{aligned}\delta K &= K_{\mu\nu}\delta\gamma^{\mu\nu} - \gamma^{\mu\nu} n_\alpha \delta\Gamma^\alpha{}_{\mu\nu} + \gamma^{\mu\nu}\nabla_\nu \delta n_\mu \\ &= -K^{\mu\nu}\delta g_{\mu\nu} - \gamma^{\mu\nu} n^\alpha \left(\nabla_{(\mu}\delta g_{\nu)\alpha} - \frac{1}{2}\nabla_\alpha \delta g_{\mu\nu}\right) + \frac{1}{2} K n^\alpha n^\beta \delta g_{\alpha\beta}. \end{aligned} \tag{1.85}$$

These expressions can be combined and inserted into the integrand of the surface term of (1.83), yielding

$$\begin{aligned}\frac{1}{2} n_\alpha v^\alpha + \delta K + \frac{1}{2} K \gamma^{\mu\nu}\delta\gamma_{\mu\nu} = &-\frac{1}{2}\gamma^{\mu\nu} n^\alpha \nabla_{(\mu}\delta g_{\nu)\alpha} \\ &+\frac{1}{2}(K n^\mu n^\nu + K\gamma^{\mu\nu} - 2K^{\mu\nu})\delta g_{\mu\nu}. \end{aligned} \tag{1.86}$$

Thus, the variation of the Einstein-Hilbert action with a Gibbons-Hawking term yields

$$2\delta S = \int_M G_{\mu\nu}\delta g^{\mu\nu} - \int_{\partial M}\left(\gamma^{\mu\nu} n^\alpha \nabla_{(\mu}\delta g_{\nu)\alpha} - (K n^\mu n^\nu + K\gamma^{\mu\nu} - 2K^{\mu\nu})\delta g_{\mu\nu}\right). \tag{1.87}$$

All we have to impose now is that the metric does not vary on the boundary: $\delta g_{\mu\nu}|_{(\partial M)} = 0$. This immediately removes all but the first term in the surface integral. A corollary of this condition is that the transverse derivative term vanishes, which completely removes the surface integral. Thus, after imposing $\delta g_{\mu\nu}|_{(\partial M)} = 0$, the variation of the Einstein-Hilbert action with the Gibbons-Hawking term (1.82) yields

$$\delta S = \frac{1}{2}\int_M G_{\mu\nu}\delta g^{\mu\nu}. \tag{1.88}$$

The data $\delta g_{\mu\nu}|_{(\partial M)} = 0$, which follows from GR with the Gibbons-Hawking term, is a smaller amount of data than that required from GR alone, to produce the same field equations. Most of the time the Gibbons-Hawking term is implicitly assumed to be present when the variational principle is used to compute field equations.

1.3 Cosmology

Over the past century the view of the universe that we find ourselves in has exploded in scale. One hundred years ago, humanity thought that the Milky Way Galaxy was alone in the universe. In 1916 Albert Einstein published his theory of gravity, General

Relativity, which revolutionized our understanding of the nature of spacetime and provided an explanation of various anomalies observed in the orbit of Mercury. In 1925 came the realization that the observed "nebulae" were in fact galaxies in their own right. In 1929, Edwin Hubble observed the recession speeds of galaxies as a function of their distance away from us, providing the rather startling conclusion that the universe is expanding. Over the next 70 years the existence of dark matter was inferred and the cosmic microwave background was detected. In 1998 it was discovered that the universe is actually apparently accelerating in its expansion.

Cosmology is the grandiose study of the universe as a whole, and is tasked with the attempt to provide an understanding of the origin of the universe, how the constituents of the universe affect its geometry and evolution, and how structures form in the universe. The universe is an incredibly complicated object whose constituents vary in size from the quantum mechanically small to the unimaginably large. At first sight it would appear a fruitless task to attempt to construct a metric for the universe which is capable of incorporating information about objects of such a wide range of scales. Indeed, this task is fruitless. However, if assertions are made about the statistical nature in which the content of the universe is distributed on the largest scales, a metric is able to be constructed and used to solve the field equations. The assertion which makes this possible is simply that we are not in a special place in the universe, meaning that the content of the universe is homogeneous and isotropic on the very largest scales. These are the scales where the individual constituents of the universe (galaxies, clusters etc.) become so small they are able to be approximated as being the elements of a homogeneous isotropic fluid. We then speak of a cosmological fluid, where the fluid is a mixture of different components each having different gravitational properties.

Of course, modeling the universe as a homogeneous isotropic fluid is an approximation which is well motivated on statistical grounds. By analogy, a sponge is homogeneous if one zooms out far enough, and one certainly does not care about the motion of individual molecules of water if one wants to understand the gross flow of water. However, if we want to model the universe on slightly smaller scales and understand how structures in the universe form and evolve, the fluid description breaks down and we must provide an alternative description. We now imagine that the localized constituents of the universe are immersed within the cosmological fluid and that the dynamics of the fluid will effect the dynamics of these localized constituents; for instance, a different background dynamic will produce a different distribution of the constituents.

We now provide a brief overview of how the cosmological background is modeled within the framework of General Relativity, and how the different constituents in the universe affect its geometry. We then go on to show how to deal with small perturbations about this background due to structure in the universe.

1.3.1 The Cosmological Background

The cosmological principle states that we do not inhabit a special place in the universe; the implication is that the universe is homogeneous and isotropic. Homogeneity implies an invariance of the metric under translation, and isotropy an invariance under rotation. The Friedmann-Robertson-Walker (FRW) metric is constructed under these principles to describe the background geometry of the universe, and is given by

$$\mathrm{d}s^2 = -\mathrm{d}t^2 + a^2(t)\left(\mathrm{d}x^2 + \mathrm{d}y^2 + \mathrm{d}z^2\right), \tag{1.89}$$

where for simplicity we have assumed that the spatial sections have zero curvature. The only "free" function in this metric is the scale factor, $a(t)$: once that is specified, the global geometry of the universe is specified. In General Relativity, the evolution of the scale factor is set by Einstein's field equations. It is common to write down the FRW metric in conformal time via the coordinate transformation $\mathrm{d}t = a\mathrm{d}\tau$. This provides a metric which is conformally flat, where the conformal factor is the scale factor: $g_{\mu\nu} = a^2(\tau)\eta_{\mu\nu}$. The FRW metric in conformal time is

$$\mathrm{d}s^2 = a^2(\tau)\left(- \mathrm{d}\tau^2 + \mathrm{d}x^2 + \mathrm{d}y^2 + \mathrm{d}z^2\right). \tag{1.90}$$

The FRW metric (1.89) is used to construct geometrical quantities, such as the Christoffel symbols, $\Gamma^{\alpha}{}_{\mu\nu}$, the Ricci scalar, R, and the components of the Einstein tensor, $G^{\mu}{}_{\nu}$. Using (1.89) the non-zero components of these quantities are

$$\Gamma^{i}{}_{j0} = H\delta^{i}{}_{j}, \qquad \Gamma^{0}{}_{ij} = a^2 H\delta_{ij}, \tag{1.91a}$$

$$R = 6\left(H^2 + \frac{\ddot{a}}{a}\right), \qquad G^{0}{}_{0} = 3H^2, \qquad G^{i}{}_{j} = -\left(H^2 + 2\frac{\ddot{a}}{a}\right)\delta^{i}{}_{j}, \tag{1.91b}$$

where the Hubble parameter is defined as $H \equiv \dot{a}/a$, and an overdot denotes derivative with respect to the coordinate time t. One can immediately notice that the Ricci scalar has a singularity at $a(t) = 0$: this is called the big bang. The trace of the extrinsic curvature (1.36) is given by $K = 3H$.

Since the Einstein tensor for the FRW metric is diagonal, so too must be the energy-momentum tensor. The content of the universe is modeled as a perfect fluid, whose energy-momentum tensor is given by

$$T^{\mu}{}_{\nu} = \mathrm{diag}(\rho, P, P, P), \tag{1.92}$$

where $\rho = \rho(t)$ is the total energy density and $P = P(t)$ the total pressure of the "cosmological fluid". Equating the components of the Einstein tensor to the relevant component of the energy-momentum tensor yields the Einstein equations in an FRW background:

$$H^2 = \frac{8\pi G}{3}\rho, \qquad \frac{\ddot{a}}{a} = -\frac{4\pi G}{3}\left(\rho + 3P\right). \tag{1.93a}$$

The first of these equations is the Friedmann equation and the second is the Raychaudhuri equation. The final equation we must calculate in the FRW background is the conservation equation, $\nabla_\mu T^\mu{}_\nu = 0$, which yields

$$\dot{\rho} = -3H(\rho + P). \tag{1.93b}$$

The cosmological Eqs. (1.93) are not yet closed: there is no rule for the evolution for the pressure P. The usual way to close the equations is via an equation of state,

$$P = w\rho, \tag{1.94}$$

which specifies the pressure of a substance in terms of its density. The equation of state of baryonic matter, radiation and a cosmological constant are respectively

$$w_{\rm m} = 0, \qquad w_{\rm r} = \frac{1}{3}, \qquad w_\Lambda = -1. \tag{1.95}$$

When the universe is dominated by different components, the evolution of the scale factor, $a(t)$, is different. To obtain the different behaviour, first the fluid Eq. (1.93b) is integrated to obtain the dependance of the energy density on scale factor when the universe is dominated by a component with a particular equation of state, yielding

$$\rho(a) \propto a^{-3(1+w)}. \tag{1.96}$$

We now use (1.96) in the Friedmann equation to provide an equation describing how the scale factor evolves when the universe is dominated by a component with equation of state w:

$$a(t) \propto t^{\frac{2}{3(1+w)}}. \tag{1.97}$$

The density and scale factor evolution for a universe dominated by matter, radiation and a cosmological constant is

$$w_{\rm m} = 0 \qquad \Rightarrow \qquad \rho_{\rm m} \propto a^{-3}, \qquad a \propto t^{2/3}, \tag{1.98a}$$

$$w_r = \frac{1}{3} \qquad \Rightarrow \qquad \rho_{\rm r} \propto a^{-4}, \qquad a \propto t^{1/2}, \tag{1.98b}$$

$$w_\Lambda = -1 \quad \Rightarrow \quad \rho_\Lambda \propto \text{const}, \quad a \propto e^{Ht}. \tag{1.98c}$$

By defining the critical density of the universe, $\rho_c \equiv 3H_0^2/8\pi G$ to be the density required to produce a flat universe, and the density fractions, $\Omega_X \equiv \rho_X/\rho_c$, the Friedmann equation can be written as

$$\left(\frac{H}{H_0}\right)^2 = \frac{\Omega_r}{a^4} + \frac{\Omega_m}{a^3} + \Omega_\Lambda. \tag{1.99}$$

This can be used in conjunction for the equation of a light-ray (radial null geodesic), $dt = a(t)dr$, to obtain the distance r to an object at scale factor a,

$$r(a) = \frac{1}{H_0}\int_a^1 \frac{da}{\sqrt{\Omega_r + a\Omega_m + a^4\Omega_\Lambda}}. \tag{1.100}$$

By measuring the angle θ subtended by an object of physical length l, the angular diameter distance is given by

$$d_A(a) = ar(a) = \frac{l}{\theta}. \tag{1.101}$$

So, by measuring θ for objects of known intrinsic lengths, one can obtain information about the cosmological parameters. It is common to use redshift $z \equiv 1/a - 1$.

1.3.2 Cosmological Perturbations

The FRW metric (1.89) is a model for the geometry of the universe on the largest possible scales, where the content of the universe was taken to be a homogeneous and isotropic fluid. What this means, however, is that the metric is not capable of encoding information about the gravitational effects of localized structures in the universe. To be able to do this the gravitational field corresponding to this structure is assumed to be a small perturbation to the metric. There are many classic texts and reviews on cosmological perturbation theory in the literature, e.g. [10, 25–29].

The metric perturbed about a conformally flat FRW background is written as

$$ds^2 = a^2(t)(\eta_{\mu\nu} + h_{\mu\nu})dx^\mu dx^\nu. \tag{1.102}$$

The perturbations $h_{\mu\nu}$ can be space and time dependent. We parameterize the $h_{\mu\nu}$ as

$$ds^2 = a^2(\tau)\Big[-(1+2\Phi)d\tau^2 + 2N_i dx^i d\tau + (\delta_{ij} + h_{ij})dx^i dx^j\Big], \tag{1.103}$$

where $\Phi = \Phi(\tau, \mathbf{x})$ is interpreted as the Newtonian gravitational potential, $N_i = N_i(\tau, \mathbf{x})$ as the lapse function and $h_{ij} = h_{ij}(\tau, \mathbf{x})$ as the spatial metric perturbations. Because the metric is perturbed, all geometrical quantities computed from the metric will also become perturbed from their value in an FRW background, as given in (1.60). In General Relativity, not all of the components of the metric (1.103) satisfy dynamical equations of motion and some of the components of the metric are merely constraints (this is nicely illustrated in the ADM formalism [30]). These correspond to gauge freedom in the theory, and so it is usual to make coordinate system choices to remove this freedom; there are four components which can be removed. Two of the popular choices are the *synchronous gauge* $\Phi = 0$, $N_i = 0$ and the *conformal Newtonian gauge* $N_i = 0$, $h_{ij} = -2\Psi\delta_{ij}$. In the conformal Newtonian gauge only scalar perturbations to the metric can be studied (the perturbations are in the form of two gravitational potentials, Φ, Ψ).

Armed with the parameterization of the perturbed metric (1.103) we must provide field equations governing the metric perturbations, and how the dynamics of the perturbations are determined by the content (which must also be perturbed). In General Relativity these governing equations are given by Einstein's field equations expanded to linear order in the perturbations to the metric,

$$\delta G_{\mu\nu} = 8\pi G \delta T_{\mu\nu}. \tag{1.104}$$

The metric (1.103) is used to calculate the components of the perturbed Einstein tensor, δG^{μ}_{ν}; we are working in conformal coordinates so over-dots are derivatives with respect to conformal time and $\mathcal{H} \equiv \dot{a}/a$ is the conformal time Hubble parameter. In the synchronous gauge, the components of $\delta G^{\mu}{}_{\nu}$ are [31]

$$a^2\delta G^0{}_0 = -\mathcal{H}\dot{h} + \frac{1}{2}\nabla^2 h - \frac{1}{2}\partial_i\partial_j h^{ij}, \tag{1.105a}$$

$$2a^2\delta G^0{}_i = \partial_i\dot{h} - \partial_j\dot{h}^j{}_i, \tag{1.105b}$$

$$\begin{aligned} 2a^2\delta G^i{}_j = &-\left[\ddot{h} + 2\mathcal{H}\dot{h} - \nabla^2 h + \partial_k\partial_l h^{kl}\right]\delta^i{}_j \\ &+\ddot{h}^i{}_j + 2\mathcal{H}\dot{h}^i{}_j - \nabla^2 h^i{}_j + \partial^i\partial_k h^k{}_j + \partial_j\partial_k h^{ik} - \partial^i\partial_j h. \end{aligned} \tag{1.105c}$$

In the conformal Newtonian gauge the components of $\delta G^{\mu}{}_{\nu}$ are [32, 33]

$$a^2\delta G^0{}_0 = 2\left[-\nabla^2\Psi + 3\mathcal{H}(\dot{\Psi} + \mathcal{H}\Phi)\right], \tag{1.106a}$$

$$a^2\delta G^0{}_i = -2\nabla_i(\dot{\Psi} + \mathcal{H}\Phi), \tag{1.106b}$$

$$a^2\delta G^i{}_j = 2\Big[\ddot{\Psi} + (\mathcal{H}^2 + 2\dot{\mathcal{H}})\Phi + (\dot{\Phi} + 2\dot{\Psi})\mathcal{H}\Big]\delta^i{}_j + \Big[\nabla^i\nabla_j - \delta^i{}_j\nabla^2\Big](\Psi - \Phi). \tag{1.106c}$$

It is convenient to write the components of the perturbed energy momentum tensor as

$$\delta T_{\mu\nu} = \delta\rho u_\mu u_\nu + \delta P \gamma_{\mu\nu} + 2(\rho + P)v_{(\mu}u_{\nu)} + P\Pi_{\mu\nu}, \tag{1.107}$$

where u_μ is a time-like unit vector $\gamma_{\mu\nu}$ a space-like tensor, v_μ a space-like vector and $\Pi_{\mu\nu}$ a spatial transverse-traceless tensor. These vectors and tensors satisfy

$$u^\mu u_\mu = -1, \quad v^\mu u_\mu = 0, \quad u^\mu\gamma_{\mu\nu} = 0, \quad \Pi^\mu{}_\mu = 0, \quad u^\mu\Pi_{\mu\nu} = 0. \tag{1.108}$$

The component $\delta\rho$ is interpreted as the perturbed density, δP as the perturbed pressure, v_μ as the velocity field and $\Pi_{\mu\nu}$ as the shear tensor. It is then usual to define the density contrast, $\delta \equiv \delta\rho/\rho$.

To be able to solve the perturbed gravitational field equations the equations of motion governing the perturbed sources, $\delta T^\mu{}_\nu$, must be provided. The relevant equation of motion is the perturbed conservation equation, $\delta(\nabla_\mu T^\mu{}_\nu) = 0$, which expands to yield

$$\nabla_\mu\delta T^\mu{}_\nu + \delta\Gamma^\mu{}_{\mu\alpha}T^\alpha{}_\nu - \delta\Gamma^\alpha{}_{\mu\nu}T^\mu{}_\alpha = 0. \tag{1.109}$$

Using (1.103) to parameterize the perturbed metric and (1.107) to parameterize the perturbed content, the time and space parts of the perturbed conservation equation respectively yield

$$\dot{\delta} = -(1 + w)\left(\nabla_j v^j + \frac{1}{2}\dot{h}\right) - 3\mathcal{H}\left(\frac{\delta P}{\delta\rho} - w\right)\delta, \tag{1.110a}$$

$$\dot{v}_i = -\mathcal{H}(1 - 3w)v_i + \big(\partial_i\Phi - \mathcal{H}N_i\big) + \frac{1}{\rho(1 + w)}\partial_i\delta P - \frac{w}{1 + w}\nabla_j\Pi^j{}_i, \tag{1.110b}$$

where we have taken $\dot{w} = 0$ for simplicity. As noted for the background cosmological equations, the perturbed equations are not closed evolution equations until δP, $\Pi^i{}_j$ are specified by, for example, equations of state. It is common to write these evolution equations in Fourier space. The scalar Fourier decomposition is

$$\mathrm{i}k_j v^j = -(\rho + P)k^2\theta, \qquad \mathrm{i}k^j\partial_k\delta P = -k^2\delta P, \tag{1.111a}$$

$$ik^j \partial_i \Pi^i{}_j = -k_i k^j \Pi^i{}_j, \qquad \Pi^i{}_j = (\hat{k}^i \hat{k}_j - \tfrac{1}{3}\delta^i{}_j)\Pi^S, \tag{1.111b}$$

where $k_i = k\hat{k}_i$. The perturbed conservation equations for scalar perturbations in Fourier space in the synchronous gauge are

$$\dot{\delta} = -(1+w)\left(-k^2\theta + \frac{1}{2}\dot{h}\right) - 3\mathcal{H}w\Gamma, \tag{1.112a}$$

$$\dot{\theta} = -\mathcal{H}(1-3w)\theta - \frac{w}{1+w}\left[\delta + \Gamma - \frac{2}{3}\Pi^S\right], \tag{1.112b}$$

and in the Conformal Newtonian gauge,

$$\dot{\hat{\delta}} = -(1+w)\left(-k^2\hat{\theta} - 3\dot{\phi}\right) - 3\mathcal{H}w\Gamma, \tag{1.113a}$$

$$\dot{\hat{\theta}} = -\mathcal{H}(1-3w)\hat{\theta} - \psi - \frac{w}{1+w}\left[\hat{\delta} + \Gamma - \frac{2}{3}\Pi\right]. \tag{1.113b}$$

1.3.3 The Dark Side

By observing different cosmological and astrophysical systems (such as galaxy rotation curves, the cosmic microwave background, distances to supernovae, gravitational lensing and structure formation) and analyzing the data within the standard cosmological framework (GR to provide the gravitational theory and an FRW metric for the geometry), observers and theorists have come to the rather startling conclusion that about 96 % of the universe is comprised of two forms of matter of which we have absolutely zero comprehension: *dark matter* and *dark energy* [34–36] (see also [37, 38]). This has rather profound implications for our understanding of a cosmological model. Dark matter was introduced to fix galaxy rotation curves and for structure formation, whilst dark energy was introduced as an explanation for the origin of the apparent observed acceleration of the universe. Within the framework of the standard GR + FRW cosmological paradigm, acceleration $\ddot{a} > 0$ only occurs whenever the energy density and pressure of the content satisfy $\rho + 3P < 0$, which requires the equation of state $w < -\frac{1}{3}$.

Dark matter has the same gravitating properties as Baryonic matter, in that it has negligible pressure, but it differs in that dark matter does not appear to interact with electromagnetic fields. Thus, our only hint as to the existence of dark matter comes from its gravitational effects.

Dark energy is rather more alien in nature, and apparently needs to be included as the currently dominant gravitating species in the content of the universe. The observation of apparent acceleration appears to suggest that dark energy must be a substance with equation of state $w_{\rm de} \approx -1$. No known substance has such an equation of state. In fact, the substance with this equation of state is highly exotic because it acts as some sort of "anti-gravity", gravitationally repelling objects.

The important point to realize is that when the observational data are analyzed through a specific gravitational and cosmological theory, 96 % of the universe is required to be invented. This raises some serious and fundamental questions:

- What is the dark energy and dark matter?
- Is it appropriate to use the FRW metric on cosmological scales?
- Is it appropriate to use GR on cosmological scales?

The way to "understand" these problems is to be clear about what it is that is assumed to be *a priori* true, and what it is that is deduced from observations given what is assumed to be true.

One possibility is that we could assume that GR is the appropriate gravitational theory and FRW is an accurate geometrical model for cosmological scales, in which case we must start to ask "what is the dark matter and the dark energy?". There are many theories in the literature, for example cosmological constant, quintessence, elastic dark energy, cold dark matter, warm dark matter and axions. The full list of theories is large and has been extensively studied in the literature (see, e.g. [39–43]). There are a few popular theories in the literature which provide a substance with an equation of state $w < -\frac{1}{3}$. The simplest is the cosmological constant, Λ, whose equation of state is $w_\Lambda = -1$; there is no variation in time or space of the energy density or equation of state of Λ. The simplest alternative to Λ is the dynamical minimally coupled homogeneous "quintessence" scalar field ϕ [44]. The Lagrangian density and energy-momentum tensor of quintessence is

$$\mathcal{L}_{(\phi)} = \frac{1}{2}\nabla_\mu\phi\nabla^\mu\phi - V(\phi), \qquad T^{\mu\nu}_{(\phi)} = \nabla^\mu\phi\nabla^\nu\phi - g^{\mu\nu}\mathcal{L}, \tag{1.114}$$

where $V = V(\phi)$ is the potential function. Because the scalar field is homogeneous, we have $\phi = \phi(t)$ and the energy density, $\rho_{(\phi)}$, and pressure, $P_{(\phi)}$, become

$$\rho_{(\phi)} = \frac{1}{2}\dot{\phi}^2 + V, \qquad P_{(\phi)} = \frac{1}{2}\dot{\phi}^2 - V, \tag{1.115}$$

which allows the value of the equation of state parameter of the quintessence field to be computed,

$$w_\phi = \frac{P_{(\phi)}}{\rho_{(\phi)}} = \frac{\frac{1}{2}\dot{\phi}^2 - V}{\frac{1}{2}\dot{\phi}^2 + V} \tag{1.116}$$

If the content of the universe were to be dominated by the quintessence scalar field then the acceleration condition $\rho_\phi + 3P_\phi < 0$ becomes $V > \dot{\phi}^2$. That is, whenever the kinetic energy of the scalar field is less than its potential energy the quintessence field can drive an acceleration of the universe. A simple generalization of the quintessence field is k-essence [45], The Lagrangian density and energy-momentum tensor of k-essence is given by

$$\mathcal{L} = \mathcal{L}(\mathcal{X}, \phi), \qquad T_{\mu\nu} = \mathcal{L}_{,\mathcal{X}} \nabla_\mu \phi \nabla_\nu \phi - g_{\mu\nu} \mathcal{L}, \tag{1.117}$$

where the kinetic term, $\mathcal{X} \equiv \frac{1}{2} \nabla^\mu \phi \nabla_\mu \phi$. For a homogeneous scalar field, the energy density and pressure are given by

$$\rho_K = \mathcal{L}_{,\mathcal{X}} \dot{\phi}^2 - \mathcal{L}, \qquad P_K = -\mathcal{L}. \tag{1.118}$$

An alternative idea is to still assume that GR is appropriate, but question the validity of applying the FRW metric on cosmological scales, and begin to study inhomogeneous cosmologies [46–50]. We must understand whether or not the effect of localized matter distributions (such as galaxies, or clusters of galaxies) could produce something which we would interpret as being cosmological acceleration if we were ignorant of these matter distributions on cosmological scales. The fact remains that by using the FRW metric we are ignorant of the matter distributions. There are a growing number of studies in the literature on inhomogeneous universe (see, e.g. [51–56]), but this field is nowhere near as developed as the study of modified gravity or dark energy theories.

The third step is to question the validity of GR on cosmological scales. It is the purpose of the rest of this thesis to discuss and unpack this point.

1.4 Modified Gravity

There is a surge in interest in gravity theories that are, in some way, different from General Relativity. These theories are collectively known as *modified gravities*. Studying modified gravities can be motivated in a few different ways. First of all, there is "pure academic interest": attempting to understand the structure of different types of field equations to build a full picture of how gravity works in the broadest sense. Secondly, the discovery of the dark side could be explained by the existence of some new gravitational theory on large scales. We will now simplistically elucidate on a historical precedent for this second point.

A few hundred years ago, humanity had almost a complete ignorance of how gravity worked, or how to determine the force a body felt due to a distribution of mass. Over time a series of earth based experiments began to be envisaged and constructed which enabled an empirical law to be written down which links the force due to to gravity, F, of two bodies separated by a distance r. This is of course the inverse square law: $F \propto 1/r^2$. Careful experimentation also enabled the constant of proportionality to be determined: $F = GMm/r^2$. The constant G is Newton's gravitational constant, M and m are the masses of the two bodies separated by a distance r, and F is the force of gravity between the bodies, pulling the bodies together. This equation encapsulates the essence of Newton's gravitational theory.

Once Newtonian gravity was formulated, more earth based experiments were constructed to test Newton's gravitational theory. When all the experiments returned the same results and agreed with the predictions from Newtonian gravity, experimenters and theorists grew in confidence that the theory is correct *in the physical scenarios in which the law was empirically deduced.* This confidence was then extended and Newtonian gravity was used to generate predictions for physical scenarios in which the law was not empirically deduced. For instance, the orbits of the planets in the Solar System or the behaviour of the paths of light next to the Sun.

Over time, theoretical predictions were extracted from Newtonian gravity and tested to high precision. As of the mid-1800s, the results can be crudely summarized as follows: the observed orbits of all planets *except two* agreed precisely with the predictions from Newtonian gravity. The closest planet to the Sun, Mercury, had a "wobble" in its orbit which could not be explained by applying Newton's gravitational theory to the Mercury-Sun system, and the orbit of Uranus had orbital discrepancies.

One of the explanations for these discrepancies was to invent two extra planets. Vulcan was predicted to exist close to the Sun, with the properties of Vulcan being chosen such that a Newtonian calculation of the Sun-Mercury-Vulcan system returned an answer which explained the observed wobble of Mercury. Neptune was predicted to exist in the vicinity of Uranus, whose properties were also chosen in such a way as to fix observed discrepancies.

Neptune was subsequently searched for and found, but Vulcan was not found (and indeed does not exist). The reason that Vulcan does not exist is actually because it was the expectation that Newton's understanding of gravity holds so close to the Sun which was false. Newtonian gravity was deduced on the earth which is a low-curvature environment, and fails in high-curvature environments such as those around the Sun. Einstein constructed his theory of General Relativity and used it to predict the dynamics of the Sun-Mercury system (where Newtonian gravity had failed). The GR prediction matched observation spectacularly well (and still provided the same predictions for the other planets as Newtonian gravity), without requiring Vulcan to be invented. The key was that Newtonian gravity was not designed to be applied in extreme gravitational fields, whereas GR was.

When we assume that the universe is comprised of matter we understand (photons, baryons, neutrinos etc.), and that the cosmological principle holds, the predictions of GR and observations are completely incompatible. Only when we *impose* that dark matter and dark energy is the dominant contribution to the matter content do the predictions and experimental observations become compatible.

This story provides a powerful philosophical point: applying the Newtonian gravitational theory to a system it was never tested in had the consequence of requiring an entire planet to be invented. Today, applying GR to cosmological scales has meant that 96 % of the universe must be invented. The analogy should now be obvious:

perhaps GR is not the gravitational theory that should be applied on cosmological scales. Instead, perhaps some new "modified" gravity theory should.

Applying the framework of General Relativity to a system as massive as a galaxy, or indeed to the whole universe, is entirely an extrapolation of the assumption of the validity of GR as *the* gravitational theory. The discrepancy between predictions from the GR and the "standard matter" theoretical model and experimental observations has motivated a search for new gravitational theories. Preferably, this new theory will produce predictions about our universe which are compatible with experimental observations without the need for introducing dark energy (or at least, dark energy will not need to be the dominant matter contribution). However, it is necessary that this new gravitational theory has Newtonian and GR limits so that in the regimes where we know GR holds, the new gravitational theory satisfies current constraints from e.g. tests of the equivalence principle via torsion balances [57] or lunar laser ranging [58, 59], frame dragging around the earth [60] or in the strong field limit around binary pulsar systems [61].

It is worth recapping that GR provides a specific set of equations for determining the components of the metric, $g_{\mu\nu}$, from a matter content, through Einstein's field equations,

$$G_{\mu\nu} = 8\pi G T_{\mu\nu}. \tag{1.119}$$

This field equation can be derived from the Einstein-Hilbert action,

$$S_{\rm GR} = \int d^4x \sqrt{-g}\mathcal{L}_{\rm EH} + S_{\rm m}[g_{\mu\nu}, \chi], \qquad \mathcal{L}_{\rm EH} = R, \tag{1.120}$$

where R is the Ricci scalar and $S_{\rm m}$ is the action of the matter fields χ. The popular way to construct a modified gravity theory is to provide extra terms or extra fields to the gravitational Lagrangian density. In general we will use an action

$$S = \int d^4x \sqrt{-g}\,\mathcal{L}, \tag{1.121}$$

where $\mathcal{L}$ is the gravitational Lagrangian density, and contains geometrical terms (such as the Ricci scalar), any extra gravitational mediators and all matter fields (which are usually collected into a matter Lagrangian $\mathcal{L}_{\rm m}$). Modified gravity theories provide alternative field equations, meaning that the metric will respond differently to the same matter content. As remarked in Sect. 1.2.4, a well posed variational principle requires some specification of the behaviour of the theory on the boundary (e.g. the Gibbons-Hawking term in GR). In the literature it is usually "assumed" that such a boundary term exists; only very recently [62] were the explicit forms of the boundary conditions worked out for general scalar-tensor theories.

1.4.1 A Catalogue of Modified Gravity Theories

There are a large number of modified gravity theories in the literature but here we will provide a brief overview of some popular modified gravity theories; for a recent extensive review see [63]. Most modified gravity theories can be classified broadly into two categories: (a) only the metric is used to mediate gravity and (b) extra mediators of gravity are introduced; this classification is not entirely unambiguous (for example, metric-only theories can be rewritten as scalar-tensor theories), but it is useful for our purposes of providing a brief review.

1.4.1.1 Tensor Theories

To begin our brief review of modified gravity theories, we will consider theories which only use the metric to mediate gravity: these are called *tensor theories*. The Lagrangian of a tensor theory is of the general form

$$\mathcal{L} = \mathcal{L}(g_{\mu\nu}, \partial_\alpha g_{\mu\nu}, \partial_\alpha \partial_\beta g_{\mu\nu}, \ldots). \tag{1.122}$$

The simplest tensor theory is the cosmological constant [64–66], whereby the gravitational Lagrangian density is

$$\mathcal{L} = R - 16\pi G \mathcal{L}_{\mathrm{m}} - \Lambda, \tag{1.123}$$

whose gravitational field equations are

$$G_{\mu\nu} = 8\pi G T_{\mu\nu} + \Lambda g_{\mu\nu}. \tag{1.124}$$

A simple way to extend the gravitational Lagrangian is the manner in which the Ricci scalar appears (in the Einstein-Hilbert action the Ricci scalar appears linearly). These are called $f(R)$ theories [67–69]. The gravitational action is modified to include an arbitrary function of the Ricci scalar:

$$\mathcal{L} = R + f(R) - 16\pi G \mathcal{L}_{\mathrm{m}}, \tag{1.125}$$

whose gravitational field equations are

$$G_{\mu\nu} = 8\pi G T_{\mu\nu} + \left(R_{\mu\nu} + g_{\mu\nu} \nabla^\alpha \nabla_\alpha - \nabla_\mu \nabla_\nu\right) f' - \frac{1}{2} f g_{\mu\nu}, \tag{1.126}$$

where $f' \equiv \mathrm{d}f/\mathrm{d}R$. One can then begin to envisage a generalization of such gravitational theories, for example where arbitrary functions of curvature invariants appear,

$$\mathcal{L} = R + f(R, R^{\mu\nu} R_{\mu\nu}, R^{\mu\nu\alpha\beta} R_{\mu\nu\alpha\beta}) - 16\pi G \mathcal{L}_{\mathrm{m}}. \tag{1.127}$$

A further example of these types of theories are the Gauss-Bonnet gravities [70–73], where the gravitational Lagrangian contains the Gauss-Bonnet term,

$$\mathcal{L} = \mathcal{L}(R, \mathcal{G}), \qquad \mathcal{G} = R^2 - 4R^{\mu\nu}R_{\mu\nu} + R^{\mu\nu\alpha\beta}R_{\mu\nu\alpha\beta}. \tag{1.128}$$

1.4.1.2 Scalar-Tensor Theories

The next type of modified gravity theory we consider contains extra mediators of gravity. The simplest such "extra mediator" is a scalar field, ϕ, so that we talk of scalar-tensor gravity theories. The Lagrangian of a scalar-tensor theory will be of the form

$$\mathcal{L} = \mathcal{L}(g_{\mu\nu}, \partial_\alpha g_{\mu\nu}, \partial_\alpha\partial_\beta g_{\mu\nu}, \ldots, \phi, \partial_\mu\phi, \ldots). \tag{1.129}$$

A simple example of theories of this class uses a function of a dynamical scalar field to couple to the Ricci scalar in the Lagrangian,

$$\mathcal{L} = \mathfrak{a}(\phi)R - \mathfrak{b}(\phi)\nabla^\mu\phi\nabla_\mu\phi - 2V(\phi), \tag{1.130}$$

where $\mathfrak{a}$, $\mathfrak{b}$, V are arbitrary functions of the scalar field (V is the potential term). The field equations for the metric scalar field are respectively given by

$$\mathfrak{a}G_{\mu\nu} + (g_{\mu\nu}\Box - \nabla_\mu\nabla_\nu)\mathfrak{a} = 8\pi G T_{\mu\nu} + T^{(\phi)}_{\mu\nu}, \tag{1.131a}$$

$$\mathfrak{b}\Box\phi = V' - \frac{1}{2}\mathfrak{a}\nabla^\mu\phi\nabla_\mu\phi - \frac{1}{2}\mathfrak{a}'R, \tag{1.131b}$$

where for convenience we have defined

$$T^{(\phi)}_{\mu\nu} \equiv \mathfrak{b}\left[\nabla_\mu\phi\nabla_\nu\phi - \frac{1}{2}g_{\mu\nu}\left(\nabla^\alpha\phi\nabla_\alpha\phi - 2\frac{V}{\mathfrak{b}}\right)\right]. \tag{1.132}$$

Brans-Dicke theory [74] is a specific example, where the functions $\mathfrak{a}$, $\mathfrak{b}$ are taken to be

$$\mathfrak{a}(\phi) = \phi, \qquad \mathfrak{b}(\phi) = \frac{\omega(\phi)}{\phi}. \tag{1.133}$$

There has been a recent surge in interest of theories whose Lagrangians contain two derivatives of a scalar field, but whose field equations are at most of second order (naively, a Lagrangian with two derivatives has fourth order field equations). An example is the kinetic gravity braiding theory [75–77],

$$\mathcal{L}_{\rm KGB} = K(\phi, \mathcal{X}) + G(\phi, \mathcal{X})\Box\phi, \tag{1.134}$$

where $\Box\phi \equiv \nabla^{\mu}\nabla_{\mu}\phi$. The theory is called "braided" because the Ricci tensor explicitly enters into the equation of motion for the scalar field in the process of making the field equations second order. Horndeski's theory [78] is constructed to be the most general scalar-tensor theory in 4D with second order field equations, and has recently been rediscovered and studied by a number of authors [63, 79–82]. The covariant galileon theory [63, 83–86], is another example of a theory whose Lagrangian contains the metric, curvature tensors, first and second derivatives of the scalar field. The theory is constructed so that the field equations contain at most second order in derivatives of the metric (the galileon and Horndeski theories have been shown to be equivalent to each other [87]). The covariant galileon Lagrangian is given by

$$\mathcal{L} = \sum_i c_i \mathcal{L}_i, \tag{1.135}$$

where c_i are dimensionless constants and the Lagrangians $\mathcal{L}_i$ are purely "kinetic" functions of the galileon scalar field π, given by

$$\mathcal{L}_1 = \frac{1}{2}\nabla_{\mu}\pi\nabla^{\mu}\pi, \qquad \mathcal{L}_2 = \frac{1}{2M^3}\nabla_{\mu}\pi\nabla^{\mu}\pi\Box\pi, \tag{1.136a}$$

$$\mathcal{L}_3 = \frac{1}{2M^6}\nabla_{\mu}\pi\nabla^{\mu}\pi\left[2(\Box\pi)^2 - 2\nabla^{\mu}\nabla^{\nu}\pi\nabla_{\mu}\nabla_{\nu}\pi - \frac{1}{2}R\nabla_{\mu}\pi\nabla^{\mu}\pi\right], \tag{1.136b}$$

$$\begin{aligned}\mathcal{L}_4 = \frac{1}{2M^9}\nabla_{\mu}\pi\nabla^{\mu}\pi\Big[&(\Box\pi)^3 - 3(\Box\pi)\nabla^{\mu}\nabla^{\nu}\pi\nabla_{\mu}\nabla_{\nu}\pi \\ &+2\nabla^{\nu}\nabla_{\mu}\pi\nabla^{\rho}\nabla_{\nu}\pi\nabla^{\mu}\nabla_{\rho}\pi - 6\nabla_{\mu}\pi\nabla^{\mu}\nabla^{\nu}\pi\nabla^{\rho}\pi G_{\nu\rho}\Big],\end{aligned} \tag{1.136c}$$

where R, $G_{\mu\nu}$ are the Ricci scalar and Einstein tensor respectively. The field equations of this theory can be found in, e.g.[84]. Recently, highly stringent constraints were placed upon the galileon theory [86, 88] appearing to suggest that the theory is strongly disfavored by cosmological data sets.

The tensor theories we discussed in Sect. 1.4.1.1 can be thought of as being non-minimal scalar-tensor theories. To see this, consider for a moment a scalar-tensor theory given by

$$S = \int d^4\sqrt{-g}\left[R + f(\psi) + f'(\psi)(R - \psi)\right], \tag{1.137}$$

where we note that the Ricci scalar appears linearly and coupled to a non-dynamical scalar field $\phi \equiv f'(\psi)$ where the scalar field ψ has some potential term, $f(\psi)$. By extremizing the variation $\delta S = 0$, one obtains the constraint $f''(\psi) = 0$ or $R = \psi$. Assuming that $f(\psi)$ is chosen such that $f''(\psi) \neq 0$, then substituting $R = \psi$ into the action (1.137) yields

$$S = \int d^4x \sqrt{-g} \left[R + f(R) \right], \tag{1.138}$$

which is indeed the action for an $F(R)$ theory. This idea can be generalized [89] for theories containing other curvature tensors. Let us write the scalar-tensor theory

$$S = \int d^4x \sqrt{-g} \Big[R + f(\phi_1, \phi_2, \phi_3) + f_1(R - \phi_1) + f_2(R^{\mu\nu} R_{\mu\nu} - \phi_2) + f_3(R^{\mu\nu\alpha\beta} R_{\mu\nu\alpha\beta} - \phi_3) \Big], \tag{1.139}$$

where $f_i \equiv \frac{\partial f}{\partial \phi_i}$. Then, if $\det \partial^2 f / \partial \phi_i \phi_j \neq 0$, the equations of motion of the ϕ_i yield a set of conditions $\phi_1 = R, \phi_2 = R^{\mu\nu} R_{\mu\nu}, \phi_3 = R^{\mu\nu\alpha\beta} R_{\mu\nu\alpha\beta}$, so that the action (1.139) is equivalent to

$$S = \int d^4x \sqrt{-g} \left[R + f(R, R^{\mu\nu} R_{\mu\nu}, R^{\mu\nu\alpha\beta} R_{\mu\nu\alpha\beta}) \right], \tag{1.140}$$

which is the action for "just" a tensor theory.

1.4.1.3 Tensor-Vector Theories

One can also introduce vector fields into the mediating sector of gravity. The Lagrangian of a general vector-tensor theory will be of the form

$$\mathcal{L} = \mathcal{L}(g_{\mu\nu}, \partial_\alpha g_{\mu\nu}, \partial_\alpha \partial_\beta g_{\mu\nu}, \ldots, A^\mu, \partial_\nu A^\mu, \ldots). \tag{1.141}$$

The most popular example of these classes of theories is the Einstein-æther theory [90–94], where a vector field A^μ is introduced into the gravitational Lagrangian and whose length is constrained to unity, with a generalized Maxwell kinetic term. The Lagrangian density for the linear Einstein-æther theory is

$$S = -\frac{1}{16\pi G} \int d^4x \sqrt{-g} \left[R + K^{\mu\nu\alpha\beta} \nabla_\mu A_\alpha \nabla_\nu A_\beta + \lambda (A^\mu A_\mu - 1) \right], \tag{1.142}$$

where

$$K^{\mu\nu\alpha\beta} = c_1 g^{\mu\nu} g^{\alpha\beta} + c_2 g^{\mu\alpha} g^{\nu\beta} + c_3 g^{\mu\beta} g^{\nu\alpha} + c_4 A^\mu A^\nu g^{\alpha\beta} \tag{1.143}$$

represents a generalization of the Maxwell kinetic term. The $\{c_i\}$ are constants and λ is a Lagrange multiplier whose role is to enforce the unit time-like constraint.

1.4.1.4 Tensor-Vector-Scalar Theories

A gravitational theory containing scalar, vector and tensor fields can be written as

$$\mathcal{L} = \mathcal{L}(g_{\mu\nu}, \partial_\alpha g_{\mu\nu}, \ldots, \phi, \partial_\mu \phi, \ldots, A^\mu, \partial_\nu A^\mu, \ldots). \tag{1.144}$$

By far the most popular modified gravity theory containing tensor, vector and scalar fields is TeVeS [95–100]. In the simplest version of TeVeS the action is given by

$$S = 16\pi G S_{\mathrm{m}}[g_{\mu\nu}, \chi, \nabla_\mu \chi] + \int \mathrm{d}^4 x \sqrt{-\tilde{g}} \Big[\mathcal{L}_{\mathrm{g}} + \mathcal{L}_{\mathrm{A}} + \mathcal{L}_\phi \Big]. \tag{1.145}$$

There are two metrics: the matter sector only couples to $g_{\mu\nu}$, and the gravitational sector is constructed from $\tilde{g}_{\mu\nu}$, which is related to the matter metric via a "disformal" transformation,

$$g_{\mu\nu} = e^{-2\phi}(\tilde{g}_{\mu\nu} + A_\mu A_\nu) - e^{2\phi} A_\mu A_\nu. \tag{1.146}$$

The Lagrangians that construct the gravitational sector are given by

$$\mathcal{L}_{\mathrm{g}} = \tilde{R}, \tag{1.147a}$$

$$\mathcal{L}_{\mathrm{A}} = -\frac{1}{2} k F^{\mu\nu} F_{\mu\nu} + \lambda (A^\mu A_\mu + 1), \tag{1.147b}$$

$$\mathcal{L}_\phi = -\mu(\tilde{g}^{\mu\nu} - A^\mu A^\nu) \tilde{\nabla}_\mu \phi \tilde{\nabla}_\nu \phi - V(\mu) \tag{1.147c}$$

TeVeS has been shown [101, 102] to be equivalent to a vector-tensor theory where the vector field is not of unit norm (which is the case in æther theories).

1.4.2 Einstein & Jordan Frames

It is possible to find an equivalence between a set of theories where matter is universally coupled to the metric but the Ricci scalar is explicitly coupled to an "extra" scalar field (i.e. the Lagrangian contains a term of the form ϕR, so called scalar-tensor theories) and a set of theories where the scalar field and Ricci scalar are minimally coupled (i.e. there are no longer any terms of the form ϕR), but matter couples to a different metric to that generated by the gravitational field equations. The frame in which matter universally couples to the gravitating metric is called the *Jordan frame* and the frame in which matter couples to a different metric is called the *Einstein frame*. The actions in these two frames take on the schematic form

$$S_{\text{Jordan}} = \int d^4x\sqrt{-g}\,\phi R + S_{\text{m}}[g_{\mu\nu};\chi], \tag{1.148}$$

$$S_{\text{Einstein}} = \int d^4x\sqrt{-g}\left[R + \frac{1}{2}(\nabla\phi)^2\right] + S_{\text{m}}[A^2(\phi)g_{\mu\nu};\chi]. \tag{1.149}$$

Clearly, performing calculations with the Einstein frame action (1.149) will be much simpler than calculations with the Jordan frame action (1.148) because the Ricci scalar appears only linearly and uncoupled to any extra fields in the Einstein frame action. However, theories usually "present themselves" in Jordan frame form (for example, $F(R)$ theories can be shown to be equivalent to a theory in the Jordan frame with non-minimal coupling between the Ricci scalar and a scalar field). So, we would like to obtain an understanding as to how to transform between the Jordan and Einstein frames. The key to understanding lies through conformal transformation of the metric.

This gives rise to the chameleon mechanism [103–105]. The coupling term causes the bare potential that the scalar field feels to be shifted, $V_{\text{eff}}(\phi) = V_{\text{bare}}(\phi) + \rho A(\phi)$, where ρ is the energy density of a non-relativistic source. This causes a screening of the modification of the gravity theory in the vicinity of the source: what this means is that experiments which are designed to test deviations from GR in the vicinity of a planet, for example, will still detect GR.

We will show how to transfer between the Jordan (1.148) and Einstein (1.149) frames. We will start our calculation by stating some useful identities for conformal transformation (see e.g. [7, 19, 106]). Performing a conformal transformation $g_{\mu\nu} \to \tilde{g}_{\mu\nu} = \Omega^2 g_{\mu\nu}$ in n-dimensions, the measure and Ricci scalar transform as

$$\sqrt{-g} \to \sqrt{-\tilde{g}} = \sqrt{-g}\,\Omega^n, \tag{1.150}$$

$$R \to \tilde{R} = \Omega^{-2}\left[R - 2(n-1)\frac{\Box\Omega}{\Omega} - (n-1)(n-4)\frac{1}{\Omega^2}\nabla^{\mu}\Omega\nabla_{\mu}\Omega\right] \tag{1.151}$$

Specifically, in $n = 4$-dimensions,

$$\sqrt{-\tilde{g}} = \sqrt{-g}\,\Omega^4, \qquad \tilde{R} = \Omega^{-2}\left[R - 6\Omega^{-1}\Box\Omega\right]. \tag{1.152}$$

We now use these relations to perform a conformal transformation upon a theory containing a coupling between a scalar field ϕ and the Ricci scalar R in $n = 4$,

$$\int d^4x\sqrt{-g}\,\phi R \rightarrow \int d^4x\sqrt{-\tilde{g}}\,\tilde{\phi}\tilde{R} = \int d^4x\sqrt{-g}\,\phi\Omega^2\left(R - 6\Omega^{-1}\Box\Omega\right), \quad (1.153)$$

and if we choose the conformal factor Ω to be related to the scalar field ϕ such that $\Omega^2\phi = 1$ then

$$\begin{aligned}\int d^4x\sqrt{-g}\,\phi R &\rightarrow \int d^4x\sqrt{-g}\left(R - 6\Omega^{-1}\Box\Omega\right) \\ &= \int d^4x\sqrt{-g}\left(R - \frac{6}{\Omega^2}\nabla^\mu\Omega\nabla_\mu\Omega - 6\nabla^\mu(\Omega^{-1}\nabla_\mu\Omega)\right).\end{aligned} \quad (1.154)$$

The final term is a total derivative and is usually ignored (it is only a surface term), leaving

$$\int d^4x\sqrt{-g}\,\phi R \rightarrow \int d^4x\sqrt{-g}\left(R - \frac{6}{\Omega^2}\nabla^\mu\Omega\nabla_\mu\Omega\right). \quad (1.155)$$

If we now write the scalar field as $\Omega = e^{\frac{1}{\sqrt{12}}\psi}$ (i.e. $\psi = -\sqrt{3}\ln\phi$) then

$$\int d^4x\sqrt{-g}\,\phi R \rightarrow \int d^4x\sqrt{-g}\left(R - \frac{1}{2}\nabla^\mu\psi\nabla_\mu\psi\right). \quad (1.156)$$

What we have done, therefore, is show that a theory whose Lagrangian contains an explicit coupling between a scalar field ϕ and the Ricci scalar is equivalent to a theory without that coupling, but with the introduction of a canonical kinetic term for the scalar field. That is, we have transformed from a scalar-tensor theory with explicit coupling to a minimally coupled scalar-tensor theory.

1.5 Summary

The current status of gravitational physics can summarized as follows: the spacetime arena appears to be influenced by the gravitating content of the universe, and GR provides a specific set of field equations for how the spacetime is determined for a given content. However, observational data are completely incompatible with the predictions from GR with the metric of a homogeneous isotropic universe, unless dark energy is introduced. Modifying the gravitational physics away from GR is one way out of this situation, although one of the major obstacles that modified gravity theories face is that all theories must posses a Newtonian limit compatible with all Solar System tests to pass current observational constraints.

It is fair to say, therefore, that we do not yet have a complete theoretical picture that is consistent with the data and which answers the simple question "*how does gravity work?*"

References

1. A. Einstein, *The Meaning of Relativity* (Routledge, 1952)
2. L. Landau, E. Lifshitz, *The Classical Theory of Fields* (Addison Wesley Press, Cambridge, 1951)
3. S. Weinberg, *Gravitation and Cosmology: Principles and Applications of the General Theory of Relativity* (John Wiley, New York, 1972)
4. S. Hawking, G. Ellis, *The Large Scale Structure of Space-Time. Cambridge Monographs on Mathematical Physics* (Cambridge University Press, New York, 1973)
5. E.W. Kolb, M.S. Turner, The early Universe. Front. Phys. **69**, 1–547 (1990)
6. P.A.M. Dirac, *General Theory of Relativity* (Princeton University Press, UK, 1996)
7. R.M. Wald, *General Relativity* (The University of Chicago Press, Chicago, 1984)
8. D.F. Lawden, *Introduction to Tensor Calculus, Relativity and Cosmology* (2002)
9. M. Trodden, S.M. Carroll, *TASI Lectures: Introduction to Cosmology* (University of Colorado, Boulder, 2004), astro-ph/0401547
10. S. Weinberg, *Cosmology* (Oxford University Press, USA, 2008)
11. P. Peter, J.-P. Uzan, *Primordial Cosmology* (Oxford University Press, Oxford, 2009)
12. T. Frankel, *The Geometry of Physics: An Introduction* (Cambridge University Press, Cambridge, 1997)
13. G. Lemaître, Expansion of the universe, a homogeneous universe of constant mass and increasing radius accounting for the radial velocity of extra-galactic nebulae. Monthly Not. R. Astron. Soc. **91**, 483–490 (1931)
14. K. Schwarzschild, On the gravitational field of a mass point according to Einstein's theory. Sitzungsber. Preuss. Akad. Wiss. Berlin (Math. Phys.) **1916**, 189–196 (1916) [physics/9905030]
15. F. Kottler, ber die physikalischen grundlagen der einsteinschen gravitationstheorie. Ann. Phys. **361**(14), 401–462 (1918)
16. G.C. McVittie, An example of gravitational collapse in general relativity. Astrophys. J. **143**, 682 (1966)
17. N. Kaloper, M. Kleban, D. Martin, McVittie's legacy: black holes in an expanding universe. Phys. Rev. **D81**, 104044 (2010) [arXiv:1003.4777]
18. R.P. Kerr, Gravitational field of a spinning mass as an example of algebraically special metrics. Phys. Rev. Lett. **11**, 237–238 (1963)
19. E. Dyer, K. Hinterbichler, Boundary terms, variational principles and higher derivative modified gravity. Phys. Rev. **D79**, 024028 (2009) [arXiv:0809.4033]
20. G. Gibbons, S. Hawking, Action integrals and partition functions in quantum gravity. Phys. Rev. **D15**, 2752–2756 (1977)
21. S.W. Hawking, G.T. Horowitz, The gravitational hamiltonian, action, entropy and surface terms. Class. Quant. Grav. **13**, 1487–1498 (1996) [gr-qc/9501014]
22. H.A. Chamblin, H.S. Reall, Dynamic dilatonic domain walls. Nucl. Phys. **B562**, 133–157 (1999) [hep-th/9903225]
23. A. Mennim, R.A. Battye, Cosmological expansion on a dilatonic brane-world. Class. Quant. Grav. **18**, 2171–2194 (2001) [hep-th/0008192]

24. J. York, W. James, Role of conformal three geometry in the dynamics of gravitation. Phys. Rev. Lett. **28**, 1082–1085 (1972)
25. G.F.R. Ellis, M. Bruni, Covariant and gauge-invariant approach to cosmological density fluctuations. Phys. Rev. D **40**, 1804–1818 (1989)
26. H. Kodama, M. Sasaki, Cosmological perturbation theory. Prog. Theor. Phys. Suppl. **78**, 1–166 (1984)
27. V.F. Mukhanov, H. Feldman, R.H. Brandenberger, Theory of cosmological perturbations. Part 1. Classical perturbations. Part 2. Quantum theory of perturbations. Part 3. Extensions. Phys. Rept. **215**, 203–333 (1992)
28. R. Durrer, Cosmological perturbation theory. Lect. Notes Phys. **653**, 31–70 (2004) [astro-ph/0402129]
29. K.A. Malik, D.R. Matravers, A concise introduction to perturbation theory in cosmology. Class. Quant. Grav. **25**, 193001 (2008) [arXiv:0804.3276]
30. R. L. Arnowitt, S. Deser, C. W. Misner, The dynamics of general relativity. gr-qc/0405109
31. R.A. Battye, A. Moss, Cosmological perturbations in elastic dark energy models. Phys. Rev. **D76**, 023005 (2007) [astro-ph/0703744]
32. V.F. Mukhanov, H.A. Feldman, R.H. Brandenberger, Theory of cosmological perturbations. Phys. Rep. **215**(5–6), 203–333 (1992)
33. C.-P. Ma, E. Bertschinger, Cosmological perturbation theory in the synchronous versus conformal Newtonian gauge. Astrophys. J. **426**, L57 (1994) [astro-ph/9401007]
34. Supernova Cosmology Project Collaboration, S. Perlmutter et al., Measurements of omega and lambda from 42 high redshift supernovae. Astrophys. J. **517**, 565–586 (1999) [astro-ph/9812133]. The Supernova Cosmology Project
35. Supernova Search Team Collaboration, A.G. Riess et al., Observational evidence from supernovae for an accelerating universe and a cosmological constant. Astron. J. **116**, 1009–1038 (1998) [astro-ph/9805201]
36. A.G. Riess et al., BVRI light curves for 22 type ia supernovae. Astron. J. **117**, 707–724 (1999) [astro-ph/9810291]
37. J. Frieman, M. Turner, D. Huterer, Dark energy and the accelerating universe. Ann. Rev. Astron. Astrophys. **46**, 385–432 (2008) [arXiv:0803.0982]
38. P. Astier and R. Pain, Observational evidence of the accelerated expansion of the universe, arXiv:1204.5493
39. R. Durrer and R. Maartens, Dark energy and modified gravity, arXiv:0811.4132
40. E.J. Copeland, M. Sami, S. Tsujikawa, Dynamics of dark energy. Int. J. Mod. Phys. **D15**, 1753–1936 (2006) [hep-th/0603057]
41. L. Amendola, S. Tsujikawa, *Dark Energy: Theory and Observations* (Cambridge University Press, Cambridge, 2010)
42. R. Durrer, What do we really know about Dark Energy?, arXiv:1103.5331
43. M. Kunz, The phenomenological approach to modeling the dark energy, arXiv:1204.5482
44. R. Caldwell, R. Dave, P.J. Steinhardt, Cosmological imprint of an energy component with general equation of state. Phys. Rev. Lett. **80**, 1582–1585 (1998) [astro-ph/9708069]
45. T. Chiba, T. Okabe, M. Yamaguchi, Kinetically driven quintessence. Phys. Rev. **D62**, 023511 (2000) [astro-ph/9912463]
46. C. Clarkson, G. Ellis, A. Faltenbacher, R. Maartens, O. Umeh, et al., (Mis-)Interpreting supernovae observations in a lumpy universe, arXiv:1109.2484
47. R. Maartens, Is the universe homogeneous?. Phil. Trans. Roy. Soc. Lond. **A369**, 5115–5137 (2011) [arXiv:1104.1300]
48. P. Bull, T. Clifton, P.G. Ferreira, The kSZ effect as a test of general radial inhomogeneity in LTB cosmology. Phys. Rev. **D85**, 024002 (2012) [arXiv:1108.2222]
49. T. Clifton, P.G. Ferreira, K. O'Donnell, An improved treatment of optics in the lindquist-wheeler models. Phys. Rev. **D85**, 023502 (2012) [arXiv:1110.3191], p. 7, 5 figures
50. N. Meures, M. Bruni, Redshift and distances in a ΛCDM cosmology with non-linear inhomogeneities. Mon. Not. Roy. Astron. Soc. **419**, 1937 (2012) [arXiv:1107.4433]
51. G.F. Ellis, Inhomogeneity effects in Cosmology, arXiv:1103.2335

52. S. Rasanen, Backreaction: Directions of progress, Class. Quant. Grav. **28**, 164008 (2011) [arXiv:1102.0408]
53. C. Clarkson, O. Umeh, Is backreaction really small within concordance cosmology?. Class. Quant. Grav. **28**, 164010 (2011) [arXiv:1105.1886]
54. T. Clifton, Cosmology without averaging. Class. Quant. Grav. **28**, 164011 (2011) [arXiv:1005.0788]
55. K. Bolejko, M.-N. Celerier, A. Krasinski, Inhomogeneous cosmological models: Exact solutions and their applications. Class. Quant. Grav. **28**, 164002 (2011) [arXiv:1102.1449]
56. V. Marra, A. Notari, Observational constraints on inhomogeneous cosmological models without dark energy. Class. Quant. Grav. **28**, 164004 (2011) [arXiv:1102.1015]
57. S. Schlamminger, K.-Y. Choi, T. Wagner, J. Gundlach, E. Adelberger, Test of the equivalence principle using a rotating torsion balance. Phys. Rev. Lett. **100**, 041101 (2008) [arXiv:0712.0607]
58. J.G. Williams, S.G. Turyshev, D.H. Boggs, Lunar laser ranging tests of the equivalence principle with the earth and moon. Int. J. Mod. Phys. **D18**, 1129–1175 (2009) [gr-qc/0507083]
59. J.G. Williams, S.G. Turyshev, D. Boggs, Lunar laser ranging tests of the equivalence principle. Class. Quant. Grav. **29**, 184004 (2012) [arXiv:1203.2150]
60. I. Ciufolini, A. Paolozzi, E.C. Pavlis, J.C. Ries, R. Koenig, R.A. Matzner, G. Sindoni, H. Neumayer, Towards a one percent measurement of frame dragging by spin with satellite laser ranging to LAGEOS, LAGEOS 2 and LARES and GRACE gravity models. Space Sci. Rev. **148**, 71–104 (2009)
61. M. Kramer, I.H. Stairs, R. Manchester, M. McLaughlin, A. Lyne, et al., Tests of general relativity from timing the double pulsar. Sci. **314**, 97–102 (2006) [astro-ph/0609417]
62. A. Padilla, V. Sivanesan, Boundary Terms and Junction Conditions for Generalized Scalar-Tensor Theories, arXiv:1206.1258
63. T. Clifton, P.G. Ferreira, A. Padilla, C. Skordis, Modified gravity and cosmology. Phys. Rept. **513**, 1–189 (2012) [arXiv:1106.2476]
64. S. Weinberg, The cosmological constant problem. Rev. Mod. Phys. **61**, 1–23 (1989)
65. T. Padmanabhan, Cosmological constant: The weight of the vacuum. Phys. Rept. **380**, 235–320 (2003) [hep-th/0212290]
66. J. Martin, Everything You Always Wanted To Know About The Cosmological Constant Problem (But Were Afraid To Ask), arXiv:1205.3365
67. H.-J. Schmidt, Fourth order gravity: Equations, history, and applications to cosmology. eConf **C0602061**, 12 (2006) [gr-qc/0602017]
68. T.P. Sotiriou, V. Faraoni, f(R) Theories Of Gravity. Rev. Mod. Phys. **82**, 451–497 (2010) [arXiv:0805.1726]
69. A. De Felice, S. Tsujikawa, f(R) theories. Living Rev. Rel. **13**, 3 (2010) [arXiv:1002.4928]
70. S. Nojiri, S.D. Odintsov, M. Sasaki, Gauss-Bonnet dark energy. Phys. Rev. **D71**, 123509 (2005) [hep-th/0504052]
71. B.M. Carter, I.P. Neupane, Towards inflation and dark energy cosmologies from modified Gauss-Bonnet theory. JCAP **0606**, 004 (2006) [hep-th/0512262]
72. L. Amendola, C. Charmousis, S.C. Davis, Constraints on Gauss-Bonnet gravity in dark energy cosmologies. JCAP **0612**, 020 (2006) [hep-th/0506137]
73. L. Amendola, C. Charmousis, S.C. Davis, Solar system constraints on Gauss-Bonnet mediated dark energy. JCAP **0710**, 004 (2007) [arXiv:0704.0175]
74. C. Brans, R.H. Dicke, Mach's principle and a relativistic theory of gravitation. Phys. Rev. **124**, 925–935 (1961)
75. A. Anisimov, E. Babichev, A. Vikman, B-inflation. JCAP **0506**, 006 (2005) [astro-ph/0504560]
76. C. Deffayet, O. Pujolas, I. Sawicki, A. Vikman, Imperfect dark energy from kinetic gravity braiding. JCAP **1010**, 026 (2010) [arXiv:1008.0048]
77. O. Pujolas, I. Sawicki, A. Vikman, The imperfect fluid behind kinetic gravity braiding. JHEP **11**, 156 (2011) [arXiv:1103.5360]

78. G.W. Horndeski, Second-order scalar-tensor field equations in a four-dimensional space. Int. J. Theor. Phys. **10**, 363–384 (1974)
79. C. Charmousis, E.J. Copeland, A. Padilla, P.M. Saffin, General second order scalar-tensor theory, self tuning, and the Fab Four. Phys. Rev. Lett. **108**, 051101 (2012) [arXiv:1106.2000]
80. C. Charmousis, E.J. Copeland, A. Padilla, P.M. Saffin, Self-tuning and the derivation of a class of scalar-tensor theories. Phys. Rev. **D85**, 104040 (2012) [arXiv:1112.4866]
81. A. De Felice, T. Kobayashi, S. Tsujikawa, Effective gravitational couplings for cosmological perturbations in the most general scalar-tensor theories with second-order field equations. Phys. Lett. **B706**, 123–133 (2011) [arXiv:1108.4242]
82. A. De Felice, S. Tsujikawa, Conditions for the cosmological viability of the most general scalar-tensor theories and their applications to extended Galileon dark energy models. JCAP **1202**, 007 (2012) [arXiv:1110.3878]
83. A. Nicolis, R. Rattazzi, E. Trincherini, The Galileon as a local modification of gravity. Phys. Rev. **D79**, 064036 (2009) [arXiv:0811.2197]
84. S. Appleby, E.V. Linder, The paths of gravity in Galileon cosmology. JCAP **1203**, 043 (2012) [arXiv:1112.1981]
85. C. de Rham, Galileons in the Sky, arXiv:1204.5492
86. S.A. Appleby, E.V. Linder, Galileons on Trial, arXiv:1204.4314
87. T. Kobayashi, M. Yamaguchi, J. Yokoyama, Generalized g-inflation: Inflation with the most general second-order field equations. Prog. Theor. Phys. **126**, 511–529 (2011) [arXiv:1105.5723]
88. A. Barreira, B. Li, C. Baugh, S. Pascoli, Linear perturbations in Galileon gravity models, arXiv:1208.0600
89. T. Chiba, Generalized gravity and ghost. JCAP **0503**, 008 (2005) [gr-qc/0502070]
90. C. Eling, T. Jacobson, D. Mattingly, Einstein-aether theory, gr-qc/0410001
91. T. Jacobson, D. Mattingly, Einstein-aether waves. Phys. Rev. **D70**, 024003 (2004) [gr-qc/0402005]
92. A. Tartaglia, N. Radicella, Vector field theories in cosmology. Phys. Rev. **D76**, 083501 (2007) [arXiv:0708.0675]
93. T.G. Zlosnik, P.G. Ferreira, G.D. Starkman, Modifying gravity with the Aether: an alternative to dark matter. Phys. Rev. **D75**, 044017 (2007) [astro-ph/0607411]
94. C. Armendariz-Picon, A. Diez-Tejedor, Aether unleashed. JCAP **0912**, 018 (2009) [arXiv:0904.0809]
95. J.D. Bekenstein, Relativistic gravitation theory for the MOND paradigm. Phys. Rev. **D70**, 083509 (2004) [astro-ph/0403694]
96. T.G. Zlosnik, P.G. Ferreira, G.D. Starkman, The vector-tensor nature of Bekenstein's relativistic theory of modified gravity. Phys. Rev. **D74**, 044037 (2006) [gr-qc/0606039]
97. C. Skordis, D.F. Mota, P.G. Ferreira, C. Boehm, Large scale structure in Bekenstein's theory of relativistic modified newtonian dynamics. Phys. Rev. Lett. **96**, 011301 (2006) [astro-ph/0505519]
98. C. Skordis, TeVeS cosmology : Covariant formalism for the background evolution and linear perturbation theory. Phys. Rev. **D74**, 103513 (2006) [astro-ph/0511591]
99. C. Skordis, Generalizing TeVeS cosmology. Phys. Rev. **D77**, 123502 (2008) [arXiv:0801.1985]
100. I. Ferreras, N. Mavromatos, M. Sakellariadou, M.F. Yusaf, Confronting MOND and TeVeS with strong gravitational lensing over galactic scales: An extended survey, arXiv:1205.4880
101. T. Zlosnik, P. Ferreira, G.D. Starkman, The vector-tensor nature of Bekenstein's relativistic theory of modified gravity. Phys. Rev. **D74**, 044037 (2006) [gr-qc/0606039]
102. C. Skordis, The tensor-vector-scalar theory and its cosmology. Class. Quant. Grav. **26**, 143001 (2009) [arXiv:0903.3602]
103. J. Khoury, A. Weltman, Chameleon fields: Awaiting surprises for tests of gravity in space. Phys. Rev. Lett. **93**, 171104 (2004) [astro-ph/0309300]
104. J. Khoury, A. Weltman, Chameleon cosmology. Phys. Rev. **D69**, 044026 (2004) [astro-ph/0309411]

105. P. Brax, A.-C. Davis, B. Li, H.A. Winther, A unified description of screened modified gravity. Phys. Rev. **D86**, 044015 (2012) [arXiv:1203.4812]
106. V. Faraoni, E. Gunzig, P. Nardone, Conformal transformations in classical gravitational theories and in cosmology. Fund. Cosmic Phys. **20**, 121 (1999) [gr-qc/9811047]

Chapter 2
The Effective Action Formalism for Cosmological Perturbations

2.1 Introduction

The standard model of cosmology uses General Relativity (GR) to describe gravitational interactions, an homogeneous/isotropic FRW metric to describe the geometry and matter content of cold dark matter (CDM)/photons/baryons to describe its constituents. Observations of the cosmic microwave background, supernovae, baryon acoustic oscillations, gravitational lensing and structure formation point to the existence of an additional component dubbed "dark energy", or a modification to gravity, which needs to be introduced to explain the the observed acceleration [1–5].

The simplest explanation is a cosmological constant, Λ, and the standard paradigm is the ΛCDM model. However, there is still considerable flexibility for the explanation to be something radically different. In general, we can model all possible theories as an extra "dark sector" component to the stress-energy-momentum tensor. The structure of the gravitational field equations means that this extra component can be used to model either "exotic matter" with an equation of state $P/\rho < -\frac{1}{3}$ or a modification to GR (i.e. modifying exactly how gravity responds to the presence of matter). Constructing viable models of modified gravity has become an important task with the discovery of the acceleration of the Universe; some modified gravity models may also be able to account for observations which otherwise require dark matter.

One way to model the dark sector is "Lagrangian engineering": write down ever more complicated new theories with a view of constraining their parameters and free functions to fit observation with the hope that self-accelerating solutions can be found. Theories where explicit forms of dark energy are written down also fall into this category. They include TeVeS [6, 7], Einstein-æther [8], Brans-Dicke [9], Horndeski [10–12] and $F(R)$ gravities [13, 14], quintessence [15, 16], k-essence [17, 18] and Gallileons [19]. This is by no means an exhaustive list, and we have made no mention of the plethora of higher dimensional theories. The reader is directed to the recent extensive review of modified gravity theories [20].

J. Pearson, *Generalized Perturbations in Modified Gravity and Dark Energy*, Springer Theses, DOI: 10.1007/978-3-319-01210-0_2,

Given this proliferation of modified gravity and dark energy models, it would be a good idea to construct a generic way of parameterising deviations from the GR+ΛCDM picture and various suggestions have been made [21–33] to do this for perturbations. This approach is called the "Parameterized-Post-Friedmannian" (PPF) framework, in analogy to the well established Parameterized-Post-Newtonian (PPN) framework which was invented for Solar System tests of General Relativity [34]. However, as we describe below, to date no generic approach has been proposed which has a physical basis.

In this thesis we describe a new way of parameterizing perturbations in the dark sector requiring, as an assumption, knowledge of the field content. We do not assume a specific Lagrangian density, but we are able to model the possible effects on observations by using an effective action to compute the possible perturbations to the gravitational field equations. This is done by limiting the action to terms which are quadratic in the perturbed field content which is sufficient to model linearized perturbations, and assuming that the spatial sections are isotropic.

Our theories will be completely general allowing for all possible degrees of freedom. Initially we do not impose reparametrization, or gauge, invariance. This is something which we would expect of a fundamental theory of dark energy, but not necessarily one for which the field content is just a coarse grained description. We will find that this can lead to an phenomological vector degree of freedom, ξ^{μ}. In the elastic dark energy theory [35–38], which can be used to describe the effects of a dark energy component composed of a topological defect lattice, ξ^{μ} represents a perturbation of the elastic medium from its equilibrium. We will see that the imposition of reparametrization invariance substantially reduces the number of free functions.

We note that many authors have consider possible dark energy theories which are effective Lagrangians in the traditional sense, that is, the terms in the Lagrangian represent an expansion of field operators which are suppressed at low energies [39–41]. Our approach is sufficiently similar to this approach to share the epitaph "effective action", but it is completely different in many ways. It is completely classical and is in no sense an expansion in energy scales. Moreover, it is just an effective action for the perturbations, and in no sense represents the full field theory of the dark energy.

2.2 Approaches to Parameterizing Dark Sector Perturbations

In this section we will provide a brief review of current approaches to studying generalized gravitational theories, concluding with a short discussion on the generalities of our approach.

2.2.1 Parameterized Post-Friedmannian Approach

A popular way to parameterize the dark sector takes an "observational" perspective. One can modify the equations governing the predictions of the Newtonian gravitational potential Φ and shear σ by introducing extra functions space and time into the relevant equations and then parametrizing these extra functions in an *ad hoc* fashion. Since it is possible to explicitly observe Φ and σ via the evolution structure and gravitational shear [23, 28, 29, 31, 42] (see also the more recent papers [43–45]), one can then compare them with the predictions of particular *ad hoc* choice and determine constraints on the deviation of a particular parameter from its value in General Relativity.

One way of doing this is by modifying the Poisson and gravitational slip equations, introducing two scale—and time-dependent functions, $Q = Q(k, a)$ and $R = R(k, a)$; k is the wavenumber in a Fourier expansion and a the scale factor. The Poisson and gravitational-slip equations then become

$$k^2\Phi = -4\pi G Q a^2 \rho \Delta, \qquad \Psi - R\Phi = -12\pi G Q a^2 \rho(1+w)\sigma, \tag{2.1}$$

where $\Delta \equiv \delta + 3H\theta(1+w)$ is the comoving density perturbation, $\delta \equiv \delta\rho/\rho$ the density contrast, θ the velocity divergence field, $w = P/\rho$ the equation of state and σ is the anisotropic stress. When these equations are derived in GR one finds that $Q(k, a) = R(k, a) = 1$. So if, by comparison to data, either of these parameters are shown to be inconsistent with unity, then deviations from GR can be established. In [32, 46] it was shown that the two functions Q, R are not necessarily independent: they can be linked by the perturbed Bianchi identity, depending on the structure of the underlying theory.

2.2.2 *Generalized Gravitational Field Equations*

Another way to investigate the dark sector takes a more theoretical standpoint, and is based on a more consistent modification of the governing field equations. The method stems from the fact that any modified gravity theory or model of dark energy can be encapsulated by writing the *generalized gravitational field equations*

$$G_{\mu\nu} = 8\pi G T_{\mu\nu} + U_{\mu\nu}, \tag{2.2}$$

where $G_{\mu\nu}$ is the Einstein tensor calculated from the spacetime metric, $T_{\mu\nu}$ is the energy-momentum tensor of all *known* species (radiation, Baryons, CDM etc) and $U_{\mu\nu}$ is a tensor which contains all *unknown* contributions to the gravitational field equations, which we call the *dark energy-momentum tensor* [24, 25, 27].

Because the Bianchi identity automatically holds for the Einstein tensor, $\nabla_\mu G^{\mu\nu} = 0$, in the standard case where the *known* and *unknown* sectors are decoupled (that is $\nabla_\mu T^{\mu\nu} = 0$) we have the conservation law

$$\nabla_\mu U^{\mu\nu} = 0. \tag{2.3}$$

This represents a constraint equation on the extra parameters and functions that may appear in a parameterization of the dark sector at the level of the background.

At perturbed order, the parameterization of $\delta U^{\mu\nu}$ is constrained by the perturbed conservation law

$$\delta(\nabla_\mu U^{\mu\nu}) = 0. \tag{2.4}$$

The shortcoming of this approach is that one must supply the components of $\delta U^{\mu\nu}$. Skordis [27] does this by expanding the components $\delta U^\mu{}_\nu$ in terms of pseudo derivative operators acting upon gauge invariant combinations of metric perturbations, by imposing the principles that (a) the field equations remain at most second order and (b) the equations are gauge-form invariant. A particular form of these components were considered in [27]:

$$-a^2\delta U^0{}_0 = \frac{1}{a}\mathcal{A}\hat{\Phi}, \quad -a^2\delta U^0{}_i = \nabla_i\left(\frac{1}{a^2}\mathcal{B}\hat{\Phi}\right), \quad a^2\delta U^i{}_i = \mathcal{C}_1\hat{\Phi} + \mathcal{C}_2\dot{\hat{\Phi}} + \mathcal{C}_3\hat{\Psi}, \tag{2.5a}$$

$$a^2\left[\delta U^i{}_j - \frac{1}{3}\delta^i_j\delta U^k{}_k\right] = (\nabla^i\nabla_j - \frac{1}{3}\delta_{ij}\nabla^2)(\mathcal{D}_1\hat{\Phi} + \mathcal{D}_2\dot{\hat{\Phi}} + \mathcal{D}_3\hat{\Psi}), \tag{2.5b}$$

where $\mathcal{O} = \{\mathcal{A}, \mathcal{B}, \mathcal{C}_i, \mathcal{D}_i\}$ is a set of pseudo differential operators and $\{\hat{\Phi}, \hat{\Psi}\}$ are gauge invariant combinations of perturbed metric variables. The possible form that the elements of $\mathcal{O}$ can take is constrained by the perturbed Bianchi identity. For instance, it was shown that $\mathcal{C}_3 = \mathcal{D}_3 = 0$ is one of the sufficient consistency relations. A generalized version of this method can be found in [20, 32].

This scheme provides a way to compute and constrain observables without ever having to write down an explicit theory for the dark sector. There appears to be, however, a weakness in the current formulation of this strategy: there does not seem to be a physically obvious way to interpret the $\mathcal{O}$; for example, if one were to find that $\mathcal{C}_3 = 0$ is "required" for consistency with observational data, what does that impose physically upon the system? It is exactly this issue we address in this paper.

2.2.3 *Effective Action Approach*

The generalized gravitational field Eq. (2.2) can be constructed from an action

$$S = \int d^4x\,\sqrt{-g}\left[R + 16\pi G\mathcal{L}_{\mathrm{m}} - 2\mathcal{L}_{\mathrm{d}}\right]. \tag{2.6}$$

The matter Lagrangian $\mathcal{L}_{\mathrm{m}}$ contains all known matter fields (e.g. baryons, photons) and is used to construct the known energy momentum tensor $T^{\mu\nu}$, and the dark sector Lagrangian $\mathcal{L}_{\mathrm{d}}$ contains all "unknown" contributions to the gravitational sector, and will be used to construct the dark energy momentum tensor $U^{\mu\nu}$. One can define

$$T^{\mu\nu} \equiv \frac{2}{\sqrt{-g}}\frac{\delta}{\delta g_{\mu\nu}}(\sqrt{-g}\mathcal{L}_{\mathrm{m}}), \qquad U^{\mu\nu} \equiv -\frac{2}{\sqrt{-g}}\frac{\delta}{\delta g_{\mu\nu}}(\sqrt{-g}\mathcal{L}_{\mathrm{d}}). \tag{2.7}$$

The dark sector Lagrangian may contain known fields in an unknown configuration or extra fields, but of course we do not know *a priori* what the dark sector Lagrangian density is.

Two simple examples are (i) a slowly-rolling minimally coupled scalar field parameterized by a potential, $V(\phi)$, and (ii) a modified gravity model parameterized by a free function of the Ricci scalar, $F(R)$. There are restrictions on the form of both of these functions to achieve acceleration, but once they have been applied there is still considerable freedom in the choices of $V(\phi)$ and $F(R)$ and wide ranges of behaviour of the expansion history, $a(t)$, can be arranged for particular choices of the functions. One would expect this to be the case in any self consistent dark energy model compatible with FRW metric and therefore it might seem reasonable to make the assumption that the dark stress-energy-momentum tensor $U_{\mu\nu} = \rho u_\mu u_\nu + P\gamma_{\mu\nu}$ where $w(a) = P/\rho$ is in 1–1 correspondence with $a(t)$. The important question, which we are concerned with, is how to parametrize the perturbations $\delta U^\mu{}_\nu$ in a general way based on some general physical principle. In this way our approach is similar to that discussed in Sect. 2.2.2

The overall ethos which we advocate is to write down an effective action, inspired by the approach that is taken in particle physics (see, e.g. [47]) where, for example, the most general modifications to the standard model are written down for a given field content that are compatible with some assumed symmetry/symmetries. Then all the free coefficients are constrained by experiment. In our case, we will specify the field content of the dark sector, for example, scalar or vector fields, and write down a general quadratic Lagrangian density for the perturbed field variables which is sufficient to generate equations of motion for linearized perturbations. We will also make the assumption that the spatial sections are isotropic which substantially reduces the number of free coefficients. See Fig. 2.1 for a schematic of our philosophy.

2.3 Formalism

2.3.1 *Second Order Lagrangian*

The underlying principle behind our method is to write down a effective Lagrangian for perturbed field variables. If our theory is constructed from a set of field variables $\{X^{(A)}\}$, then we write each field variable as a linearized perturbation about some background value,

$$X^{(A)} = \bar{X}^{(A)} + \delta X^{(A)}. \tag{2.8}$$

The action for the perturbed field variables $\{\delta X^{(A)}\}$ is computed by integrating a Lagrangian which is quadratic in the perturbed field variables. If there are "N" perturbed field variables, the effective Lagrangian for the perturbed field variables is given by

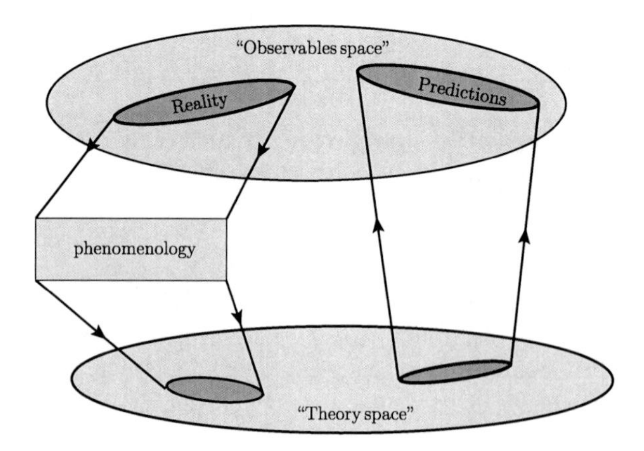

Fig. 2.1 Schematic depiction of our effective action approach. The "traditional" *modus operandi* is to pick a theory from "theory space" and obtain the possible observational predictions from that theory; this requires an educated guess as to what the theory that describes reality should be. The *modus operandi* of phenomenological field theory is to determine which points in theory space are consistent with observations of reality: this will automatically rule out theories which are incompatible with observations without ever needing to consider them. The focus of this thesis is to advocate the use of phenomenological field theory in the determination of the underlying gravitational theory to be used on cosmological scales

$$\mathcal{L}_{\{2\}}(\delta X^{(\mathrm{C})}) = \sum_{\mathrm{A}=1}^{\mathrm{N}} \sum_{\mathrm{B}=1}^{\mathrm{N}} \mathsf{G}_{\mathrm{AB}} \delta X^{(\mathrm{A})} \delta X^{(\mathrm{B})}, \tag{2.9}$$

where $\mathsf{G}_{\mathrm{AB}} = \mathsf{G}_{\mathrm{AB}}(\bar{X}^{(\mathrm{C})})$ is a set of arbitrary functions only depending on the background field variables; clearly, $\mathsf{G}_{\mathrm{AB}} = \mathsf{G}_{\mathrm{BA}}$. To obtain the equation of motion of the perturbed field variables $\{\delta X^{(\mathrm{A})}\}$ we must induce some variation in the $\{\delta X^{(\mathrm{A})}\}$ and subsequently demand that $\mathcal{L}_{\{2\}}$ is independent of these variations. If we vary the perturbed field variables with a variational operator $\hat{\delta}$,

$$\delta X^{(\mathrm{A})} \to \delta X^{(\mathrm{A})} + \hat{\delta}(\delta X^{(\mathrm{A})}), \tag{2.10}$$

then the effective Lagrangian will vary according to $\mathcal{L}_{\{2\}} \to \mathcal{L}_{\{2\}} + \hat{\delta}\mathcal{L}_{\{2\}}$, where

$$\hat{\delta}\mathcal{L}_{\{2\}} = 2 \sum_{\mathrm{A}=1}^{\mathrm{N}} \sum_{\mathrm{B}=1}^{\mathrm{N}} \mathsf{G}_{\mathrm{AB}} \delta X^{(\mathrm{A})} \hat{\delta}(\delta X^{(\mathrm{B})}). \tag{2.11}$$

The demand that the effective Lagrangian is independent of these variations is the statement that

$$\frac{\hat{\delta}}{\hat{\delta}(\delta X^{(\mathrm{B})})} \mathcal{L}_{\{2\}} = 0, \tag{2.12}$$

that is,

$$\sum_{A=1}^{N}\sum_{B=1}^{N} \mathsf{G}_{AB}\delta X^{(A)} = 0. \tag{2.13}$$

These equations provide the equations of motion of the perturbed field variables. We will now show how to obtain the effective action for perturbations by directly perturbing the background action.

We will consider an action of the form

$$S = \int d^4x\, \sqrt{-g}\mathcal{L}, \tag{2.14}$$

where g is the determinant of the spacetime metric, $g_{\mu\nu}$, and $\mathcal{L}$ is the Lagrangian, which contains all fields in the theory. It will be useful to write the first and second variations of the action as

$$\delta S = \int d^4x\, \sqrt{-g}\Diamond\mathcal{L}, \qquad \delta^2 S = \int d^4x\, \sqrt{-g}\Diamond^2\mathcal{L}, \tag{2.15}$$

where "$\Diamond$" is a useful measure-weighted pseudo-operator introduced in [48, 49], and is defined by

$$\Diamond^n\mathcal{L} \equiv \frac{1}{\sqrt{-g}}\delta^n(\sqrt{-g}\mathcal{L}). \tag{2.16}$$

We will only consider first perturbations of the field content of a theory. For the action (2.14) we can use the well known result

$$\frac{1}{\sqrt{-g}}\delta\sqrt{-g} = -\frac{1}{2}g_{\mu\nu}\delta g^{\mu\nu} = +\frac{1}{2}g^{\mu\nu}\delta g_{\mu\nu}, \tag{2.17}$$

to show that to quadratic order in the perturbations that the integrands in (2.15) are given by

$$\Diamond\mathcal{L} = \delta\mathcal{L} + \frac{1}{2}\mathcal{L}g^{\mu\nu}\delta g_{\mu\nu}, \tag{2.18a}$$

$$\Diamond^2\mathcal{L} = \delta^2\mathcal{L} + g^{\mu\nu}\delta g_{\mu\nu}\delta\mathcal{L} + \frac{1}{4}\mathcal{L}\left(g^{\mu\nu}g^{\alpha\beta} - 2g^{\mu(\alpha}g^{\beta)\nu}\right)\delta g_{\mu\nu}\delta g_{\alpha\beta}. \tag{2.18b}$$

We treat the integrand of the second variation of the action, i.e. $\Diamond^2\mathcal{L}$, as the effective Lagrangian, $\mathcal{L}_{\{2\}}$, for linearized perturbations,

$$\delta^2 S = \int d^4x\, \sqrt{-g}\Diamond^2\mathcal{L} = \int d^4x \sqrt{-g}\, \mathcal{L}_{\{2\}} \tag{2.19}$$

It is called the *second order Lagrangian*. The final term of (2.18b) is an effective mass-term for the gravitational fluctuations $\delta g_{\mu\nu}$ which is always present even when the field which constitutes the dark sector does not vary, i.e. when $\delta\mathcal{L} = \delta^2\mathcal{L} = 0$.

Although we will be providing various explicit examples later on in the Thesis, we will briefly discuss how to write down $\mathcal{L}_{\{2\}}$ once the field content has been specified. If the field content is $\{X, Y\}$, then we write $\mathcal{L} = \mathcal{L}(X, Y)$, and then $\mathcal{L}_{\{2\}}$ is written down by writing all quadratic interactions of the perturbed fields with appropriate coefficients,

$$\mathcal{L}_{\{2\}} = A(t)\delta X\delta X + B(t)\delta X\delta Y + C(t)\delta Y\delta Y. \tag{2.20}$$

Notice that we have moved from having complete ignorance of how the fields X, Y combine to construct the Lagrangian density $\mathcal{L}$ to only requiring 3 "background" functions, $A(t), B(t), C(t)$ to be able to write $\mathcal{L}_{\{2\}}$ down. Typically, we would expect these functions to be specified in terms of the scale factor $a(t)$.

The theories we consider contribute to the gravitational field equations via the dark energy-momentum tensor, $U^{\mu\nu}$, which we define in the usual way, (2.7). The indices on the dark energy momentum tensor are symmetric by construction,

$$U_{\mu\nu} = U_{\nu\mu} = U_{(\mu\nu)}, \tag{2.21}$$

where tensor indices are symmetrised as $A_{(\mu\nu)} = \frac{1}{2}\big(A_{\mu\nu} + A_{\nu\mu}\big)$. The dark energy-momentum tensor above can be directly perturbed to give

$$\delta U^{\mu\nu} = -\frac{1}{2}\left[\sum_{\mathrm{A}}\left(\delta X^{(\mathrm{A})}\frac{1}{\sqrt{-g}}\frac{\delta}{\delta X^{(\mathrm{A})}}\frac{\delta}{\delta g_{\mu\nu}}(\sqrt{-g}\mathcal{L})\right) + U^{\mu\nu}g^{\alpha\beta}\delta g_{\alpha\beta}\right], \tag{2.22}$$

where $\{\delta X^{(\mathrm{A})}\}$ are the perturbed field variables. This can be written in a more succinct way by using the second order Lagrangian,

$$\delta U^{\mu\nu} = -\frac{1}{2}\left[4\frac{\partial(\mathcal{L}_{\{2\}})}{\partial(\delta g_{\mu\nu})} + U^{\mu\nu}g^{\alpha\beta}\delta g_{\alpha\beta}\right]. \tag{2.23}$$

Therefore, to obtain the gravitational contribution at perturbed order, due to our effective Lagrangian for perturbed field variables, one must compute the derivative of the second order Lagrangian with respect to the perturbed metric.

The equations of motion for a field X and its perturbation δX are found by regarding $\mathcal{L}$ and $\mathcal{L}_{\{2\}}$ as the relevant Lagrangian densities. Explicitly, the equations of motion for the field X and its perturbation, δX, are respectively given by

$$\partial_\mu\left(\frac{\partial\mathcal{L}}{\partial(\partial_\mu X)}\right) - \frac{\partial\mathcal{L}}{\partial X} = 0, \qquad \partial_\mu\left(\frac{\partial(\mathcal{L}_{\{2\}})}{\partial(\partial_\mu\delta X)}\right) - \frac{\partial(\mathcal{L}_{\{2\}})}{\partial\delta X} = 0. \tag{2.24}$$

The equations of motion governing the perturbation to the metric, $\delta g_{\mu\nu}$, are given by the perturbed gravitational field equations,

$$\delta G_{\mu\nu} = 8\pi G \delta T_{\mu\nu} + \delta U_{\mu\nu}. \tag{2.25}$$

The perturbed conservation law for the dark energy-momentum tensor is

$$\delta(\nabla_\mu U^{\mu\nu}) = 0, \tag{2.26}$$

which can be written as

$$\nabla_\mu \delta U^{\mu\nu} + \frac{1}{2}\Big[U^{\mu\nu} g^{\alpha\beta} - U^{\alpha\beta} g^{\mu\nu} + 2g^{\nu\beta} U^{\alpha\mu} \Big] \nabla_\mu \delta g_{\alpha\beta} = 0. \tag{2.27}$$

2.3.2 *Isotropic* $(3+1)$ *Decomposition*

One of the things it will be useful for us to do is to impose isotropy of spatial sections on the background spacetime. The motivation for doing this is that our goal is to study perturbations about an FRW background. After imposing isotropy we are able to use an isotropic $(3+1)$ decomposition to significantly simplify expressions. It is also possible to include anisotropic backgrounds as described in [37].

We foliate the 4D spacetime by 3D surfaces orthogonal to a time-like vector u_μ, which is normalized via

$$u^\mu u_\mu = -1. \tag{2.28a}$$

This induces an embedding of a 3D surface in a 4D space. The 4D metric is $g_{\mu\nu}$ and the 3D metric is $\gamma_{\mu\nu}$, and they are related by

$$\gamma_{\mu\nu} = g_{\mu\nu} + u_\mu u_\nu. \tag{2.28b}$$

The foliation implies that the time-like vector is orthogonal to the 3D metric,

$$u^\mu \gamma_{\mu\nu} = 0. \tag{2.28c}$$

The foliation induces a symmetric extrinsic curvature,

$$K_{\mu\nu} \equiv \nabla_\mu u_\nu, \tag{2.28d}$$

which is entirely spatial,

$$u^\mu K_{\mu\nu} = 0, \qquad K \equiv K^\mu{}_\mu = \gamma^{\mu\nu} K_{\mu\nu}. \tag{2.28e}$$

This can be used to deduce that

$$\nabla_\mu \gamma_{\alpha\beta} = 2K_{\mu(\alpha} u_{\beta)}, \qquad \nabla_\mu \gamma^{\mu\nu} = K u^\nu \tag{2.28f}$$

A common application of the $(3+1)$ decomposition is to write down the only energy-momentum tensor compatible with the globally isotropic FRW metric,

$$T_{\mu\nu} = \rho u_\mu u_\nu + P \gamma_{\mu\nu}. \tag{2.29}$$

There are only two "coefficients" used in the decomposition of the energy-momentum tensor: the energy-density ρ and pressure P,

$$\rho = u^\mu u^\nu T_{\mu\nu}, \qquad P = \frac{1}{3}\gamma^{\mu\nu} T_{\mu\nu}. \tag{2.30}$$

Writing a tensor as a sum over combinations of u^μ and $\gamma_{\mu\nu}$ defines the isotropic $(3+1)$ decomposition. We will now show how to decompose tensors of higher rank. For example, an isotropic vector is completely decomposed as

$$A^\mu = A u^\mu, \tag{2.31}$$

where $A = A(t)$. Notice that before we imposed isotropy upon A^μ we would need 4 functions to specify all "free" components of A^μ; by imposing isotropy we have reduced the number of "free" functions from $4 \to 1$. A symmetric rank-2 isotropic tensor is completely decomposed as

$$B_{\mu\nu} = B_1 u_\mu u_\nu + B_2 \gamma_{\mu\nu} = B_{\nu\mu}, \tag{2.32}$$

where $B_1 = B_1(t)$, $B_2 = B_2(t)$. The time-like part of $B_{\mu\nu}$ is B_1 and the space-like part is B_2. A rank-3 tensor symmetric in its second two indices is completely decomposed as

$$C_{\lambda\mu\nu} = C_1 u_\lambda \gamma_{\mu\nu} + C_2 u_\lambda u_\mu u_\nu + C_3 \gamma_{\lambda(\mu} u_{\nu)} = C_{\lambda\nu\mu}. \tag{2.33}$$

This formalism can also be used to construct tensors which are entirely spatial. For example, a rank-4 tensor defined as

$$D_{\mu\nu\alpha\beta} = D_1 \gamma_{\mu\nu}\gamma_{\alpha\beta} + D_2 \gamma_{\mu(\alpha}\gamma_{\beta)\nu}, \tag{2.34}$$

is entirely spatial, a fact which is manifested by $u^\mu D_{\mu\nu\alpha\beta} = 0$, after one notes the symmetries in the indices $D_{\mu\nu\alpha\beta} = D_{(\mu\nu)(\alpha\beta)} = D_{\alpha\beta\mu\nu}$.

The coefficients which appear in an isotropic decomposition can only have time-like derivatives. For the coefficients B_1, B_2 in (2.32) we have

$$\nabla_\mu B_1 = -\dot{B}_1 u_\mu, \qquad \nabla_\mu B_2 = -\dot{B}_2 u_\mu, \tag{2.35}$$

where an overdot is used to denote differentiation in the direction of the time-like vector: $\dot{X} \equiv u^{\mu}\nabla_{\mu}X$.

2.3.3 Perturbation Theory

We will be making substantial use of perturbation theory in this paper, and so here we will take the time to concrete the notation and terminology we use. A large portion of the technology we are about to discuss was developed, amongst other things, to model relativistic elastic materials [35, 37, 50–60]; we will recapitulate the ideas and bring the technology into the language of perturbation theory to be used with a gravitational theory.

A quantity Q is perturbed about a background value, $\bar{Q}$, as $Q = \bar{Q} + \delta Q$. For example, the metric perturbed about a background $\bar{g}_{\mu\nu}$ is written as

$$g_{\mu\nu} = \bar{g}_{\mu\nu} + \delta g_{\mu\nu}. \tag{2.36}$$

It is important to realize that the operation of index raising and lowering does not commute with the variation. For example, $\delta g_{\mu\nu} = -g_{\mu(\alpha}g_{\beta)\nu}\delta g^{\alpha\beta}$ for the metric and $\delta(\nabla^{\mu}\phi) = g^{\mu\nu}\nabla_{\nu}\delta\phi + \delta g^{\mu\nu}\nabla_{\nu}\phi$ for the derivative of a scalar field ϕ.

Consider a quantity which is perturbed about some background value, $Q(t, \mathbf{x}) = \bar{Q}(t) + \delta Q(t, \mathbf{x})$. We can then employ two classes of coordinate system to follow the perturbation δQ through evolution; time evolution can be thought of as Lie-dragging a quantity along a time-like vector, u^{μ}, to "carve out" the world-line of the perturbation, i.e. operating on a quantity with $£_{u}$. The first is where the density of the perturbations remains fixed (i.e. the coordinate system evolves to comove with the perturbations); this is a *Lagrangian* system. In the second, the coordinate system is fixed by some means (such as knowledge of the background geometry) and the density of the perturbations changes; this is an *Eulerian* system. We write perturbations in the Lagrangian system as δ_{L} and perturbations in the Eulerian system as δ_{E}. Evidently, a coordinate transformation can be used to transfer between the two systems, $x^{\mu} \rightarrow x^{\mu} + \xi^{\mu}$. The Eulerian and Lagrangian variations are linked by

$$\delta_{\mathrm{L}} = \delta_{\mathrm{E}} + £_{\xi}, \tag{2.37}$$

where $£_{\xi}$ is the Lie derivative along the gauge field ξ^{μ}. This setup is schematically depicted in Fig. 2.2.

We set the gauge field ξ^{μ} and time-like vector u_{μ} to be mutually orthogonal,

$$\xi^{\mu}u_{\mu} = 0. \tag{2.38}$$

This is because the time-like transformations which the component ξ^{0} could induce are world-line preserving, and are redundant when u^{μ} is present (which is inherently

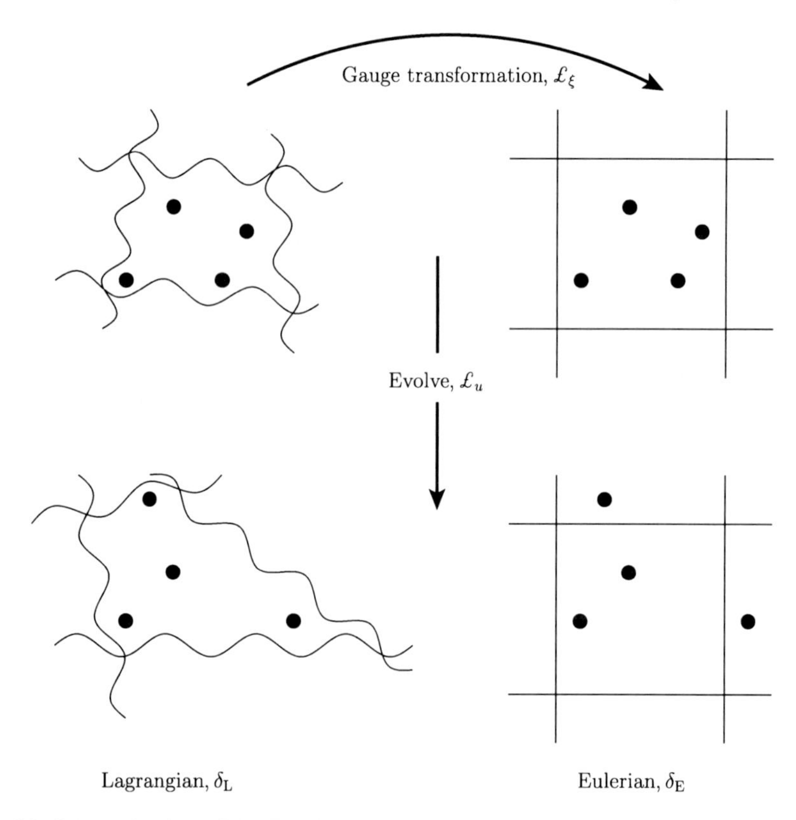

Fig. 2.2 Schematic view of the Eulerian and Lagrangian coordinate systems. The Lagrangian system can be said to be *comoving*, and the Eulerian system as being *fixed*. The Lagrangian system retains the density of a field, whereas the Eulerian system does not. This is schematically depicted by the "grid square" becoming deformed in the Lagrangian system on the *left*, to accommodate the movement of "particles" upon evolution in time. The grid square in the Eulerian system has remained fixed, meaning that the number of particles in a given square changes upon evolution. In cosmology we are perturbing against a *fixed* background: the FRW background, however calculations are often easier to perform in a comoving system. This means that physical relevance is taken from equations perturbed according to a Eulerian scheme

a world-line preserving evolution). See Fig. 2.3 for a schematic view illustrating this point. This is a choice which we discuss the ramifications of in Chap. 6.

There is an important question which arises: which perturbation scheme should we use to derive cosmologically relevant results, i.e. which δ should we use: δ_{E} or δ_{L}? In cosmological perturbation theory a quantity is perturbed from its value in a *fixed* (or known) background (such as its value in an FRW background). Therefore, equations should be perturbed relative to a fixed background, and so we should employ the Eulerian scheme.

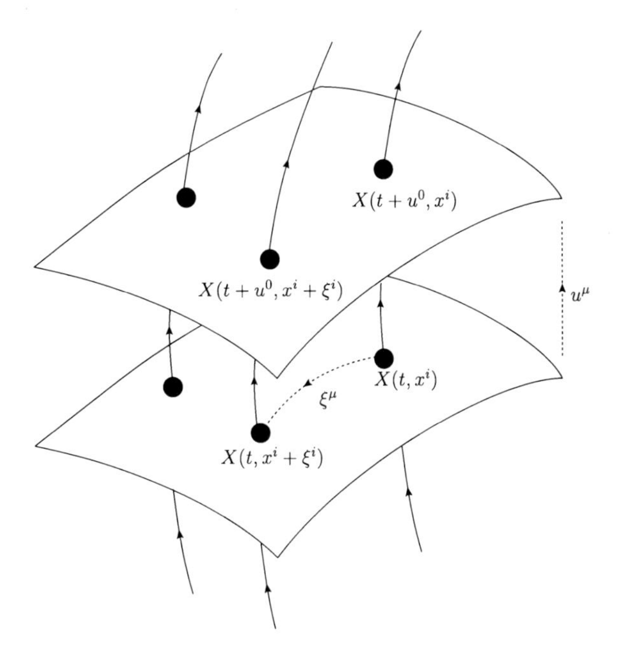

Fig. 2.3 Schematic view of the foliation and evolution, with three example world-lines drawn on, each piercing two 3D surfaces; u^μ is a time-like vector satisfying $u^\mu u_\mu = -1$. A quantity X on a surface with spacetime location (t, x^i) can be transformed into a quantity on the same surface but at a different location by transforming the coordinate on the surface, $x^i \rightarrow x^i + \xi^i$. This is a diffeomorphism which drags one world-line into another. If the time coordinate is transformed $t \rightarrow t + u^0$ then the quantity is evaluated on a different 3d surface, but on the same world-line. Thus, if we were to have a transformation $x^\mu \rightarrow x^\mu + \xi^\mu + \chi u^\mu$, where χ is an arbitrary scalar field, the time-like part of ξ^μ is redundant. Hence, we are free to set $\xi^\mu u_\mu = 0$, fixing the time-like part of the diffeomorphism field to be zero. So, we should have the interpretation that ξ^μ moves between world-lines and u^μ moves along world-lines

The equation of motion governing the metric perturbations is

$$\delta_{\rm E} G_{\mu\nu} = 8\pi G \delta_{\rm E} T_{\mu\nu} + \delta_{\rm E} U_{\mu\nu}, \tag{2.39}$$

and the perturbed conservation law that should be solved is the one evaluated in a Eulerian system,

$$\delta_{\rm E}\left(\nabla_\mu U^{\mu\nu}\right) = 0, \tag{2.40}$$

which can be written as

$$\nabla_\mu \delta_{\rm E} U^{\mu\nu} + \frac{1}{2}\left[U^{\mu\nu} g^{\alpha\beta} - U^{\alpha\beta} g^{\mu\nu} + 2g^{\nu\beta} U^{\alpha\mu}\right] \nabla_\mu \delta_{\rm E} g_{\alpha\beta} = 0. \tag{2.41}$$

If the Lagrangian variation of the dark energy-momentum tensor is the quantity that is supplied, (i.e. $\delta_{\rm L} U^{\mu\nu}$ is given), then one must be careful to use (2.37), to obtain the Eulerian perturbed quantity,

$$\delta_{\rm E} U^{\mu\nu} = \delta_{\rm L} U^{\mu\nu} - \xi^{\alpha} \nabla_{\alpha} U^{\mu\nu} + 2U^{\alpha(\mu} \nabla_{\alpha} \xi^{\nu)}. \tag{2.42}$$

Furthermore, to obtain the components of the mixed Eulerian perturbed dark energy-momentum tensor, one must use

$$\delta_{\rm E} U^{\mu}{}_{\nu} = g_{\alpha\nu} \delta_{\rm E} U^{\mu\alpha} + U^{\mu\alpha} \delta_{\rm E} g_{\nu\alpha}. \tag{2.43}$$

The Lagrangian and Eulerian perturbations of the metric are linked by

$$\delta_{\rm E} g_{\mu\nu} = \delta_{\rm L} g_{\mu\nu} - 2\nabla_{(\mu} \xi_{\nu)}. \tag{2.44}$$

For a vector field A^{μ} one finds that

$$\delta_{\rm E} A^{\mu} = \delta_{\rm L} A^{\mu} - \xi^{\alpha} \nabla_{\alpha} A^{\mu} + A^{\alpha} \nabla_{\alpha} \xi^{\mu}. \tag{2.45}$$

As final explicit example, the Eulerian and Lagrangian variations of a scalar field ϕ are linked via

$$\delta_{\rm E} \phi = \delta_{\rm L} \phi - \xi^{\mu} \nabla_{\mu} \phi. \tag{2.46}$$

An interesting lemma is that if $\nabla_{\mu} \phi \propto \dot{\phi} u_{\mu}$ then by (2.38) we find that the Eulerian and Lagrangian variations of a scalar field are identical, $\delta_{\rm E} \phi = \delta_{\rm L} \phi$. This means that a diffeomorphism does not change the perturbations of the scalar field; this is a consequence of the background field being homogeneous.

References

1. Supernova Cosmology Project Collaboration, S. Perlmutter et al. Measurements of Omega and Lambda from 42 high redshift supernovae. Astrophys. J. **517**, 565–586 (1999). [astro-ph/9812133]. The Supernova Cosmology Project
2. Supernova Search Team Collaboration, A. G. Riess et al. Observational evidence from supernovae for an accelerating universe and a cosmological constant. Astron. J. **116**, 1009–1038 (1998).[astro-ph/9805201]
3. A.G. Riess et al. BVRI light curves for 22 type Ia supernovae. Astron. J. **117**, 707–724 (1999). [astro-ph/9810291]
4. F. Zwicky, Die Rotverschiebung von extragalaktischen Nebeln. Helv. Phys. Acta. **6**, 110–127 (1933)
5. E. Komatsu et al. Seven-year wilkinson microwave anisotropy probe (wmap) observations: cosmological interpretation. Astrophys. J. Supplement Ser. **192**(218), (2011)
6. C. Skordis, The tensor-vector-scalar theory and its cosmology. Class. Quant. Grav. **26**, 143001 (2009). [arXiv:0903.3602]
7. J.D. Bekenstein, Relativistic gravitation theory for the MOND paradigm. Phys. Rev. **D70**, 083509 (2004). [astro-ph/0403694]
8. T.G. Zlosnik, P.G. Ferreira, G.D. Starkman, Modifying gravity with the Aether: an alternative to Dark Matter. Phys. Rev. **D75**, 044017 (2007). [astro-ph/0607411]

9. C. Brans, R.H. Dicke, Mach's principle and a relativistic theory of gravitation. Phys. Rev. **124**, 925–935 (1961)
10. G.W. Horndeski, Second-order scalar-tensor field equations in a four-dimensional space. Int. J. Theor. Phys. **10**, 363–384 (1974)
11. C. Charmousis, E.J. Copeland, A. Padilla, P.M. Saffin, General second order scalar-tensor theory, self tuning, and the Fab Four. Phys. Rev. Lett. **108**, 051101 (2012). [arXiv:1106.2000]
12. T. Kobayashi, M. Yamaguchi, J. Yokoyama, Generalized G-inflation: inflation with the most general second-order field equations. Prog. Theor. Phys. **126**, 511–529 (2011). [arXiv:1105.5723]
13. S. Capozziello, S. Carloni, A. Troisi, Quintessence without scalar fields. Recent Res. Dev. Astron. Astrophys. **1**, 625 (2003). [astro-ph/0303041]
14. S.M. Carroll, V. Duvvuri, M. Trodden, M.S. Turner, Is cosmic speed—up due to new gravitational physics?. Phys. Rev. **D70**, 043528 (2004). [astro-ph/0306438]
15. E.J. Copeland, M. Sami, S. Tsujikawa, Dynamics of dark energy. Int. J. Mod. Phys. **D15**, 1753–1936 (2006). [hep-th/0603057]
16. Y. Fujii, Origin of the gravitational constant and particle masses in scale invariant scalar-tensor theory. Phys. Rev. **D26**, 2580 (1982)
17. C. Armendariz-Picon, T. Damour, V.F. Mukhanov, k-Inflation. Phys. Lett. **B458**, 209–218 (1999). [hep-th/9904075]
18. C. Armendariz-Picon, V.F. Mukhanov, P.J. Steinhardt, Essentials of k essence. Phys. Rev. **D63**, 103510 (2001). [astro-ph/0006373]
19. A. Nicolis, R. Rattazzi, E. Trincherini, The Galileon as a local modification of gravity. Phys. Rev. **D79**, 064036 (2009). [arXiv:0811.2197]
20. T. Clifton, P.G. Ferreira, A. Padilla, C. Skordis, Modified gravity and cosmology. Phys. Rept. **513**, 1–189 (2012). [arXiv:1106.2476]
21. W. Hu, Structure formation with generalized dark matter. Astrophys. J.**506**, 485–494 (1998). [astro-ph/9801234]
22. J. Weller, A.M. Lewis, Large scale cosmic microwave background anisotropies and dark energy. Mon. Not. Roy. Astron. Soc. **346**, 987–993 (2003). [astro-ph/0307104]
23. R. Bean, O. Dore, Probing dark energy perturbations: the dark energy equation of state and speed of sound as measured by WMAP. Phys. Rev. **D69**, 083503 (2004). [astro-ph/0307100]
24. W. Hu, I. Sawicki, A parameterized post-friedmann framework for modified gravity. Phys. Rev. **D76**, 104043 (2007). [arXiv:0708.1190]
25. W. Hu, Parametrized post-friedmann signatures of acceleration in the CMB. Phys. Rev. **D77**, 103524 (2008). [arXiv:0801.2433]
26. L. Amendola, M. Kunz, D. Sapone, Measuring the dark side (with weak lensing). JCAP **0804**, 013 (2008). [arXiv:0704.2421]
27. C. Skordis, Consistent cosmological modifications to the Einstein equations. Phys. Rev. **D79**, 123527 (2009). [arXiv:0806.1238]
28. R. Bean, M. Tangmatitham, Current constraints on the cosmic growth history. Phys. Rev. **D81**, 083534 (2010). [arXiv:1002.4197]
29. L. Pogosian, A. Silvestri, K. Koyama, G.-B. Zhao, How to optimally parametrize deviations from General Relativity in the evolution of cosmological perturbations? Phys. Rev. **D81**, 104023 (2010). [arXiv:1002.2382]
30. S.A. Appleby, J. Weller, Parameterizing scalar-tensor theories for cosmological probes. JCAP **1012**, 006 (2010). [arXiv:1008.2693]
31. A. Hojjati, L. Pogosian, G.-B. Zhao, Testing gravity with CAMB and CosmoMC. JCAP **1108**, 005 (2011). [arXiv:1106.4543]
32. T. Baker, P.G. Ferreira, C. Skordis, J. Zuntz, Towards a fully consistent parameterization of modified gravity. Phys. Rev. **D84**, 124018 (2011). [arXiv:1107.0491]
33. T. Baker, Phi Zeta Delta: growth of perturbations in parameterized gravity for an Einstein-de Sitter Universe. Phys. Rev. **D85**, 044020 (2012). [arXiv:1111.3947]
34. C.M. Will, Theory and experiment in gravitational physics. (Cambridge University Press, Cambridge, 1993)

35. R.A. Battye, A. Moss, Cosmological perturbations in elastic dark energy models. Phys. Rev. **D76**, 023005 (2007). [astro-ph/0703744]
36. R.A. Battye, A. Moss, Constraints on the solid dark universe model. JCAP **0506**, 001 (2005). [astro-ph/0503033]
37. R.A. Battye, A. Moss, Anisotropic perturbations due to dark energy. Phys. Rev. **D74**, 041301 (2006). [astro-ph/0602377]
38. R. Battye, A. Moss, Anisotropic dark energy and CMB anomalies. Phys. Rev. **D80**, 023531 (2009). [arXiv:0905.3403]
39. S. Weinberg, Effective field theory for inflation. Phys. Rev. **D77**, 123541 (2008). [arXiv:0804.4291]
40. C. Cheung, P. Creminelli, A.L. Fitzpatrick, J. Kaplan, L. Senatore, The effective field theory of inflation. JHEP **03**, 014 (2008). [arXiv:0709.0293]
41. P. Creminelli, G. D'Amico, J. Norena, F. Vernizzi, The effective theory of Quintessence: the w¡-1 side unveiled. JCAP **0902**, 018 (2009). [arXiv:0811.0827]
42. R. Caldwell, A. Cooray, A. Melchiorri, Constraints on a new post-general relativity cosmological parameter. Phys. Rev. **D76**, 023507 (2007). [astro-ph/0703375]
43. J.N. Dossett, M. Ishak, J. Moldenhauer, Testing general relativity at cosmological scales: implementation and parameter correlations. Phys. Rev. **D84**, 123001 (2011). [arXiv:1109.4583]
44. D. Kirk, I. Laszlo, S. Bridle, R. Bean, Optimising cosmic shear surveys to measure modifications to gravity on cosmic scales. arXiv:1109.4536
45. I. Laszlo, R. Bean, D. Kirk, S. Bridle, Disentangling dark energy and cosmic tests of gravity from weak lensing systematics. arXiv:1109.4535
46. J. Zuntz, T. Baker, P. Ferreira, C. Skordis, Ambiguous tests of general relativity on cosmological scales. JCAP **1206**, 032 (2012). [arXiv:1110.3830]
47. T.D. Lee, A theory of spontaneous *t* violation. Phys. Rev. **D8**, 1226–1239 (1973)
48. B. Carter, Equations of motion of a stiff geodynamic string or higher brane. Class. Quant. Grav. **11**(11), 2677 (1994)
49. R.A. Battye, B. Carter, Second order Lagrangian and symplectic current for gravitationally perturbed Dirac-Goto-Nambu strings and branes. Class. Quant. Grav. **17**, 3325–3334 (2000). [hep-th/9811075]
50. B. Carter, H. Quintana, Foundations of general relativistic high-pressure elasticity theory. Proc. R. Soci. Lond. Math. Phys. Sci. **331**(1584), 57–83 (1972)
51. B. Carter, Elastic perturbation theory in general relativity and a variation principle for a rotating solid star. Commun. Math. Phys. **30**, 261–286 (1973)
52. B. Carter, Speed of sound in a high-pressure general-relativistic solid. Phys. Rev. **D7**, 1590–1593 (1973)
53. J.L. Friedman, B.F. Schutz, Erratum: on the stability of relativistic systems. Astrophys. J. **200**, 204–220 (1975)
54. B. Carter, H. Quintana, Gravitational and acoustic waves in an elastic medium. Phys. Rev. **D16**(1016), 2928–2938 (1977)
55. B. Carter, Rheometric structure theory, convective differentiation and continuum electrodynamics. Proc. R. Soci. Lond. Math. Phys. Sci. **372**(1749), 169–200 (1980)
56. L.D. Landau, E.M. Lifshitz, *Theory of Elasticity*, 3rd edn. (Pergamon Press, Oxford, 1986)
57. M. Bucher, D.N. Spergel, Is the dark matter a solid?. Phys. Rev. **D60**, 043505 (1999). [astro-ph/9812022]
58. B. Carter, Interaction of gravitational waves with an elastic solid medium. gr-qc/0102113
59. T. Azeyanagi, M. Fukuma, H. Kawai, K. Yoshida, Universal description of viscoelasticity with foliation preserving diffeomorphisms. Phys. Lett. **B681**, 290–295 (2009). [arXiv:0907.0656]
60. M. Fukuma, Y. Sakatani, Relativistic viscoelastic fluid mechanics. Phys. Rev. **E84**, 026316 (2011). [arXiv:1104.1416]

Chapter 3
Metric Only and First Order Scalar Field Theory

3.1 Introduction

We will now apply the formalism we developed in Chap. 2 to the two simplest possible theories: first where the field content of the dark sector only contains the metric (and no derivatives thereof) and secondly where the field content of the dark sector contains a scalar field, its first derivative and the metric. We will not *a priori* impose any theoretical structure upon the field content, but we will show how this can be done. We will conclude this chapter by providing explicit formulae for the components of the dark energy momentum tensor that source the perturbed gravitational field equations for these simple theories.

We remind that the Lagrangian perturbed metric is calculated from the Eulerian perturbed metric via

$$\delta_{\rm L} g_{\mu\nu} = \delta_{\rm E} g_{\mu\nu} + 2\nabla_{(\mu}\xi_{\nu)}, \tag{3.1}$$

and that the dark energy momentum tensor is constructed from the Lagrangian for perturbations via

$$\delta_{\rm L} U^{\mu\nu} = -\frac{1}{2}\left[4\frac{\hat{\delta}}{\hat{\delta}\delta_{\rm L} g_{\mu\nu}}\mathcal{L}_{\{2\}} + U^{\mu\nu} g^{\alpha\beta}\delta_{\rm L} g_{\alpha\beta}\right]. \tag{3.2}$$

3.2 No Extra Fields: $\mathcal{L} = \mathcal{L}(g_{\mu\nu})$

Our first and simplest example is where the dark sector does not contain any extra fields: only the metric is present, albeit in an arbitrary combination. In this section we do not allow the dark sector to contain derivatives of the metric—this is discussed in a subsequent chapter. One of the aims of this section is to build an intuition for

J. Pearson, *Generalized Perturbations in Modified Gravity and Dark Energy*, Springer Theses, DOI: 10.1007/978-3-319-01210-0_3,

understanding how to write down perturbative quantities and how to decompose tensors which arise in the perturbative equations. The Lagrangian density we will consider in this section is of the form

$$\mathcal{L} = \mathcal{L}(g_{\mu\nu}). \tag{3.3}$$

Thse theories contain the cosmological constant, elastic dark energy [1, 2], linearized massive gravity theories [3–11], and will also include more general theories that have not been previously considered. The Lagrangian for perturbations for this field content is simply

$$\mathcal{L}_{\{2\}} = \frac{1}{8}\mathcal{W}^{\mu\nu\alpha\beta}\delta_{\mathrm{L}}g_{\mu\nu}\delta_{\mathrm{L}}g_{\alpha\beta}. \tag{3.4}$$

The rank-4 tensor $\mathcal{W}_{\mu\nu\alpha\beta}$ is only a function of background quantities, and is therefore manifestly gauge invariant. By inspecting (3.4) it follows that the tensor $\mathcal{W}_{\alpha\beta\mu\nu}$ enjoys the following symmetries,

$$\mathcal{W}_{\alpha\beta\mu\nu} = \mathcal{W}_{(\alpha\beta)(\mu\nu)} = \mathcal{W}_{\mu\nu\alpha\beta}. \tag{3.5}$$

This theory is very simple in the sense that functional derivatives of $\mathcal{L}_{\{2\}}$ can be easily "read off". This will not be the case in more complicated theories, and so here we will take some care to be methodical about how we compute a functional derivative. Varying the Lagrangian perturbation to the metric $\delta_{\mathrm{L}}g_{\mu\nu} \rightarrow \delta_{\mathrm{L}}g_{\mu\nu} + \hat{\delta}\delta_{\mathrm{L}}g_{\mu\nu}$ induces a response $\mathcal{L}_{\{2\}} \rightarrow \mathcal{L}_{\{2\}} + \hat{\delta}\mathcal{L}_{\{2\}}$, where one can easily compute

$$\hat{\delta}\mathcal{L}_{\{2\}} = \frac{1}{4}\mathcal{W}^{\mu\nu\alpha\beta}\delta_{\mathrm{L}}g_{\alpha\beta}\hat{\delta}\delta_{\mathrm{L}}g_{\mu\nu}. \tag{3.6}$$

Thus, the functional derivative of $\mathcal{L}_{\{2\}}$ with respect to the perturbed metric is

$$\frac{\hat{\delta}}{\hat{\delta}\delta_{\mathrm{L}}g_{\mu\nu}}\mathcal{L}_{\{2\}} = \frac{1}{4}\mathcal{W}^{\mu\nu\alpha\beta}\delta_{\mathrm{L}}g_{\alpha\beta}. \tag{3.7}$$

We can use (3.2), (3.4) and (3.7) to find that the perturbations to the dark energy momentum tensor are given by

$$\delta_{\mathrm{L}}U^{\mu\nu} = -\frac{1}{2}\left\{\mathcal{W}^{\alpha\beta\mu\nu} + g^{\alpha\beta}U^{\mu\nu}\right\}\delta_{\mathrm{L}}g_{\alpha\beta}. \tag{3.8}$$

Because it is the Lagrangian variation which appears above we must convert to Eulerian variations to obtain cosmologically relevant perturbations. By using (2.42) and (2.44) in (3.8) we obtain

$$\delta_{\mathrm{E}}U^{\mu\nu} = -\frac{1}{2}\left\{\mathcal{W}^{\alpha\beta\mu\nu} + g^{\alpha\beta}U^{\mu\nu}\right\}\left(\delta_{\mathrm{E}}g_{\alpha\beta} + 2\nabla_{(\alpha}\xi_{\beta)}\right) - \xi^{\alpha}\nabla_{\alpha}U^{\mu\nu} + 2U^{\alpha(\mu}\nabla_{\alpha}\xi^{\nu)}. \tag{3.9}$$

To find the equation of motion of the vector field, ξ^{μ}, we must compute the Eulerian perturbed Bianchi identity. Substituting (3.9) into (2.41) we obtain

$$2\Big[L^{\mu\alpha\beta\nu}\Big]\nabla_{\mu}\nabla_{\alpha}\xi_{\beta} + 2\Big[\nabla_{\sigma}\mathcal{W}^{\sigma\nu\mu\alpha}\Big]\nabla_{\mu}\xi_{\alpha} + 2\Big[\nabla_{\mu}\nabla_{\alpha}U^{\mu\nu}\Big]\xi^{\alpha} = \delta_{\rm E}J^{\nu}, \quad (3.10a)$$

where, for convenience, we defined

$$L^{\mu\alpha\beta\nu} \equiv \mathcal{W}^{\mu\nu\alpha\beta} + g^{\alpha\beta}U^{\mu\nu} - 2U^{\alpha(\mu}g^{\nu)\beta}, \quad (3.10b)$$

and where the perturbed source term, $\delta_{\rm E}J^{\nu}$, is given by

$$\delta_{\rm E}J^{\nu} \equiv \Big[2g^{\nu\beta}U^{\alpha\mu} - U^{\alpha\beta}g^{\mu\nu} - \mathcal{W}^{\mu\nu\alpha\beta}\Big]\nabla_{\mu}\delta_{\rm E}g_{\alpha\beta} - \Big[\nabla_{\mu}\mathcal{W}^{\mu\nu\alpha\beta}\Big]\delta_{\rm E}g_{\alpha\beta}. \quad (3.10c)$$

Here we observe that the metric perturbations $\delta_{\rm E}g_{\mu\nu}$ and the diffeomorphism field ξ^{μ} are intimately linked: one cannot consistently set either to zero (in general). Equation (3.10a) is the constraint equation for any parameters/functions that appear in a parameterization of the dark sector, under the rather general assumption that $\mathcal{L} = \mathcal{L}(g_{\mu\nu})$; the only freedom that remains is how to construct $\mathcal{W}_{\mu\nu\alpha\beta}$ out of background quantities. In Sect. 3.4 we will provide the components of the equation of motion for a perturbed FRW spacetime.

The only way to write the tensors $U_{\mu\nu}$, $\mathcal{W}_{\alpha\beta\mu\nu}$ with an isotropic $(3+1)$ decomposition which respects the symmetries (3.5) is

$$U_{\mu\nu} = \rho u_{\mu}u_{\nu} + P\gamma_{\mu\nu}, \quad (3.11a)$$

$$\begin{aligned}\mathcal{W}_{\mu\nu\alpha\beta} &= A_{\mathcal{W}}u_{\mu}u_{\nu}u_{\alpha}u_{\beta} + B_{\mathcal{W}}\big(\gamma_{\mu\nu}u_{\alpha}u_{\beta} + \gamma_{\alpha\beta}u_{\mu}u_{\nu}\big)\\ &\quad + 2C_{\mathcal{W}}\big(\gamma_{\mu(\alpha}u_{\beta)}u_{\nu} + \gamma_{\nu(\alpha}u_{\beta)}u_{\mu}\big) + \mathcal{E}_{\mu\nu\alpha\beta}, \end{aligned} \quad (3.11b)$$

where $\mathcal{E}_{\mu\nu\alpha\beta}$ respects the same symmetries as $\mathcal{W}_{\mu\nu\alpha\beta}$, satisfies $u^{\mu}\mathcal{E}_{\mu\nu\alpha\beta} = 0$ (i.e. $\mathcal{E}_{\mu\nu\alpha\beta}$ is entirely spatial) and is given by

$$\mathcal{E}_{\mu\nu\alpha\beta} = D_{\mathcal{W}}\gamma_{\mu\nu}\gamma_{\alpha\beta} + 2E_{\mathcal{W}}\gamma_{\mu(\alpha}\gamma_{\beta)\nu}. \quad (3.11c)$$

A concrete example of a theory which only contains the metric is the elastic dark energy theory [1, 2] where one can find that the coefficients in terms of physical quantities such as energy density ρ, pressure P, bulk β and shear moduli μ are given by

$$A_{\mathcal{W}} = -\rho, \qquad B_{\mathcal{W}} = P, \qquad C_{\mathcal{W}} = -P, \quad (3.12a)$$

$$D_{\mathcal{W}} = \beta - P - \frac{2}{3}\mu, \qquad E_{\mathcal{W}} = \mu + P, \quad (3.12b)$$

where the bulk modulus is defined via $\beta \equiv (\rho + P)\frac{\mathrm{d}P}{\mathrm{d}\rho}$, and the pressure and shear modulus are functions of the density $P = P(\rho)$, $\mu = \mu(\rho)$ (e.g. one way to choose these functional dependancies is with an "equation of state", w and $\hat{\mu}$, so that $P = w\rho$, $\mu = \hat{\mu}\rho$).

We will now briefly study the structure of the dark sector Lagrangian (3.4). Inserting (3.1) into the Lagrangian (3.4) yields

$$\mathcal{L}_{\{2\}} = \frac{1}{8}\mathcal{W}^{\mu\nu\alpha\beta}\delta_{\mathrm{E}}g_{\mu\nu}\delta_{\mathrm{E}}g_{\alpha\beta} + \frac{1}{2}\mathcal{W}^{\mu\nu\alpha\beta}\delta_{\mathrm{E}}g_{\mu\nu}\nabla_{(\mu}\xi_{\nu)} + \frac{1}{2}\mathcal{W}^{\mu\nu\alpha\beta}\nabla_{(\mu}\xi_{\nu)}\nabla_{(\alpha}\xi_{\beta)}. \quad (3.13)$$

It is now manifest that the theory describes a vector field interacting with the perturbed metric. The role of the vector field can be thought of as being "artificial" in the sense that it is there to restore gauge invariance. An alternative interpretation is that the vector field is a genuine field, which is the case in elastic dark energy, where the vector field describes the deformations of an elastic substance. To further understand of the structure of the Lagrangian, we parameterize the Lagrangian perturbations to the metric as

$$\delta_{\mathrm{L}}g_{\mu\nu} = 2\Phi u_{\mu}u_{\nu} + 2N_{(\mu}u_{\nu)} + \bar{H}_{\alpha\beta}\gamma^{\alpha}{}_{\mu}\gamma^{\beta}{}_{\nu}, \quad (3.14)$$

where $N^{\mu}u_{\mu} = 0$. Inserting (3.14) into the Lagrangian (3.4), along with the $(3+1)$ decomposition of $\mathcal{W}$ (3.11b), we obtain the following expression for $\mathcal{L}_{\{2\}}$ in terms of the Lagrangian perturbed metric:

$$8\mathcal{L}_{\{2\}} = 4A_{\mathcal{W}}\Phi^2 + 4B_{\mathcal{W}}\bar{H}\Phi + 2C_{\mathcal{W}}N^{\alpha}N_{\alpha} + D_{\mathcal{W}}\bar{H}^2 + 2E_{\mathcal{W}}\bar{H}^{\alpha\beta}\bar{H}_{\alpha\beta}, \quad (3.15)$$

where $\bar{H} = \bar{H}^{i}{}_{i}$. Thus, we see that the functions $A_{\mathcal{W}}, \dots, E_{\mathcal{W}}$ parameterize the interactions between the perturbed metric components. This is more commonly studied in the content of massive gravity theories, where the $A_{\mathcal{W}}, \dots, E_{\mathcal{W}}$ play the roles of masses of the gravitons. We further study this connection to massive gravity theories in Chap. 6.

To summarize: we have written down a Lagrangian for perturbations (3.4) under the simple assumption that only the metric is present in the dark sector. We then computed the Lagrangian (3.8) and Eulerian (3.9) perturbation to the dark energy momentum tensor, which sources the perturbed gravitational field equations. These formulae enabled us to write down the perturbed conservation Eq. (3.10a) which must be solved in tandem with the gravitational field equations. We showed that the theory actively interacts a vector field ξ^{μ} with the Eulerian perturbed metric. By employing a $(3+1)$ decomposition of the background tensors, we identified that a maximum of just five functions are required to specify the Lagrangian for perturbations in the dark sector when no extra fields are present. These five functions are

$$X = \left\{A_{\mathcal{W}}, B_{\mathcal{W}}, C_{\mathcal{W}}, D_{\mathcal{W}}, E_{\mathcal{W}}\right\} \quad (3.16)$$

where each function only depends on background quantities.

3.3 Scalar Field Theory: $\mathcal{L} = \mathcal{L}(g_{\mu\nu}, \phi, \nabla_\mu\phi)$

The second example is when the dark sector contains an arbitrary combination of scalar field ϕ, the first derivative of the field $\nabla_\mu\phi$, and the metric $g_{\mu\nu}$. This encompasses scalar field theories such as quintessence and k-essence, but we could also encompass a range of other possible theories.

For a theory with field content

$$\mathcal{L} = \mathcal{L}(g_{\mu\nu}, \phi, \nabla_\mu\phi) \tag{3.17}$$

the Lagrangian for perturbations is given by

$$\begin{aligned}\mathcal{L}_{\{2\}} &= \mathcal{A}(\delta_{\rm L}\phi)^2 + \mathcal{B}^\mu\delta_{\rm L}\phi\nabla_\mu\delta_{\rm L}\phi + \frac{1}{2}\mathcal{C}^{\mu\nu}\nabla_\mu\delta_{\rm L}\phi\nabla_\nu\delta_{\rm L}\phi \\ &+\frac{1}{4}\left[\mathcal{Y}^{\alpha\mu\nu}\nabla_\alpha\delta_{\rm L}\phi\delta_{\rm L}g_{\mu\nu} + \mathcal{V}^{\mu\nu}\delta_{\rm L}\phi\delta_{\rm L}g_{\mu\nu} + \frac{1}{2}\mathcal{W}^{\mu\nu\alpha\beta}\delta_{\rm L}g_{\mu\nu}\delta_{\rm L}g_{\alpha\beta}\right].\end{aligned} \tag{3.18}$$

It is useful to regard the quantities $\mathcal{A}, \ldots, \mathcal{W}^{\mu\nu\alpha\beta}$ as *coupling coefficients*, since they dictate the coupling between various perturbed field variables in the Lagrangian for perturbations. The coupling coefficients above comprise: one scalar $\mathcal{A}$, one vector $\mathcal{B}^\mu$, two rank-2 tensors $\mathcal{C}^{\mu\nu}$, $\mathcal{V}^{\mu\nu}$, one rank-3 tensor, $\mathcal{Y}_{\alpha\mu\nu}$ and one rank-4 tensor $\mathcal{W}_{\mu\nu\alpha\beta}$, all of which are only functions of background quantities and are therefore gauge invariant. At first sight these are all independent quantities, but we will show later that the conservation and Euler-Lagrange equations can be used to link the quantities. By inspecting (3.18) these tensors enjoy the following symmetries,

$$\mathcal{C}^{\mu\nu} = \mathcal{C}^{(\mu\nu)}, \qquad \mathcal{Y}^{\alpha\mu\nu} = \mathcal{Y}^{\alpha(\mu\nu)}, \qquad \mathcal{V}^{\mu\nu} = \mathcal{V}^{(\mu\nu)}, \tag{3.19a}$$

$$\mathcal{W}^{\alpha\beta\mu\nu} = \mathcal{W}^{(\alpha\beta)(\mu\nu)} = \mathcal{W}^{\mu\nu\alpha\beta}. \tag{3.19b}$$

In what follows we will assume $\gamma^\mu{}_\nu\nabla_\mu\phi = 0$ (alternatively this can be stated as $\xi^\mu\nabla_\mu\phi = 0$), so that $\delta_{\rm E}\phi = \delta_{\rm L}\phi$. This is the covariant statement that $\nabla_\mu\phi$ is entirely time-like, while using the fact that the diffeomorphism is entirely space-like. Therefore, since the Eulerian and Lagrangian perturbations of a scalar field are identical we will not distinguish between them and we will write $\delta\phi \equiv \delta_{\rm E}\phi = \delta_{\rm L}\phi$.

We now want to apply the variational principle to this Lagrangian for perturbations. This means that we must vary the perturbed field variables, and we will eventually require that $\mathcal{L}_{\{2\}}$ is independent of these varied perturbed fields. Importantly, realize that these virtual variations will not affect the unperturbed field variables, ϕ, $g_{\mu\nu}$. The virtual variations are given by

$$\delta\phi \to \delta\phi + \hat{\delta}\delta\phi, \qquad \delta_{\rm L}g_{\mu\nu} \to \delta_{\rm L}g_{\mu\nu} + \hat{\delta}\delta_{\rm L}g_{\mu\nu}. \tag{3.20}$$

The Lagrangian will respond with a virtual variation, $\mathcal{L}_{\{2\}} \to \mathcal{L}_{\{2\}} + \hat{\delta}\hat{\delta}\mathcal{L}_{\{2\}}$ where

$$\hat{\delta}\mathcal{L}_{\{2\}} = A\hat{\delta}\delta\phi + B^{\mu}\nabla_{\mu}\hat{\delta}\delta\phi + C^{\mu\nu}\hat{\delta}\delta_{\mathrm{L}}g_{\mu\nu}, \tag{3.21}$$

where we defined

$$A \equiv \left[2\mathcal{A} + \mathcal{B}^{\mu}\nabla_{\mu}\right]\delta\phi + \left[\frac{1}{4}\mathcal{V}^{\mu\nu}\right]\delta_{\mathrm{L}}g_{\mu\nu}, \tag{3.22a}$$

$$B^{\mu} \equiv \left[\mathcal{B}^{\mu} + \mathcal{C}^{\mu\nu}\nabla_{\nu}\right]\delta\phi + \left[\frac{1}{4}\mathcal{Y}^{\mu\alpha\beta}\right]\delta_{\mathrm{L}}g_{\alpha\beta}, \tag{3.22b}$$

$$4C^{\mu\nu} \equiv \left[\mathcal{V}^{\mu\nu} + \mathcal{Y}^{\alpha\mu\nu}\nabla_{\alpha}\right]\delta\phi + \left[\mathcal{W}^{\mu\nu\alpha\beta}\right]\delta_{\mathrm{L}}g_{\alpha\beta}. \tag{3.22c}$$

After isolating the total derivative in (3.21), the variation of $\mathcal{L}_{\{2\}}$ can be compactly written as

$$\hat{\delta}\mathcal{L}_{\{2\}} = \left[A - \nabla_{\mu}B^{\mu}\right]\hat{\delta}\delta\phi + C^{\mu\nu}\hat{\delta}\delta_{\mathrm{L}}g_{\mu\nu} + \nabla_{\mu}\left[B^{\mu}\hat{\delta}\delta\phi\right]. \tag{3.23}$$

We can now read off the functional derivatives of $\mathcal{L}_{\{2\}}$ with respect to the perturbed field variables:

$$\frac{\hat{\delta}}{\hat{\delta}\delta\phi}\mathcal{L}_{\{2\}} = A - \nabla_{\mu}B^{\mu}, \tag{3.24a}$$

$$\frac{\hat{\delta}}{\hat{\delta}\delta_{\mathrm{L}}g_{\mu\nu}}\mathcal{L}_{\{2\}} = C^{\mu\nu}. \tag{3.24b}$$

The variation also contains a total derivative, which we write as

$$\vartheta^{\mu} = B^{\mu}\hat{\delta}\delta\phi. \tag{3.24c}$$

Equation (3.24a) will provide us with the equation of motion of the $\delta\phi$-field, (3.24b) will provide us with $\delta U^{\mu\nu}$ and (3.24c) provides the condition that $\hat{\delta}\delta\phi$ must vanish on the boundary for a well-posed variational principle to be used.

3.3.1 Perturbed Euler–Lagrange Equation

The equation of motion of the perturbed scalar field, $\delta\phi$, is found by demanding that (3.24a) vanishes. This yields

$$\nabla_{\mu}B^{\mu} = A. \tag{3.25}$$

Substituting the definitions of A, B from (3.22a) into (3.25) yields

$$\mathcal{C}^{\mu\nu}\nabla_\mu\nabla_\nu\delta\phi + \big(\nabla_\mu\mathcal{C}^{\mu\nu}\big)\nabla_\nu\delta\phi + \big(\nabla_\mu\mathcal{B}^\mu - 2\mathcal{A}\big)\delta\phi = \delta_{\mathrm{E}}S, \tag{3.26}$$

where the "perturbed source" $\delta_{\mathrm{E}}S$ is given by

$$\delta_{\mathrm{E}}S \equiv \frac{1}{4}\Big[\big(\mathcal{V}^{\alpha\beta} - \nabla_\mu\mathcal{Y}^{\mu\alpha\beta}\big)\delta_{\mathrm{L}}g_{\alpha\beta} - \mathcal{Y}^{\mu\alpha\beta}\nabla_\mu\delta_{\mathrm{L}}g_{\alpha\beta}\Big], \tag{3.27}$$

where $\delta_{\mathrm{L}}g_{\alpha\beta} = \delta_{\mathrm{E}}g_{\alpha\beta} + 2\nabla_{(\alpha}\xi_{\beta)}$. We note that $\mathcal{C}^{\mu\nu}$ plays the role of an "effective metric", due to its resemblance to the corresponding term in the perturbed Klein-Gordon equation, namely $g^{\mu\nu}\nabla_\mu\nabla_\nu\delta\phi$, and there is also an effective mass of the $\delta\phi$-field, $M^2_{\mathrm{eff}} = \nabla_\mu\mathcal{B}^\mu - 2\mathcal{A}$.

3.3.2 Perturbed Dark Energy-Momentum Tensor and Conservation Equation

The Lagrangian perturbations to the dark energy-momentum tensor are found by using (3.24b) in (3.2), which yields

$$\delta_{\mathrm{L}}U^{\mu\nu} = -\frac{1}{2}\Big\{\mathcal{V}^{\mu\nu}\delta\phi + \mathcal{Y}^{\alpha\mu\nu}\nabla_\alpha\delta\phi\Big\} - \frac{1}{2}\Big\{\mathcal{W}^{\alpha\beta\mu\nu} + g^{\alpha\beta}U^{\mu\nu}\Big\}\delta_{\mathrm{L}}g_{\alpha\beta}. \tag{3.28}$$

The Eulerian perturbed conservation law $\delta_{\mathrm{E}}(\nabla_\mu U^{\mu\nu}) = 0$ can be computed using (3.28). We obtain

$$\mathcal{Y}^{\alpha\mu\nu}\nabla_\mu\nabla_\alpha\delta\phi + \big(\mathcal{V}^{\nu\alpha} + \nabla_\mu\mathcal{Y}^{\alpha\mu\nu}\big)\nabla_\alpha\delta\phi + \nabla_\mu\mathcal{V}^{\mu\nu}\delta\phi = \delta_{\mathrm{E}}J^\nu + 2E^\nu, \tag{3.29}$$

where the sources are written as a contribution from the Eulerian perturbed metric, $\delta_{\mathrm{E}}J^\nu$, and a contribution from the ξ^μ field, and are given by

$$\delta_{\mathrm{E}}J^\nu \equiv -(\mathcal{W}^{\mu\nu\alpha\beta} + U^{\alpha\beta}g^{\mu\nu} - 2g^{\nu\beta}U^{\mu\alpha})\nabla_\mu\delta_{\mathrm{E}}g_{\alpha\beta} - (\nabla_\mu\mathcal{W}^{\mu\nu\alpha\beta})\delta_{\mathrm{E}}g_{\alpha\beta}, \tag{3.30a}$$

$$E^\nu \equiv -L^{\mu\alpha\beta\nu}\nabla_\mu\nabla_\alpha\xi_\beta - (\nabla_\mu\mathcal{W}^{\mu\nu\alpha\beta})\nabla_\alpha\xi_\beta - (\nabla_\mu\nabla_\alpha U^{\mu\nu})\xi^\alpha, \tag{3.30b}$$

where

$$L^{\mu\alpha\beta\nu} \equiv \mathcal{W}^{\mu\nu\alpha\beta} + g^{\alpha\beta}U^{\mu\nu} - 2U^{\alpha(\mu}g^{\nu)\beta}. \tag{3.30c}$$

Equation (3.29) is an evolution equation for the scalar field perturbation $\delta\phi$, sourced by the metric perturbations, $\delta_{\mathrm{E}}g_{\mu\nu}$, and the vector field, ξ^μ. The scalar field perturbation sources the equation of motion for $\delta_{\mathrm{E}}g_{\mu\nu}$, i.e. the perturbed gravitational field equations, via the components $\delta_{\mathrm{E}}U^\mu{}_\nu$. In general one cannot consistently solve the evolution equations for $\delta\phi$ independently from those for the vector field ξ^μ; in Sect.

3.3.4.2 we will show how these two fields might decouple, but the decoupling only occurs in special cases.

3.3.3 (3 + 1) Decomposition of the Coupling Coefficients

We decompose the coupling coefficients $\mathcal{A}, \mathcal{B}, \mathcal{C}, \mathcal{V}, \mathcal{Y}, \mathcal{W}$ which appear in the Lagrangian for perturbations (3.18) with an isotropic (3 + 1) split, whilst respecting the symmetries (3.19a), and obtain

$$\mathcal{A} = A_{\mathcal{A}}, \tag{3.31a}$$
$$\mathcal{B}^{\mu} = A_{\mathcal{B}} u^{\mu}, \tag{3.31b}$$
$$\mathcal{C}_{\mu\nu} = A_{\mathcal{C}} u_{\mu} u_{\nu} + B_{\mathcal{C}} \gamma_{\mu\nu}, \tag{3.31c}$$
$$\mathcal{V}_{\mu\nu} = A_{\mathcal{V}} u_{\mu} u_{\nu} + B_{\mathcal{V}} \gamma_{\mu\nu}, \tag{3.31d}$$
$$\mathcal{Y}_{\alpha\mu\nu} = A_{\mathcal{Y}} u_{\alpha} u_{\mu} u_{\nu} + B_{\mathcal{Y}} u_{\alpha} \gamma_{\mu\nu} + 2C_{\mathcal{Y}} \gamma_{\alpha(\mu} u_{\nu)}. \tag{3.31e}$$
$$\mathcal{W}_{\mu\nu\alpha\beta} = A_{\mathcal{W}} u_{\mu} u_{\nu} u_{\alpha} u_{\beta} + B_{\mathcal{W}} \big(\gamma_{\mu\nu} u_{\alpha} u_{\beta} + \gamma_{\alpha\beta} u_{\mu} u_{\nu}\big) \tag{3.31f}$$
$$+ 2C_{\mathcal{W}} \big(\gamma_{\mu(\alpha} u_{\beta)} u_{\nu} + \gamma_{\nu(\alpha} u_{\beta)} u_{\mu}\big) + D_{\mathcal{W}} \gamma_{\mu\nu} \gamma_{\alpha\beta} + 2E_{\mathcal{W}} \gamma_{\mu(\alpha} \gamma_{\beta)\nu}. \tag{3.31g}$$

There are a total of 14 "free" functions in this (3 + 1) decomposition. This is an absolute upper bound on the freedom in the theory under the assumption that the background is spatially isotropic. In the subsequent discussions we will show how the number of free functions dramatically decreases when we begin to impose simple theoretical restrictions.

3.3.4 Imposing Theoretical Restrictions

We will now show how to impose realistic theoretical restrictions upon the theory. This will aid the link between our generalized scalar field theory and explicit theories, and will also serve to motivate the reduction of freedom in the theory.

3.3.4.1 Linking Conditions

The perturbed Euler-Lagrange equation (3.26) and perturbed conservation law (3.29) are both evolution equations for $\delta\phi$, and both have apparently different coefficients, resulting in an over-determined system. We can choose to remove this apparent over-determination by "forcing" the time-like (i.e. scalar) part of the perturbed conservation law to be identical to the perturbed Euler-Lagrange equation. It is important to realize that the perturbed conservation law is a vector equation, and only one of its components can be set equal to the perturbed Euler-Lagrange equation; the other components of the vector equation will introduce a set of constraint equations.

When we contract the perturbed conservation equation with a time-like vector $\tau_\mu = \omega u_\mu$ (where $\nabla_\mu \omega = -u_\mu \dot{\omega}$), we can read off a set of conditions that link the coefficients appearing in the Euler-Lagrange equation (3.26) and the perturbed conservation equation (3.29). Doing this we obtain the linking conditions

$$\mathcal{C}^{\mu\alpha} = \tau_\nu \mathcal{Y}^{\alpha\mu\nu}, \tag{3.32a}$$

$$\nabla_\mu \mathcal{C}^{\mu\alpha} = \tau_\nu \left(\mathcal{V}^{\nu\alpha} + \nabla_\mu \mathcal{Y}^{\alpha\mu\nu}\right), \tag{3.32b}$$

$$\nabla_\mu \mathcal{B}^\mu - 2\mathcal{A} = \tau_\nu \nabla_\mu \mathcal{V}^{\mu\nu}, \tag{3.32c}$$

$$\delta_{\mathrm{E}} S = \tau_\nu \left(\delta_{\mathrm{E}} J^\nu + 2E^\nu\right). \tag{3.32d}$$

We see, therefore, that the coefficients $\{\mathcal{A}, \mathcal{B}^\mu, \mathcal{C}^{\mu\nu}\}$ and $\{\mathcal{V}^{\mu\nu}, \mathcal{Y}^{\alpha\mu\nu}, \mathcal{W}^{\alpha\beta\mu\nu}\}$ that appear in $\mathcal{L}_{\{2\}}$ (3.18) are not independent, which is now obvious from (3.32a). By differentiating (3.32a) and comparing with (3.32b) one finds that

$$u_\mu u_\nu \mathcal{Y}^{\alpha\mu\nu} \dot{\omega} - (K_{\mu\nu} \mathcal{Y}^{\alpha\mu\nu} - u_\nu \mathcal{V}^{\nu\alpha})\omega = 0, \tag{3.33}$$

where $K_{\mu\nu} = \nabla_\mu u_\nu$ is the induced extrinsic curvature and an overdot is used to denote differentiation along the time-like vector. The $(3+1)$ decomposition (3.31a) introduces some interesting structure and can be used to explicitly evaluate the linking conditions (3.32a). From (3.32a) we find that

$$A_{\mathcal{C}} = -\omega A_{\mathcal{Y}}, \qquad B_{\mathcal{C}} = -\omega C_{\mathcal{Y}}. \tag{3.34}$$

After combining (3.32a) and (3.32b) to yield (3.33) we find that

$$\dot{\omega} A_{\mathcal{Y}} - (A_{\mathcal{V}} + K B_{\mathcal{Y}})\omega = 0. \tag{3.35}$$

In a similar fashion, it follows from (3.32c) that

$$\mathcal{A} = \frac{1}{2}\left[\dot{A}_{\mathcal{B}} + \omega \dot{A}_{\mathcal{V}} + K\left(A_{\mathcal{B}} + A_{\mathcal{V}} + B_{\mathcal{V}}\right)\right], \tag{3.36}$$

where $K = K^\mu{}_\mu$.

3.3.4.2 Reparameterization Invariance

One can think of ξ^μ as being an "artificial" vector field whose role was to restore reparameterization invariance, and it would therefore be desirable to have a theory that does not require ξ^μ to be present and reparameterization invariance is manifest.

We will derive conditions that the tensors in the Lagrangian must satisfy in order for reparameterization invariance to be manifest.

We can rewrite the Lagrangian with the vector field ξ^{μ} explicitly present to show how the three fields $\{\delta_{\mathrm{E}} g_{\mu\nu}, \delta\phi, \xi^{\mu}\}$ interact and how the parameters can be arranged so that they ultimately decouple. To ease our calculation we will write $h_{\mu\nu} \equiv \delta_{\mathrm{E}} g_{\mu\nu}$; we use (2.44) to replace $\delta_{\mathrm{L}} g_{\mu\nu}$ with $h_{\mu\nu} + 2\nabla_{(\mu}\xi_{\nu)}$ in the Lagrangian (3.18). Rearranging, whilst keeping track of total derivatives yields

$$\begin{aligned}
\mathcal{L}_{\{2\}} &= \mathcal{A}(\delta\phi)^2 + \mathcal{B}^{\mu}\delta\phi\nabla_{\mu}\delta\phi + \frac{1}{2}\mathcal{C}^{\mu\nu}\nabla_{\mu}\delta\phi\nabla_{\nu}\delta\phi + \frac{1}{8}\mathcal{W}^{\mu\nu\alpha\beta}h_{\alpha\beta}h_{\mu\nu} \\
&\quad + \frac{1}{4}\Big[\mathcal{V}^{\mu\nu}\delta\phi + \mathcal{Y}^{\alpha\mu\nu}\nabla_{\alpha}\delta\phi\Big]h_{\mu\nu} - \frac{1}{2}\xi_{\nu}\Big[(\nabla_{\mu}\mathcal{W}^{\mu\nu\alpha\beta})h_{\alpha\beta} + \mathcal{W}^{\mu\nu\alpha\beta}\nabla_{\mu}h_{\alpha\beta}\Big] \\
&\quad - \frac{1}{2}\xi_{\nu}\Big[\mathcal{Y}^{\alpha\mu\nu}\nabla_{\mu}\nabla_{\alpha}\delta\phi + (\mathcal{V}^{\alpha\nu} + \nabla_{\beta}\mathcal{Y}^{\alpha\beta\nu})\nabla_{\alpha}\delta\phi + (\nabla_{\mu}\mathcal{V}^{\mu\nu})\delta\phi\Big] \\
&\quad - \frac{1}{2}\xi_{\nu}\Big[4(\nabla_{\mu}\mathcal{W}^{\mu\nu\alpha\beta})\nabla_{\alpha}\xi_{\beta} + 4\mathcal{W}^{\mu\nu\alpha\beta}\nabla_{\mu}\nabla_{\alpha}\xi_{\beta}\Big] \\
&\quad + \frac{1}{2}\nabla_{\alpha}\Big[\xi_{\beta}\big(\mathcal{Y}^{\mu\alpha\beta}\nabla_{\mu}\delta\phi + \mathcal{V}^{\alpha\beta}\delta\phi + \mathcal{W}^{\mu\nu\alpha\beta}h_{\mu\nu} + 2\mathcal{W}^{\mu\nu\alpha\beta}\nabla_{\mu}\xi_{\nu}\big)\Big].
\end{aligned} \tag{3.37}$$

To enable us to identify the "free" and "interaction" Lagrangians, we note that (3.37) can be written schematically as

$$\begin{aligned}
\mathcal{L}_{\{2\}} &= \mathcal{L}^{\mathrm{A}}_{\{2\}}[\delta\phi] + \mathcal{L}^{\mathrm{B}}_{\{2\}}[h_{\mu\nu}] + \mathcal{L}^{\mathrm{C}}_{\{2\}}[\xi^{\alpha}] + \mathcal{L}^{\mathrm{D}}_{\{2\}}[h_{\mu\nu}, \delta\phi] \\
&\quad + \mathcal{L}^{\mathrm{E}}_{\{2\}}[h_{\mu\nu}, \xi^{\alpha}] + \mathcal{L}^{\mathrm{F}}_{\{2\}}[\delta\phi, \xi^{\alpha}] + \nabla_{\alpha}\mathcal{S}^{\alpha},
\end{aligned} \tag{3.38}$$

where $\mathcal{L}_{\{2\}}$ of a single field variable represents the self-interaction of that field, and of two fields represents the interaction between the two fields. The final line of (3.37) is a pure surface term, and will not contribute to the dynamics, and thus does not require consideration in what we are about to discuss. However, if we note the definition of $\mathcal{S}^{\mu}$ and compare to the perturbed energy momentum tensor (3.28), we find that $\mathcal{S}^{\mu} = -\xi_{\nu}(\delta_{\mathrm{L}} U^{\mu\nu} + \frac{1}{2}U^{\mu\nu}g^{\alpha\beta}\delta_{\mathrm{L}} g_{\alpha\beta})$.

Notice that the perturbed scalar field, $\delta\phi$, and vector field ξ_{μ} are coupled in the Lagrangian, and only decouple when their interaction Lagrangian, $\mathcal{L}^{\mathrm{F}}_{\{2\}}$, vanishes. This will remove the *direct* coupling but they may remain *indirectly* coupled if the interaction Lagrangian for the perturbed metric and vector field remains non-zero (i.e. if $\mathcal{L}^{\mathrm{E}}_{\{2\}} \neq 0$), since the perturbed metric and scalar field will remain coupled, $\mathcal{L}^{\mathrm{D}}_{\{2\}} \neq 0$. So, the interaction Lagrangian between the vector field and perturbed scalar field vanishes, i.e. $\mathcal{L}^{\mathrm{F}}_{\{2\}} = 0$, when

$$\xi_{\nu}\Big[\mathcal{Y}^{\alpha\mu\nu}\nabla_{\mu}\nabla_{\alpha}\delta\phi + (\mathcal{V}^{\alpha\nu} + \nabla_{\beta}\mathcal{Y}^{\alpha\beta\nu})\nabla_{\alpha}\delta\phi + (\nabla_{\mu}\mathcal{V}^{\mu\nu})\delta\phi\Big] = 0. \tag{3.39}$$

For arbitrary values of the perturbed scalar field and vector field, this is satisfied by the covariant conditions

$$\xi_\nu \mathcal{Y}^{\alpha\mu\nu} = 0, \qquad \xi_\nu(\mathcal{V}^{\alpha\nu} + \nabla_\beta \mathcal{Y}^{\alpha\beta\nu}) = 0, \qquad \xi_\nu \nabla_\mu \mathcal{V}^{\mu\nu} = 0. \tag{3.40}$$

To find the decoupling conditions for the perturbed metric we realize that because $\delta_{\mathrm{E}}(\nabla_\mu U^{\mu\nu}) = 0$,

$$\xi_\nu \delta_{\mathrm{E}}(\nabla_\mu U^{\mu\nu}) = 0 \tag{3.41}$$

is an identity. If we contract (3.29) with ξ_μ and use (3.39) then

$$\xi_\nu \delta_{\mathrm{E}} J^\nu + 2\xi_\nu E^\nu = 0, \tag{3.42}$$

where $\delta_{\mathrm{E}} J^\mu$ and E^μ are given respectively by (3.30a). Inserting these definitions of $\delta_{\mathrm{E}} J^\nu$, E^ν into (3.42) yields

$$\begin{aligned}&\xi_\nu(\mathcal{W}^{\mu\nu\alpha\beta} + U^{\alpha\beta} g^{\mu\nu} - 2g^{\nu\beta} U^{\mu\alpha})\nabla_\mu h_{\alpha\beta} + (\xi_\nu \nabla_\mu \mathcal{W}^{\mu\nu\alpha\beta}) h_{\alpha\beta} \\ &+ 2\xi_\nu L^{\mu\alpha\beta\nu} \nabla_\mu \nabla_\alpha \xi_\beta + 2(\xi_\nu \nabla_\mu \mathcal{W}^{\mu\nu\alpha\beta}) \nabla_\alpha \xi_\beta + 2(\xi_\nu \nabla_\mu \nabla_\alpha U^{\mu\nu})\xi^\alpha = 0.\end{aligned} \tag{3.43}$$

For arbitrary values of $h_{\mu\nu}$, the decoupling of ξ^μ from $h_{\mu\nu}$ occurs when the coefficients of $\nabla_\mu h_{\alpha\beta}$, $h_{\alpha\beta}$ vanish, which occurs when the covariant conditions

$$\xi_\nu(\mathcal{W}^{\mu\nu\alpha\beta} + U^{\alpha\beta} g^{\mu\nu} - 2g^{\nu\beta} U^{\mu\alpha}) = 0, \qquad \xi_\nu \nabla_\mu \mathcal{W}^{\mu\nu\alpha\beta} = 0 \tag{3.44}$$

are satisfied.

Inserting the (3+1) decomposition (3.31a) into (3.40) and (3.44) allows us to evaluate the decoupling conditions. This yields

$$\xi_\nu \mathcal{Y}^{\alpha\mu\nu} = B_{\mathcal{Y}} u^\alpha \xi^\mu + C_{\mathcal{Y}} u^\mu \xi^\alpha, \tag{3.45a}$$

$$\xi_\nu(\mathcal{V}^{\alpha\nu} + \nabla_\beta \mathcal{Y}^{\alpha\beta\nu}) = \left[\dot{C}_{\mathcal{Y}} + C_{\mathcal{Y}} K + B_{\mathcal{V}}\right]\xi^\alpha + \left[B_{\mathcal{Y}} + C_{\mathcal{Y}}\right]\xi_\nu K^{\alpha\nu}, \tag{3.45b}$$

$$\xi_\nu \nabla_\mu \mathcal{V}^{\mu\nu} = \left[\dot{A}_{\mathcal{V}} + (A_{\mathcal{V}} + B_{\mathcal{V}})K\right]\xi_\nu u^\nu = 0, \tag{3.45c}$$

$$\begin{aligned}(B_{\mathcal{W}} + \rho)\xi^\mu u^\alpha u^\beta + 2(C_{\mathcal{W}} - \rho)\xi^{(\alpha} u^{\beta)} u^\mu + (D_{\mathcal{W}} + P)\xi^\mu \gamma^{\alpha\beta} \\ + 2(E_{\mathcal{W}} - P)\xi^{(\alpha} \gamma^{\beta)\mu} = 0,\end{aligned} \tag{3.46a}$$

$$2\left[\dot{C}_{\mathcal{W}} + K(C_{\mathcal{W}} + E_{\mathcal{W}})\right] u^{(\alpha} \xi^{\beta)} + 2\left[B_{\mathcal{W}} + C_{\mathcal{W}} + D_{\mathcal{W}} + E_{\mathcal{W}}\right] \xi^\mu K^{(\alpha}{}_\mu u^{\beta)} = 0. \tag{3.46b}$$

Note that (3.45c) gives us no information since $u^\mu \xi_\mu = 0$. Hence, we conclude that the decoupling conditions (3.45a, 3.46a) are satisfied by the parameter choices

$$\dot{C}_{\mathcal{Y}} + C_{\mathcal{Y}} K + B_{\mathcal{V}} = 0, \qquad B_{\mathcal{Y}} = -C_{\mathcal{Y}}, \tag{3.47a}$$

$$B_{\mathcal{W}} = -\rho, \qquad C_{\mathcal{W}} = \rho, \qquad D_{\mathcal{W}} = -P, \qquad E_{\mathcal{W}} = P, \tag{3.47b}$$

$$\dot{C}_{\mathcal{W}} + K(C_{\mathcal{W}} + E_{\mathcal{W}}) = 0, \qquad B_{\mathcal{W}} + C_{\mathcal{W}} + D_{\mathcal{W}} + E_{\mathcal{W}} = 0. \tag{3.47c}$$

When (3.47b) is used the first condition of (3.47c) becomes

$$\dot{\rho} + K(\rho + P) = 0, \tag{3.48}$$

and the second is satisfied identically.

3.3.5 Scalar Fields $\mathcal{L} = \mathcal{L}(\mathcal{X}, \phi)$

Section. 5.2 we provide the explicit calculation for computing $\mathcal{L}_{\{2\}}$ and $\delta_{\mathrm{L}} U^{\mu\nu}$ in a kinetic scalar field theory $\mathcal{L} = \mathcal{L}(\mathcal{X}, \phi)$, where $\mathcal{X} = -\frac{1}{2} g^{\mu\nu} \nabla_\mu \phi \nabla_\nu \phi$ is the kinetic term of a scalar field. In Table 3.1 we give a summary of the functions that appear in the decomposition of $\delta_{\mathrm{L}} U^{\mu\nu}$ for some explicit scalar field theories; in these examples it is natural to see that upon specifying a scalar field theory, relationships between the functions and the time evolution of the functions is set. For a canonical scalar field theory, $\mathcal{L} = \mathcal{X} - V(\phi)$, the functions are given by

Table 3.1 Collection of the functions in the decomposition of $\delta_{\mathrm{L}} U^{\mu\nu}$.

Function	(a) EDE	(b) $\mathcal{L} = \mathcal{L}(\phi, \mathcal{X})$	(c) $\mathcal{L} = F(\mathcal{X})$	(d) $\mathcal{L} = \mathcal{X} - V(\phi)$
$A_{\mathcal{V}}$	0	$-2(\mathcal{L}_{,\mathcal{X}\phi}\dot{\phi}^2 - \mathcal{L}_{,\phi})$	0	$-2V'$
$B_{\mathcal{V}}$	0	$-2\mathcal{L}_{,\phi}$	0	$2V'$
$A_{\mathcal{Y}}$	0	$-2(\mathcal{L}_{,\mathcal{X}\mathcal{X}}\dot{\phi}^3 + \mathcal{L}_{,\mathcal{X}}\dot{\phi})$	$-2(F''\dot{\phi}^2 + F'\dot{\phi})$	$-2\dot{\phi}$
$B_{\mathcal{Y}}$	0	$-2\mathcal{L}_{,\mathcal{X}}\dot{\phi}$	$-2F'\dot{\phi}$	$-2\dot{\phi}$
$C_{\mathcal{Y}}$	0	$2\mathcal{L}_{,\mathcal{X}}\dot{\phi}$	$2F'\dot{\phi}$	$2\dot{\phi}$
$A_{\mathcal{W}}$	$-\rho$	$-(\mathcal{L}_{,\mathcal{X}\mathcal{X}}\dot{\phi}^4 + 2\rho + P)$	$-(F''\dot{\phi}^4 + 2\rho + P)$	$-(2\rho + P)$
$B_{\mathcal{W}}$	P	$-\rho$	$-\rho$	$-\rho$
$C_{\mathcal{W}}$	$-P$	ρ	ρ	ρ
$D_{\mathcal{W}}$	$\beta - P - \frac{2}{3}\mu$	$-P$	$-P$	$-P$
$E_{\mathcal{W}}$	$\mu + P$	P	P	P

The theories we have presented are: (a) elastic dark energy, (b) generic kinetic scalar field theory, (c) k-essence and (d) canonical scalar field theory. It is interesting to realize that the theories with $\mathcal{L} = \mathcal{L}(g_{\mu\nu})$ are subsets of theories with $\mathcal{L} = \mathcal{L}(g_{\mu\nu}, \phi, \nabla_\mu \phi)$. Comma denotes partial differentiation (e.g. $\mathcal{L}_{,\phi} = \partial\mathcal{L}/\partial\phi$), prime denotes differentiation with respect the functions single argument: $F' = \mathrm{d}F/\mathrm{d}\mathcal{X}$, $V' = \mathrm{d}V/\mathrm{d}\phi$ and an overdot denotes differentiation with respect to time; for conformal time coefficients one should replace $\dot{\phi} \rightarrow \dot{\phi}/a$

$$\rho = \frac{1}{2}\dot{\phi}^2 + V, \qquad P = \frac{1}{2}\dot{\phi}^2 - V, \tag{3.49a}$$

$$A_{\mathcal{V}} = -B_{\mathcal{V}} = -2V', \tag{3.49b}$$

$$A_{\mathcal{Y}} = B_{\mathcal{Y}} = -C_{\mathcal{Y}} = -2\dot{\phi}, \tag{3.49c}$$

$$A_{\mathcal{W}} = -(2\rho + P), \qquad B_{\mathcal{W}} = -C_{\mathcal{W}} = -\rho, \qquad D_{\mathcal{W}} = -E_{\mathcal{W}} = -P, \tag{3.49d}$$

where an overdot is understood to denote differentiation with respect to time and $V' = \mathrm{d}V/\mathrm{d}\phi$. Using (3.49a) and taking $\omega = 1/\dot{\phi}$ it transpires that (3.33) is the Klein-Gordon equation.

3.3.6 Multiple Scalar Fields

Many of our results apply to theories with multiple scalar fields (see e.g. assisted inflation [12] or multi-field dark energy [13–15]). For example, if a theory has field content given by $\mathcal{L} = \mathcal{L}(g_{\mu\nu}, \phi^{\text{A}}, \nabla_\mu\phi^{\text{A}})$, then, regardless of the details of the theory (e.g. before imposing some generalized kinetic term, $\mathcal{X}_{\text{gen}} = -\frac{1}{2}g^{\mu\nu}G_{\text{AB}}\nabla_\mu\phi^{\text{A}}\nabla_\nu\phi^{\text{B}}$), the Lagrangian for perturbations will be given by

$$\mathcal{L}_{\{2\}} = \mathcal{A}_{\text{AB}}\delta\phi^{\text{A}}\delta\phi^{\text{B}} + \mathcal{B}^{\mu}_{\text{AB}}\delta\phi^{\text{A}}\nabla_\mu\delta\phi^{\text{B}} + \frac{1}{2}\mathcal{C}^{\mu}_{\text{AB}}\nabla_\mu\delta\phi^{\text{A}}\nabla_\nu\delta\phi^{\text{B}}$$
$$+ \frac{1}{4}\left[\mathcal{Y}^{\alpha\mu\nu}_{\text{A}}\nabla_\alpha\delta\phi^{\text{A}}\delta_{\text{L}}g_{\mu\nu} + \mathcal{V}^{\mu\nu}_{\text{A}}\delta\phi^{\text{A}}\delta_{\text{L}}g_{\mu\nu} + \frac{1}{2}\mathcal{W}^{\mu\nu\alpha\beta}\delta_{\text{L}}g_{\mu\nu}\delta_{\text{L}}g_{\alpha\beta}\right], \tag{3.50}$$

where repeated field-indices denote summation:

$$\mathcal{V}^{\mu\nu}_{\text{A}}\delta\phi^{\text{A}} = \sum_{\text{A}=1}^{\text{N}}\mathcal{V}^{\mu\nu}_{\text{A}}\delta\phi^{\text{A}}. \tag{3.51}$$

All results hold, except that we would need to identify

$$A_{\mathcal{V}}\delta\phi = A_{\mathcal{V}\text{A}}\delta\phi^{\text{A}}, \qquad B_{\mathcal{V}}\delta\phi = B_{\mathcal{V}\text{A}}\delta\phi^{\text{A}}, \tag{3.52a}$$

$$A_{\mathcal{Y}}\delta\dot{\phi} = A_{\mathcal{Y}\text{A}}\delta\dot{\phi}^{\text{A}}, \qquad B_{\mathcal{Y}}\delta\dot{\phi} = B_{\mathcal{Y}\text{A}}\delta\dot{\phi}^{\text{A}}, \qquad C_{\mathcal{Y}}\delta\phi = C_{\mathcal{Y}\text{B}}\delta\phi^{\text{A}}. \tag{3.52b}$$

3.3.7 Summary

The process of identifying the time-like part of the perturbed conservation equation with the Euler-Lagrange equation has reduced the number of functions required to specify $\mathcal{L}_{\{2\}}$ from $14 \to 11$. The 11 functions are

$$\left\{A_{\mathcal{B}}, A_{\mathcal{W}}, B_{\mathcal{W}}, C_{\mathcal{W}}, D_{\mathcal{W}}, E_{\mathcal{W}}, A_{\mathcal{V}}, B_{\mathcal{V}}, A_{\mathcal{Y}}, B_{\mathcal{Y}}, C_{\mathcal{Y}}\right\}, \tag{3.53}$$

as well as the energy density ρ and pressure P of the "dark sector fluid". Also imposing reparameterization invariance reduces the number of functions from $11 \to 5$. The 5 functions are

$$\left\{A_{\mathcal{B}}, A_{\mathcal{W}}, A_{\mathcal{V}}, A_{\mathcal{Y}}, C_{\mathcal{Y}}\right\}. \tag{3.54}$$

Later on we will show that $A_{\mathcal{B}}$ does not affect the cosmological dynamics, and $A_{\mathcal{W}}$ becomes irrelevant in the synchronous gauge. This means that there are just 3 free functions left to completely specify the dark sector perturbations for a reparameterization-invariant scalar field theory.

3.4 Cosmological Perturbations

In this section we provide explicit expressions for the components of $\delta_{\mathrm{E}} U^{\mu}{}_{\nu}$ and the perturbed conservation equation specialized to the case of an FRW background. We will pay special attention to the scalar field theory, where we will again show how the vector field ξ^{μ} decouples from the equation of motion for $\delta\phi$.

We will perturb the line element about a conformally flat FRW background, and write

$$\mathrm{d}s^2 = a^2(\tau)\Big[-(1-2\Phi)\mathrm{d}\tau^2 + 2N_i\mathrm{d}x^i\mathrm{d}\tau + (\delta_{ij}+h_{ij})\mathrm{d}x^i\mathrm{d}x^j\Big]. \tag{3.55}$$

This means that we are setting the components of the Eulerian perturbed metric to

$$\delta_{\mathrm{E}} g_{00} = 2a^2(\tau)\Phi(\tau,\mathbf{x}), \quad \delta_{\mathrm{E}} g_{0i} = a^2(\tau)N_i(\tau,\mathbf{x}), \quad \delta_{\mathrm{E}} g_{ij} = a^2(\tau)h_{ij}(\tau,\mathbf{x}). \tag{3.56}$$

The time-like vector is given by $u_{\mu} = a(\tau)(-1,0,0,0)$, and we set $\xi^{\mu}u_{\mu} = 0$. All functions (3.53) are only functions of time. The background conservation equation $\nabla_{\mu}U^{\mu\nu} = 0$ becomes

$$\dot{\rho} + 3\mathcal{H}(\rho + P) = 0, \tag{3.57}$$

where an overdot denotes derivative with respect to conformal time τ and $\mathcal{H}$ is the conformal time Hubble parameter. The components of $\delta_{\mathrm{E}} U^{\mu}{}_{\nu}$ for the theory with field content $\mathcal{L} = \mathcal{L}(g_{\mu\nu}, \phi, \nabla_{\mu}\phi)$, (3.28), are given by

$$\delta_{\rm E}U^0{}_0 = (\rho + B_{\mathcal{W}})\left(\partial_k\xi^k + \frac{1}{2}h\right) + (\rho + A_{\mathcal{W}})\Phi + \frac{1}{2}\left(A_{\mathcal{V}}\delta\phi + \frac{1}{a}A_{\mathcal{Y}}\delta\dot{\phi}\right), \tag{3.58a}$$

$$\delta_{\rm E}U^i{}_0 = (C_{\mathcal{W}} - \rho)\dot{\xi}^i + (P + C_{\mathcal{W}})N^i + \frac{1}{2a}C_{\mathcal{Y}}\partial^i\delta\phi, \tag{3.58b}$$

$$\delta_{\rm E}U^0{}_i = \left(\rho - C_{\mathcal{W}}\right)\left(\dot{\xi}_i + N_i\right) - \frac{1}{2a}C_{\mathcal{Y}}\partial_i\delta\phi, \tag{3.58c}$$

$$\begin{aligned}\delta_{\rm E}U^i{}_j = &-\left\{(D_{\mathcal{W}} + P)\left(\partial_k\xi^k + \frac{1}{2}h\right) + (B_{\mathcal{W}} - P)\Phi + \frac{1}{2}\left(B_{\mathcal{V}}\delta\phi + \frac{1}{a}B_{\mathcal{Y}}\delta\dot{\phi}\right)\right\}\delta^i{}_j \\ &+ (P - E_{\mathcal{W}})\left(h^i{}_j + \partial^i\xi_j + \partial_j\xi^i\right).\end{aligned} \tag{3.58d}$$

These are the sources to the equations governing the evolution of the metric perturbations, and can be used to obtain the components of $\delta_{\rm E}U^\mu{}_\nu$ in the conformal Newtonian and synchronous gauges.

We will now work in the synchronous gauge (by setting $\Phi = N_i = 0$), and we will study the general theory $\mathcal{L} = \mathcal{L}(g_{\mu\nu}, \phi, \nabla_\mu\phi)$, which will trivially encompass the no-extra-fields case. The components $\delta_{\rm E}U^\mu{}_\nu$ become

$$\delta_{\rm E}U^0{}_0 = (\rho + B_{\mathcal{W}})\left(\partial_k\xi^k + \frac{1}{2}h\right) + \frac{1}{2}\left(A_{\mathcal{V}}\delta\phi + \frac{1}{a}A_{\mathcal{Y}}\delta\dot{\phi}\right), \tag{3.59a}$$

$$\delta_{\rm E}U^i{}_0 = (C_{\mathcal{W}} - \rho)\dot{\xi}^i + \frac{1}{2a}C_{\mathcal{Y}}\partial^i\delta\phi, \tag{3.59b}$$

$$\begin{aligned}\delta_{\rm E}U^i{}_j = &-\left\{(D_{\mathcal{W}} + P)\left(\partial_k\xi^k + \frac{1}{2}h\right) + \frac{1}{2}\left(B_{\mathcal{V}}\delta\phi + \frac{1}{a}B_{\mathcal{Y}}\delta\dot{\phi}\right)\right\}\delta^i{}_j \\ &+ (P - E_{\mathcal{W}})\left(h^i{}_j + \partial^i\xi_j + \partial_j\xi^i\right).\end{aligned} \tag{3.59c}$$

The components of the Eulerian perturbed conservation law $\delta_{\rm E}(\nabla_\mu U^{\mu\nu}) = 0$, (3.29), has a "scalar" $\nu = 0$ component and a "vector" $\nu = i$ component. The $\nu = 0$ component of the perturbed conservation law (3.29) yields

$$\begin{aligned}&A_{\mathcal{Y}}\delta\ddot{\phi} + C_{\mathcal{Y}}\nabla^2\delta\phi + \left[\dot{A}_{\mathcal{Y}} + aA_{\mathcal{V}} + (2A_{\mathcal{Y}} + 3B_{\mathcal{Y}})\mathcal{H}\right]\delta\dot{\phi} \\ &\quad + a\left[\dot{A}_{\mathcal{V}} + 3\mathcal{H}(A_{\mathcal{V}} + B_{\mathcal{V}})\right]\delta\phi = a\left[P - B_{\mathcal{W}}\right]\dot{h} - 2a\left[C_{\mathcal{W}} + B_{\mathcal{W}}\right]\partial_i\dot{\xi}^i \\ &\quad - a\left[\dot{B}_{\mathcal{W}} + \mathcal{H}(3B_{\mathcal{W}} + 3D_{\mathcal{W}} + 2E_{\mathcal{W}} - 2P)\right](h + 2\partial_i\xi^i),\end{aligned} \tag{3.60a}$$

and the $\nu = i$ component yields

$$\Big[\rho - C_{\mathcal{W}}\Big](\ddot{\xi}^i + \mathcal{H}\dot{\xi}^i) - \Big[\dot{C}_{\mathcal{W}} + 3\mathcal{H}(C_{\mathcal{W}} + P)\Big]\dot{\xi}^i$$
$$-\Big[D_{\mathcal{W}} + E_{\mathcal{W}}\Big]\partial^i\partial_k\xi^k + \Big[P - E_{\mathcal{W}}\Big]\partial_k\partial^k\xi^i$$
$$= \frac{1}{2a}\Big[B_{\mathcal{Y}} + C_{\mathcal{Y}}\Big]\partial^i\dot{\delta\phi} + \frac{1}{2a}\Big[\dot{C}_{\mathcal{Y}} + 3C_{\mathcal{Y}}\mathcal{H} + aB_{\mathcal{V}}\Big]\partial^i\delta\phi$$
$$-\Big[P - E_{\mathcal{W}}\Big]\partial^j h^i{}_j + \frac{1}{2}\Big[D_{\mathcal{W}} + P\Big]\partial^i h. \tag{3.60b}$$

We observe that the "scalar" piece (3.60a) of the perturbed conservation law represents the evolution equation for the perturbed scalar field sourced by metric perturbations and the vector field, and the "vector" piece (3.60b) constitutes an evolution equation for the vector field, sourced by the perturbed scalar field and metric perturbations. Notice that nine functions are required to be specified to be able to write down the components $\delta_{\mathrm{E}}U^{\mu}{}_{\nu}$ and the perturbed conservation law: $A_{\mathcal{V}}$, $B_{\mathcal{V}}$, $A_{\mathcal{Y}}$, $B_{\mathcal{Y}}$, $C_{\mathcal{Y}}$, $B_{\mathcal{W}}$, $C_{\mathcal{W}}$, $D_{\mathcal{W}}$, $E_{\mathcal{W}}$ (note that $A_{\mathcal{B}}$, $A_{\mathcal{W}}$ do not enter into these quantities).

We will now study the conditions under which ξ^{μ} and $\delta\phi$ decouple; this will represent a simpler subset of theories, and will provide us with another understanding of how the decoupling conditions arise. It is useful to write the perturbed conservation equation (3.60a) as

$$\mathcal{C}_1\ddot{\delta\phi} + \mathcal{C}_2\nabla^2\delta\phi + \mathcal{C}_3\dot{\delta\phi} + \mathcal{C}_4\delta\phi = \mathcal{D}_1(h + 2\partial_i\xi^i) + \mathcal{D}_2\dot{h} + \mathcal{D}_3\partial_i\dot{\xi}^i, \tag{3.61a}$$

$$\mathcal{F}_1\ddot{\xi}^i + \mathcal{F}_2\dot{\xi}^i + \mathcal{F}_3\partial^i\partial_k\xi^j + \mathcal{F}_4\partial_k\partial^k\xi^i = \mathcal{G}_1\partial^i h + \mathcal{G}_2\partial^j h^i{}_j + \mathcal{G}_3\partial^i\delta\phi + \mathcal{G}_4\partial^i\dot{\delta\phi}, \tag{3.61b}$$

where the sets of coefficients $\{\mathcal{C}_{(\mathrm{A})}, \mathcal{D}_{(\mathrm{A})}, \mathcal{F}_{(\mathrm{A})}, \mathcal{G}_{(\mathrm{A})}\}$ can be read off from (3.60a). The ξ^{μ} and $\delta\phi$ decouple when all common terms in (3.61a) vanish. This yields the conditions $\mathcal{D}_1 = \mathcal{D}_3 = 0$ and $\mathcal{G}_1 = \mathcal{G}_2 = \mathcal{G}_3 = \mathcal{G}_4 = 0$. The former decoupling condition yields

$$\dot{B}_{\mathcal{W}} + \mathcal{H}(3B_{\mathcal{W}} + 3D_{\mathcal{W}} + 2E_{\mathcal{W}} - 2P) = 0, \tag{3.62a}$$

$$C_{\mathcal{W}} = -B_{\mathcal{W}}, \tag{3.62b}$$

and the latter decoupling condition yields

$$D_{\mathcal{W}} = -P, \qquad E_{\mathcal{W}} = P, \qquad B_{\mathcal{Y}} = -C_{\mathcal{Y}}, \tag{3.62c}$$

$$\dot{C}_{\mathcal{Y}} + 3C_{\mathcal{Y}}\mathcal{H} + aB_{\mathcal{V}} = 0. \tag{3.62d}$$

We also require that

$$B_{\mathcal{W}} = -\rho, \tag{3.62e}$$

for decoupling to occur in the $\delta_{\mathrm{E}}U^{\mu}{}_{\nu}$. Combining (3.62a, 3.62c, 3.62e) yields the conservation equation: $\dot{\rho}+3\mathcal{H}(\rho+P) = 0$. These decoupling conditions are identical to those we derived covariantly, (3.47a).

Applying the decoupling conditions (3.62) to the components $\delta_{\mathrm{E}}U^{\mu}{}_{\nu}$ (3.59a) we obtain

$$\delta_{\mathrm{E}}U^{0}{}_{0} = \frac{1}{2}\left(A_{\mathcal{V}}\delta\phi + \frac{1}{a}A_{\mathcal{Y}}\delta\dot{\phi}\right), \qquad \delta_{\mathrm{E}}U^{i}{}_{0} = \frac{1}{2a}C_{\mathcal{Y}}\partial^{i}\delta\phi, \tag{3.63a}$$

$$\delta_{\mathrm{E}}U^{i}{}_{j} = \frac{1}{2a}\left((\dot{C}_{\mathcal{Y}} + 3\mathcal{H}C_{\mathcal{Y}})\delta\phi + C_{\mathcal{Y}}\delta\dot{\phi}\right), \tag{3.63b}$$

and to the perturbed conservation equation (3.60a) we obtain

$$\begin{aligned} &A_{\mathcal{Y}}\delta\ddot{\phi} + C_{\mathcal{Y}}\nabla^{2}\delta\phi + \left[\dot{A}_{\mathcal{Y}} + aA_{\mathcal{V}} + (2A_{\mathcal{Y}} - 3C_{\mathcal{Y}})\mathcal{H}\right]\delta\dot{\phi} \\ &\quad + \left[a\dot{A}_{\mathcal{V}} + 3a\mathcal{H}A_{\mathcal{V}} - 3\mathcal{H}(\dot{C}_{\mathcal{Y}} + 3\mathcal{H}C_{\mathcal{Y}})\right]\delta\phi = a(\rho + P)\dot{h}. \end{aligned} \tag{3.64}$$

We now observe that only three functions are required to be specified: $A_{\mathcal{V}}$, $A_{\mathcal{Y}}$, $C_{\mathcal{Y}}$. When the decoupling conditions are satisfied, one can consistently set $\xi^{\mu} = 0$ and reparameterization invariance is enforced. The theory is now equivalent to the theory studied in [16], where reparameterization invariance was implicitly imposed.

3.4.1 No Extra Fields

For the theory with no extra fields, we can obtain the components of $\delta_{\mathrm{E}}U^{\mu}{}_{\nu}$ and those of the perturbed conservation law by ignoring all terms with $\delta\phi$ in (3.59a) and (3.60a) respectively. The components $\delta_{\mathrm{E}}U^{\mu}{}_{\nu}$ become

$$\delta_{\mathrm{E}} U^0{}_0 = (\rho + B_{\mathcal{W}})\left(\partial_k \xi^k + \frac{1}{2}h\right), \tag{3.65a}$$

$$\delta_{\mathrm{E}} U^i{}_0 = -(\rho - C_{\mathcal{W}})\dot{\xi}^i, \tag{3.65b}$$

$$\delta_{\mathrm{E}} U^i{}_j = -(D_{\mathcal{W}} + P)\left(\partial_k \xi^k + \frac{1}{2}h\right)\delta^i{}_j + (P - E_{\mathcal{W}})(h^i{}_j + \partial^i \xi_j + \partial_j \xi^i). \tag{3.65c}$$

The $\nu = i$ component of the perturbed conservation law becomes

$$\left[\rho - C_{\mathcal{W}}\right](\ddot{\xi}^i + \mathcal{H}\dot{\xi}^i) - \left[\dot{C}_{\mathcal{W}} + 3\mathcal{H}(C_{\mathcal{W}} + P)\right]\dot{\xi}^i - \left[D_{\mathcal{W}} + E_{\mathcal{W}}\right]\partial^i \partial_k \xi^k$$
$$+\left[P - E_{\mathcal{W}}\right]\partial_k \partial^k \xi^i = \frac{1}{2}(D_{\mathcal{W}} + P)\partial^i h + (E_{\mathcal{W}} - P)\partial^k h^i{}_k, \tag{3.66}$$

and the $\nu = 0$ component of the perturbed conservation law yields

$$\left[\dot{B}_{\mathcal{W}} + \mathcal{H}(3B_{\mathcal{W}} + 3D_{\mathcal{W}} + 2E_{\mathcal{W}} - 2P)\right]\left(\partial_i \xi^i + \frac{1}{2}h\right)$$
$$+\left[C_{\mathcal{W}} + B_{\mathcal{W}}\right]\partial_i \dot{\xi}^i = \frac{1}{2}\left[B_{\mathcal{W}} - P\right]\dot{h}. \tag{3.67}$$

We can use (3.67) to obtain a set of conditions that enforces the constraint (3.67) on the $\nu = i$ component of the perturbed conservation equation. We set the coefficients of $(\partial_i \xi^i + \frac{1}{2}h)$, $\partial_i \dot{\xi}^i$, $\dot{h}$ to zero and obtain

$$B_{\mathcal{W}} = -C_{\mathcal{W}} = P, \tag{3.68a}$$

$$\dot{P} + \mathcal{H}(P + 3D_{\mathcal{W}} + 2E_{\mathcal{W}}) = 0. \tag{3.68b}$$

Applying these conditions to the components of $\delta_{\mathrm{E}} U^\mu{}_\nu$ (3.65a) we find

$$\delta_{\mathrm{E}} U^0{}_0 = (\rho + P)\left(\partial_k \xi^k + \frac{1}{2}h\right), \tag{3.69a}$$

$$\delta_{\mathrm{E}} U^i{}_0 = -(\rho + P)\dot{\xi}^i, \tag{3.69b}$$

$$\delta_{\mathrm{E}} U^i{}_j = -(D_{\mathcal{W}} + P)\left(\partial_k \xi^k + \frac{1}{2}h\right)\delta^i_j + (P - E_{\mathcal{W}})(h^i{}_j + \partial^i \xi_j + \partial_j \xi^i), \tag{3.69c}$$

and the equation of motion (3.66) becomes

$$\left[\rho + P\right]\ddot{\xi}^i + \mathcal{H}\left[\rho - 3D_{\mathcal{W}} - 2E_{\mathcal{W}}\right]\dot{\xi}^i - (D_{\mathcal{W}} + E_{\mathcal{W}})\partial^i\partial_k\xi^k - \frac{1}{2}(D_{\mathcal{W}} + P)\partial^i h$$
$$+ (P - E_{\mathcal{W}})\left[\partial_k\partial^k\xi^i + \partial^k h^i{}_k\right] = 0. \tag{3.70}$$

Hence, we see that after applying the conditions (3.68a) there are only two free coefficients which describe perturbations in the dark sector: $D_{\mathcal{W}}$ and $E_{\mathcal{W}}$. Making the choice which defines elastic dark energy, (3.12a), one finds that (3.70) agrees with the equation of motion for ξ^i given in [2].

3.4.2 Scalar Fields

As an explicit example, we can construct a theory where $g_{\mu\nu}$ and $\nabla_\mu\phi$ enter the field content by combining into the kinetic scalar $\mathcal{X} \equiv -\frac{1}{2}g^{\mu\nu}\nabla_\mu\phi\nabla_\nu\phi$, so that the field content is $\mathcal{L} = \mathcal{L}(\phi, \mathcal{X})$. In Table 3.1 we supplied the coefficients for this general kinetic scalar field theory. The decoupling conditions (3.62b, 3.62c) are trivially satisfied, the condition (3.62a) becomes

$$-\left[\dot{\rho} + 3\mathcal{H}(\rho + P)\right] = 0, \tag{3.71}$$

which is always true, and (3.62d) becomes

$$-\frac{2}{a}\left[\mathcal{L}_{,\mathcal{X}}\ddot{\phi} + 2\mathcal{H}\mathcal{L}_{,\mathcal{X}}\dot{\phi} + \mathcal{L}_{,\mathcal{X}\mathcal{X}}\dot{\phi}\dot{\mathcal{X}} + \mathcal{L}_{,\phi\mathcal{X}}\dot{\phi}^2 - a^2\mathcal{L}_{,\phi}\right] = 0, \tag{3.72}$$

which is the Euler-Lagrange equation that one can compute directly and therefore vanishes identically. What this means is that for all scalar field theories of the form $\mathcal{L} = \mathcal{L}(\phi, \mathcal{X})$, the ξ^μ and $\delta\phi$ fields decouple (i.e. one can set $\xi^\mu = 0$ consistently). The components of the perturbed dark energy-momentum tensor become

$$\delta_{\mathrm{E}}U^0{}_0 = \frac{1}{2}\left(A_{\mathcal{V}}\delta\phi + \frac{1}{a}A_{\mathcal{Y}}\dot{\delta\phi}\right), \qquad \delta_{\mathrm{E}}U^i{}_0 = \frac{1}{2a}C_{\mathcal{Y}}\partial^i\delta\phi, \tag{3.73a}$$

$$\delta_{\mathrm{E}}U^i{}_j = \frac{1}{2a}\left((\dot{C}_{\mathcal{Y}} + 3\mathcal{H}C_{\mathcal{Y}})\delta\phi + C_{\mathcal{Y}}\dot{\delta\phi}\right). \tag{3.73b}$$

The $\nu = 0$-component of the perturbed conservation law (3.60a) becomes

$$A_{\mathcal{Y}}\ddot{\delta\phi} + C_{\mathcal{Y}}\nabla^2\delta\phi + \left[\dot{A}_{\mathcal{Y}} + aA_{\mathcal{V}} + (2A_{\mathcal{Y}} - 3C_{\mathcal{Y}})\mathcal{H}\right]\dot{\delta\phi}$$
$$+ \left[a\dot{A}_{\mathcal{V}} + 3a\mathcal{H}A_{\mathcal{V}} - 3\mathcal{H}(\dot{C}_{\mathcal{Y}} + 3\mathcal{H}C_{\mathcal{Y}})\right]\delta\phi = a(\rho + P)\dot{h}, \tag{3.74}$$

which is the evolution equation for the perturbed scalar field (we have verified that this reproduces known results) and only three functions are required to be specified: $A_{\mathcal{V}}, A_{\mathcal{Y}}, C_{\mathcal{Y}}$, in addition to the equation of state, w.

3.5 Summary

In this chapter we provided two simple, but phenomenologically rich, examples illustrating how our formalism can be used. We started off with theories which satisfied almost no theoretical restrictions (only a given field content), but we showed how to impose restrictions (such as reparameterization invariance) which imposed constraints upon the "free" functions, dramatically reducing the freedom in the theory. We developed both examples to such an extent that we were able to obtain the sources to the perturbed gravitational field equations.

Our formalism can be extended to encompass theories with interactions between the known and dark sectors. These types of interactions are relevant in, for example, coupled quintessence theories and modified gravity models (see, e.g. [17–24]). At the level of the field equations, the interaction is usually described with a transfer current Q_{μ}, allowing energy-momentum to be transfered between the dark sector and the known sector fields. From the generalized gravitational field equations, $G_{\mu\nu} = 8\pi G T_{\mu\nu} + U_{\mu\nu}$, and the Bianchi identity for the Einstein tensor, $\nabla^{\mu} G_{\mu\nu} = 0$, one can immediately find that

$$\nabla^{\mu}\big(8\pi G T_{\mu\nu} + U_{\mu\nu}\big) = 0. \tag{3.75}$$

We can recast our formalism to now incorporate coupled dark energy models by introducing a transfer-current, Q_{ν}. The Bianchi identity (3.75) is equivalent to

$$\nabla^{\mu} T_{\mu\nu} = -\frac{1}{8\pi G} Q_{\nu}, \qquad \nabla^{\mu} U_{\mu\nu} = Q_{\nu}. \tag{3.76}$$

The transfer-current Q_{ν} now describes transfer of energy-momentum between the known and dark sectors. We now write down the perturbed conservation equation for this coupled dark sector,

$$\delta_{\mathrm{E}}\big(\nabla_{\mu} U^{\mu\nu}\big) = \delta_{\mathrm{E}} Q^{\nu}, \tag{3.77a}$$

which expands to give

$$\nabla_{\mu}\delta_{\mathrm{E}} U^{\mu\nu} + \frac{1}{2}\Big[U^{\mu\nu} g^{\alpha\beta} - U^{\alpha\beta} g^{\mu\nu} + 2 g^{\nu\beta} U^{\alpha\mu}\Big]\nabla_{\mu}\delta_{\mathrm{E}} g_{\alpha\beta} = \delta_{\mathrm{E}} Q^{\nu}. \tag{3.77b}$$

Using the $\delta_{\mathrm{E}} U^{\mu\nu}$ for the no extra fields case, (3.9), the coupled Eulerian perturbed conservation equation reads

$$2L^{\mu\alpha\beta\nu}\nabla_\mu\nabla_\alpha\xi_\beta + 2\left[\nabla_\sigma \mathcal{W}^{\sigma\nu\mu\alpha} + 2g^{\alpha[\mu}Q^{\nu]}\right]\nabla_\mu\xi_\alpha$$
$$+ 2\xi^\alpha\nabla_\mu\nabla_\alpha U^{\mu\nu} = \delta_{\mathrm{E}}J^\nu - Q^\nu g^{\alpha\beta}\delta_{\mathrm{E}}g_{\alpha\beta} + \delta_{\mathrm{E}}Q^\nu, \tag{3.78}$$

where $L_{\mu\alpha\beta\nu}$ and $\delta_{\mathrm{E}}J^\mu$ retain their definitions from (3.10b) and (3.10c) respectively. The existence of the transfer current induces an explicit coupling term between the scalar perturbations in the metric, $g^{\mu\nu}\delta_{\mathrm{E}}g_{\mu\nu}$, the gauge field, ξ^μ, and the energy-momentum transfer current Q_ν.

References

1. M. Bucher, D.N. Spergel, Is the dark matter a solid?. Phys. Rev. **D60**, 043505 (1999) [astro-ph/9812022]
2. R.A. Battye, A. Moss, Cosmological perturbations in elastic dark energy models. Phys. Rev. **D76**, 023005 (2007) [astro-ph/0703744]
3. M. Fierz, W. Pauli, On relativistic wave equations for particles of arbitrary spin in an electromagnetic field. Proc. Roy. Soc. Lond. **A173**, 211–232 (1939)
4. D. Boulware, S. Deser, Can gravitation have a finite range? Phys. Rev. **D6**, 3368–3382 (1972)
5. V. Rubakov, Lorentz-violating graviton masses: getting around ghosts, low strong coupling scale and VDVZ discontinuity, hep-th/0407104
6. S. Dubovsky, Phases of massive gravity, JHEP **0410**. 076 (2004) [hep-th/0409124]
7. G. Gabadadze, L. Grisa, Lorentz-violating massive gauge and gravitational fields. Phys. Lett. **B617**, 124–132 (2005) [hep-th/0412332]
8. V.A. Rubakov, P.G. Tinyakov, Infrared-modified gravities and massive gravitons. Phys. Usp. **51**, 759–792 (2008) [arXiv:0802.4379]
9. D. Blas, D. Comelli, F. Nesti, L. Pilo, Lorentz breaking massive gravity in curved space. Phys. Rev. **D80**, 044025 (2009) [arXiv:0905.1699]
10. K. Hinterbichler, Theoretical aspects of massive gravity. Rev. Mod. Phys. **84**, 671–710 (2012) [arXiv:1105.3735]
11. D. Comelli, M. Crisostomi, F. Nesti, L. Pilo, Degrees of freedom in massive gravity. Phys. Rev. **D86**, 101502 (2012). doi:10.1103/PhysRevD.86.101502 [arXiv:1204.1027]
12. A.R. Liddle, A. Mazumdar, F.E. Schunck, Assisted inflation. Phys. Rev. **D58**, 061301 (1998) [astro-ph/9804177]
13. S. Tsujikawa, General analytic formulae for attractor solutions of scalar-field dark energy models and their multi-field generalizations. Phys. Rev. **D73**, 103504 (2006) [hep-th/0601178]
14. C. van de Bruck, J.M. Weller, Quintessence dynamics with two scalar fields and mixed kinetic terms. Phys. Rev. **D80**, 123014 (2009) [arXiv:0910.1934]
15. J. Frazer, A.R. Liddle, Stability of multi-field cosmological solutions in the presence of a fluid. Phys. Rev. **D82**, 043516 (2010) [arXiv:1004.3888]
16. P. Creminelli, G. D'Amico, J. Norena, F. Vernizzi, The effective theory of quintessence: the w<-1 side unveiled, JCAP **0902**. 018 (2009) [arXiv:0811.0827]
17. J. Ellis, S. Kalara, K.A. Olive, C. Wetterich, Density-dependent couplings and astrophysical bounds on light scalar particles. Appl. Phys. B **228**(2), 264–272 (1989)
18. L. Amendola, Coupled quintessence. Phys. Rev. **D62**, 043511 (2000) [astro-ph/9908023]
19. E.J. Copeland, M. Sami, S. Tsujikawa, Dynamics of dark energy. Int. J. Mod. Phys. **D15**, 1753–1936 (2006) [hep-th/0603057]
20. L. Amendola, M. Quartin, S. Tsujikawa, I. Waga, Challenges for scaling cosmologies. Phys. Rev. **D74**, 023525 (2006) [astro-ph/0605488]

21. V. Pettorino, C. Baccigalupi, Coupled and extended quintessence: theoretical differences and structure formation. Phys. Rev. **D77**, 103003 (2008) [arXiv:0802.1086]
22. J. Valiviita, E. Majerotto, R. Maartens, Instability in interacting dark energy and dark matter fluids. JCAP **0807**, 020 (2008) [arXiv:0804.0232]
23. L. Amendola, V. Pettorino, C. Quercellini, A. Vollmer, Testing coupled dark energy with next-generation large-scale observations. Phys. Rev. **D85**, 103008 (2012) [arXiv:1111.1404]
24. V. Pettorino, L. Amendola, C. Baccigalupi, C. Quercellini, Constraints on coupled dark energy using CMB data from WMAP and SPT. Phys. Rev. **D86**, 103507 (2012). doi:10.1103/PhysRevD.86.103507 [arXiv:1207.3293]

Chapter 4
High Derivative Theories

4.1 Introduction

In this chapter we develop formal aspects of an *effective field theory for perturbations*. The original motivation for this work was to show how to provide a complete set of coherent and consistent modifications to the gravitational field equations at perturbed order for modified gravity and dark energy theories from a Lagrangian for perturbations. However, the formalism can be applied in much more diverse systems where relativistic field theories play a central role, such as massive gravities, inflation and even effective theories for quantum gravity.

Our formalism is highly algorithmic and can be succinctly summarized as follows. First one must pick a field content and then write down the Lagrangian for perturbations, $\mathcal{L}_{\{2\}}$ by writing down all possible quadratic terms in the perturbed field variables, inserting appropriately indexed coupling coefficients infront of each term. Once the Lagrangian for perturbations is written down all the "usual" techniques of field theory can be employed to find (for example) conservation laws, equations of motion, energy-momentum tensors and symmetries. The theory that one will have will be very general. Picking a symmetry of the background spacetime allows the maximal freedom in the theory to be identified. It is then possible to impose additional restrictions on the theory (e.g. reparameterization invariance) to obtain more specialized (but still very general) subsets of theories. Crucially, we will discover how couplings in the underlying Lagrangian combine to construct the field equations (this is non-trivial for high-derivative theories). The high orders of derivatives present will produce total derivatives in $\mathcal{L}_{\{2\}}$ which we isolate and keep track of.

One of the rather interesting features of our formalism is that we are not required to construct scalar quantities from the *just* the perturbed field variables. It is of course true that the Lagrangian for perturbations requires scalar quantities, but these scalar quantities can be constructed from a combination of background *and* perturbed field variables. This opens up the possibility for Lagrangians for perturbations to be written down for theories which have never been considered at the level of the background. In particular, it is easy for our formalism to naturally include Lorentz-violating theories.

J. Pearson, *Generalized Perturbations in Modified Gravity and Dark Energy*, Springer Theses, DOI: 10.1007/978-3-319-01210-0_4,

4.1.1 Field Content

The focus in this chapter is theories whose field content includes high-order derivatives of scalar fields, the metric and vector fields. We call an nth-order field theory a field theory containing at most n-derivatives of the field. The field contents of the theories we consider are of the form

$$\mathcal{L} = \mathcal{L}(g_{\mu\nu}, \partial_\alpha g_{\mu\nu}), \qquad \{3\} \tag{4.1a}$$

$$\mathcal{L} = \mathcal{L}(g_{\mu\nu}, \phi, \nabla_\mu\phi, \nabla_\mu\nabla_\nu\phi), \qquad \{10\} \tag{4.1b}$$

$$\mathcal{L} = \mathcal{L}(g_{\mu\nu}, \partial_\alpha g_{\mu\nu}, \phi, \partial_\mu\phi, \partial_\mu\partial_\nu\phi), \qquad \{15\} \tag{4.1c}$$

$$\mathcal{L} = \mathcal{L}(g_{\mu\nu}, \partial_\alpha g_{\mu\nu}, \partial_\alpha\partial_\beta g_{\mu\nu}, \phi, \partial_\mu\phi, \partial_\mu\partial_\nu\phi), \qquad \{21\}, \tag{4.1d}$$

$$\mathcal{L} = \mathcal{L}(g_{\mu\nu}, A^\mu, \nabla_\alpha A^\mu), \qquad \{6\} \tag{4.1e}$$

$$\mathcal{L} = \mathcal{L}(g_{\mu\nu}, \partial_\alpha g_{\mu\nu}, A^\mu, \partial_\alpha A^\mu), \qquad \{10\} \tag{4.1f}$$

where ϕ is a scalar field, A^μ is a vector field, $g_{\mu\nu}$ is the metric of the spacetime, ∂_μ is the partial derivative and ∇_μ the covariant derivative (satisfying $\nabla_\mu g_{\alpha\beta} = 0$). The number in braces is the number of (naively) independent coupling coefficients in $\mathcal{L}_{\{2\}}$. In Sect. 3.3.4.2 we showed that *linking conditions* can be derived which enforce the Euler–Lagrange equation to be contained within the conservation equation, which will eliminate some of the independent coefficients.

The theories (4.1a) will have some correspondence with linearized massive gravity theories (i.e. the mass and kinetic term of the graviton is generalized). The second-order scalar field theories (4.1b) and (4.1c) are related: the theories (4.1c) contain the theories (4.1b) because in the latter the derivative of the metric is constrained to enter only via the Christoffel symbols (which are inside the covariant derivatives); we can impose a *structure* upon (4.1c) to obtain (4.1b). The theories with (4.1d) would completely encompass the Galileon, Horndeski, Gauss-Bonnet, $F(R)$, and all theories with curvature tensors and second order derivatives of scalar fields. The "structure consideration" is also true of the first order vector field theories (4.1e) and (4.1f).

4.1.2 The Projectors

Before we proceed with our formalism for high derivative theories, we will introduce some useful technology and ways of thinking about the perturbed dark energy momentum tensor.

Although the perturbed dark energy-momentum tensor $\delta_{\rm E} U^{\mu}{}_{\nu}$ is complicated to construct for general models, it is fundamentally just a symmetric rank-2 tensor and therefore only has 10 components. We identify the components $\delta_{\rm E} U^{\mu}{}_{\nu}$ with the density contrast $\delta \equiv \delta\rho/\rho$, perturbed pressure δP, velocity v^{μ} and anisotropic stress $\Pi^{\mu}{}_{\nu}$ of a *generalized fluid*. This allows some sort of physical picture to be attached to the $\delta_{\rm E} U^{\mu}{}_{\nu}$. The fluid decomposition of $\delta_{\rm E} U^{\mu}{}_{\nu}$ is

$$\delta_{\rm E} U^{\mu}{}_{\nu} = \delta\rho u^{\mu} u_{\nu} + \delta P \gamma^{\mu}{}_{\nu} + 2(\rho + P) v^{(\mu} u_{\nu)} + P \Pi^{\mu}{}_{\nu}, \tag{4.2}$$

where ρ, P are the density and pressure of the "dark sector fluid", u^{μ} is a time-like unit vector; these vectors and tensors satisfy

$$u^{\mu} u_{\mu} = -1, \qquad u^{\mu} v_{\mu} = 0, \qquad \Pi^{\mu}{}_{\mu} = 0, \qquad \Pi^{\mu}{}_{\nu} u_{\mu} = 0. \tag{4.3}$$

The components can be isolated by acting on $\delta_{\rm E} U^{\mu}{}_{\nu}$ with geometrical projectors.

We obtain a perturbed fluid quantity, f, from the perturbed energy-momentum tensor by acting with a projection operator,

$$f = {\rm P}_{(f)}{}^{\mu\nu} \delta_{\rm E} U_{\mu\nu}. \tag{4.4}$$

The projectors which return the perturbed density, pressure, velocity and anisotropic stress are respectively given by

$${\rm P}_{(\delta\rho)}{}^{\mu\nu} = u^{\mu} u^{\nu}, \quad {\rm P}_{(\delta P)}{}^{\mu\nu} = \frac{1}{3}\gamma^{\mu\nu}, \quad {\rm P}_{(V)\nu}{}^{\alpha\beta} = -\frac{1}{\rho + P}\gamma^{\alpha}{}_{\nu} u^{\beta}, \tag{4.5a}$$

$${\rm P}_{(\Pi)}{}^{\alpha\beta\mu\nu} = \frac{1}{P}\left(\gamma^{\alpha\mu}\gamma^{\nu\beta} - \frac{1}{3}\gamma^{\alpha\beta}\gamma^{\mu\nu}\right). \tag{4.5b}$$

Acting the projectors (4.5) upon the Eulerian perturbed dark energy momentum tensor (4.2) yields

$${\rm P}_{(\delta\rho)}{}^{\mu\nu} \delta_{\rm E} U_{\mu\nu} = \delta\rho, \quad {\rm P}_{(\delta P)}{}^{\mu\nu} \delta_{\rm E} U_{\mu\nu} = \delta P, \tag{4.6a}$$

$${\rm P}_{(V)\mu}{}^{\alpha\beta} \delta_{\rm E} U_{\alpha\beta} = V_{\mu}, \quad {\rm P}_{(\Pi)}{}^{\alpha\beta\mu\nu} \delta_{\rm E} U_{\mu\nu} = \Pi^{\alpha\beta}. \tag{4.6b}$$

Because there are only two "fundamental" tensors, u_{μ} and $\gamma_{\mu\nu}$, the four projectors (4.5) are not all independent. For instance, the pressure perturbation projector can be written as

$$3{\rm P}_{(\delta P)}{}^{\mu\nu} = (\rho + P) u_{\alpha} {\rm P}_{(V)}{}^{\nu\mu\alpha} = -(\rho + P) u_{\sigma} {\rm P}_{(\delta\rho)}{}^{\sigma}{}_{\alpha} {\rm P}_{(V)}{}^{\nu\mu\alpha}, \tag{4.7}$$

and the anisotropic stress projector can be written as

$$\mathrm{P}_{(\Pi)}{}^{\alpha\beta\mu\nu} = \frac{(\rho+P)^2}{P} u_\sigma u_\epsilon \mathrm{P}_{(\delta\rho)}{}^{\sigma}{}_{\lambda} \mathrm{P}_{(\delta\rho)}{}^{\epsilon}{}_{\omega} \left(\mathrm{P}_{(V)}{}^{\mu\alpha\lambda} \mathrm{P}_{(V)}{}^{\nu\beta\omega} - \frac{1}{3} \mathrm{P}_{(V)}{}^{\nu\mu\lambda} \mathrm{P}_{(V)}{}^{\beta\alpha\omega} \right). \tag{4.8}$$

Interestingly, both the pressure perturbation and anisotropic stress perturbation projectors can be constructed from the density contrast and velocity projectors. Schematically:

$$\mathrm{P}_{(\delta P)} = \mathrm{P}_{(\delta P)}(\mathrm{P}_{(\delta\rho)}, \mathrm{P}_{(V)}), \quad \mathrm{P}_{(\Pi)} = \mathrm{P}_{(\Pi)}(\mathrm{P}_{(\delta\rho)}, \mathrm{P}_{(V)}). \tag{4.9}$$

This hints at the existence of a generic expression which links the perturbed pressure and anisotropic stress in terms of the density and velocity.

We will now show how these projectors act upon the dark energy momentum tensor for a first order scalar field theory. We recall that the dark energy momentum tensor for the first order scalar field theory, $\mathcal{L} = \mathcal{L}(g_{\mu\nu}, \phi, \nabla_\mu\phi)$ is given by

$$\delta_{\mathrm{E}} U^{\mu\nu} = -\frac{1}{2}\left[\mathcal{V}^{\mu\nu} + \mathcal{Y}^{\alpha\mu\nu}\nabla_\alpha\right]\delta\phi - \frac{1}{2}\left[\mathcal{W}^{\mu\nu\alpha\beta} + U^{\mu\nu}g^{\alpha\beta}\right]\delta_{\mathrm{L}} g_{\alpha\beta} - £_\xi U^{\mu\nu}, \tag{4.10}$$

and the $(3+1)$ decomposition of the tensors $g, U, \mathcal{V}, \mathcal{Y}, \mathcal{W}$ was given in (3.31). Acting the projectors (4.5) upon the scalar field sector tensors yields

$$\mathrm{P}_{(\delta\rho)}^{\mu\nu} \mathcal{V}_{\mu\nu} = A_{\mathcal{V}}, \quad \mathrm{P}_{(\delta\rho)}{}^{\mu\nu} \mathcal{Y}^{\alpha}_{\mu\nu} = A_{\mathcal{Y}} u^\alpha \tag{4.11a}$$

$$\mathrm{P}_{(V)}{}^{\alpha\mu\nu} \mathcal{V}_{\mu\nu} = 0, \quad \mathrm{P}_{(V)}{}^{\alpha\mu\nu} \mathcal{Y}_{\beta\mu\nu} = \frac{C_{\mathcal{Y}}}{\rho+P} \gamma^{\alpha}{}_{\beta}, \tag{4.11b}$$

$$\mathrm{P}_{(\delta P)}{}^{\mu\nu} \mathcal{V}_{\mu\nu} = B_{\mathcal{V}}, \quad \mathrm{P}_{(\delta P)}{}^{\mu\nu} \mathcal{Y}^{\alpha}_{\mu\nu} = B_{\mathcal{Y}} u^\alpha, \tag{4.11c}$$

$$\mathrm{P}_{(\Pi)}{}^{\alpha\beta\mu\nu} \mathcal{V}_{\mu\nu} = 0, \quad \mathrm{P}_{(\Pi)}{}^{\alpha\beta\mu\nu} \mathcal{Y}_{\rho\mu\nu} = 0. \tag{4.11d}$$

Notice that there are no contributions to the anisotropic stress from the scalar-field sector; this is a generic feature of first order scalar field theories which is not shared by higher-order scalar field theories. Acting the projectors (4.5) upon the metric sector tensors yields

$$\mathrm{P}_{(\delta\rho)}{}^{\mu\nu} \mathcal{W}_{\mu\nu\alpha\beta} = A_{\mathcal{W}} u_\alpha u_\beta + B_{\mathcal{W}} \gamma_{\alpha\beta}, \tag{4.12a}$$

$$\mathrm{P}_{(V)}{}^{\sigma\mu\nu} \mathcal{W}_{\mu\nu\alpha\beta} = \frac{C_{\mathcal{W}}}{\rho+P} 2\gamma^{\sigma}{}_{(\alpha} u_{\beta)}, \tag{4.12b}$$

$$\mathrm{P}_{(\delta P)}{}^{\mu\nu} \mathcal{W}_{\mu\nu\alpha\beta} = B_{\mathcal{W}} u_\alpha u_\beta + \left(D_{\mathcal{W}} + \frac{2}{3} E_{\mathcal{W}} \right) \gamma_{\alpha\beta}, \tag{4.12c}$$

$$\mathrm{P}_{(\Pi)}{}^{\rho\sigma\mu\nu}\mathcal{W}_{\mu\nu\alpha\beta} = \frac{2E_{\mathcal{W}}}{P}\left(\gamma^{\rho}{}_{(\alpha}\gamma^{\sigma}{}_{\beta)} - \frac{1}{3}\gamma^{\rho\sigma}\gamma_{\alpha\beta}\right). \tag{4.12d}$$

To compute the action of the projectors on the Lie derivative of the energy momentum tensor we note that

$$\begin{aligned}\pounds_{\xi}U^{\mu\nu} &= \xi^{\alpha}\nabla_{\alpha}U^{\mu\nu} - 2U^{\alpha(\mu}\nabla_{\alpha}\xi^{\nu)} \\ &= 2(\rho+P)\xi^{\alpha}K^{(\mu}{}_{\alpha}u^{\nu)} - 2\rho u^{(\mu}\dot{\xi}^{\nu)} - 2P\bar{\nabla}^{(\mu}\xi^{\nu)}\end{aligned} \tag{4.13}$$

where $\dot{X} \equiv u^{\mu}\nabla_{\mu}X$, $\bar{\nabla}_{\mu}X \equiv \gamma^{\nu}{}_{\mu}\nabla_{\nu}X$ are "time" and "space" derivatives and $K_{\mu\nu} \equiv \nabla_{\mu}u_{\nu}$ is the extrinsic curvature. Acting with the projectors (4.5) on the Lie derivative (4.13) yields

$$\mathrm{P}_{(\delta\rho)\mu\nu}\pounds_{\xi}U^{\mu\nu} = 0, \quad \mathrm{P}_{(\delta P)\mu\nu}\pounds_{\xi}U^{\mu\nu} = -\frac{2}{3}P\bar{\nabla}_{\mu}\xi^{\mu}, \tag{4.14a}$$

$$\mathrm{P}_{(V)\alpha\mu\nu}\pounds_{\xi}U^{\mu\nu} = \xi^{\mu}K_{\mu\alpha}, \quad \mathrm{P}_{(\Pi)\alpha\beta\mu\nu}\pounds_{\xi}U^{\mu\nu} = \frac{2}{3}\gamma_{\alpha\beta}\bar{\nabla}_{\mu}\xi^{\mu} - 2\bar{\nabla}_{(\alpha}\xi_{\beta)}. \tag{4.14b}$$

Putting (4.11), (4.12), (4.14) together, we obtain the following expressions for the perturbed fluid variables for a first order scalar field theory:

$$\delta\rho = -\frac{1}{2}\left[A_{\mathcal{V}}\delta\phi + A_{\mathcal{Y}}\dot{\delta\phi}\right] - \frac{1}{2}\left[(A_{\mathcal{W}}-\rho)u^{\alpha}u^{\beta} + (B_{\mathcal{W}}+\rho)\gamma^{\alpha\beta}\right]\delta_{\mathrm{L}}g_{\alpha\beta}, \tag{4.15}$$

$$\begin{aligned}\delta P = &-\frac{1}{2}\left[B_{\mathcal{V}}\delta\phi + B_{\mathcal{Y}}\dot{\delta\phi}\right] + \frac{2}{3}P\bar{\nabla}_{\mu}\xi^{\mu} \\ &-\frac{1}{2}\left[(B_{\mathcal{W}}-P)u^{\alpha}u^{\beta} + \left(D_{\mathcal{W}} + \frac{2}{3}E_{\mathcal{W}} + P\right)\gamma^{\alpha\beta}\right]\delta_{\mathrm{L}}g_{\alpha\beta},\end{aligned} \tag{4.16}$$

$$V^{\mu} = -\frac{1}{2}\frac{C_{\mathcal{Y}}}{\rho+P}\bar{\nabla}^{\mu}\delta\phi - \frac{C_{\mathcal{W}}}{\rho+P}\gamma^{\mu(\alpha}u^{\beta)}\delta_{\mathrm{L}}g_{\alpha\beta} - \xi_{\alpha}K^{\mu\alpha}, \tag{4.17}$$

$$\Pi^{\alpha\beta} = -\frac{E_{\mathcal{W}}}{P}\left(\gamma^{\mu(\alpha}\gamma^{\beta)\nu} - \frac{1}{3}\gamma^{\mu\nu}\gamma^{\alpha\beta}\right)\delta_{\mathrm{L}}g_{\mu\nu} - \frac{2}{3}\gamma^{\alpha\beta}\bar{\nabla}_{\mu}\xi^{\mu} + 2\bar{\nabla}^{(\alpha}\xi^{\beta)}. \tag{4.18}$$

We can use a similar decomposition to write the Lagrangian perturbed metric variables. We write

$$\delta_{\mathrm{L}}g^{\mu}{}_{\nu} = 2\Phi u^{\mu}u_{\nu} + H\gamma^{\mu}{}_{\nu} + (N^{\mu}u_{\nu} + N_{\nu}u^{\mu}) + H^{(\Pi)\mu}{}_{\nu}, \tag{4.19}$$

where

$$u^{u}u_{\mu} = -1, \quad u^{\mu}N_{\mu} = 0, \quad u_{\mu}H^{(\Pi)\mu}{}_{\nu} = 0, \quad \gamma^{\nu}{}_{\mu}H^{(\Pi)\mu}{}_{\nu} = 0. \tag{4.20}$$

4.2 High-Order Metric Theories

We will apply the formalism to the "next simplest" theory, whose field content is just the metric and its first derivative, $\mathcal{L} = \mathcal{L}(g_{\mu\nu}, \partial_\alpha g_{\mu\nu})$. The field content of the perturbative sector will be given by

$$\mathcal{L}_{\{2\}} = \mathcal{L}_{\{2\}}(\delta_{\mathrm{L}} g_{\mu\nu}, \nabla_\alpha \delta_{\mathrm{L}} g_{\mu\nu}). \tag{4.21}$$

By writing down all quadratic terms in this perturbed field content the Lagrangian for perturbations is given by

$$\begin{aligned} 8\mathcal{L}_{\{2\}} &= \mathcal{Q}^{\sigma\mu\nu\rho\alpha\beta} \nabla_\sigma \delta_{\mathrm{L}} g_{\mu\nu} \nabla_\rho \delta_{\mathrm{L}} g_{\alpha\beta} + 2\mathcal{P}^{\mu\nu\rho\alpha\beta} \delta_{\mathrm{L}} g_{\mu\nu} \nabla_\rho \delta_{\mathrm{L}} g_{\alpha\beta} \\ &\quad + \mathcal{W}^{\mu\nu\alpha\beta} \delta_{\mathrm{L}} g_{\mu\nu} \delta_{\mathrm{L}} g_{\alpha\beta}. \end{aligned} \tag{4.22}$$

We had to introduce three tensors, $\mathcal{Q}$, $\mathcal{P}$, $\mathcal{W}$ (the *coupling coefficients*), each of which only contains background field variables. One of our aims will be to understand how the coupling coefficients combine to construct the generalized gravitational field equations. This was a rather trivial task for the theories we studied in Chap. 3. We will find rather interesting structures appearing, and we will see that the coupling coefficients combine in a non-trivial way to construct the field equations in these more general models.

The Lagrangian (4.22) will encompass a very large variety of the linearized field theories which are studied in the context of massive gravities (for recent reviews see [1, 2]). The tensor $\mathcal{W}$ provide the "mass term" of the graviton, the tensor $\mathcal{Q}$ will provide the "kinetic term" and $\mathcal{P}$ will provide some "graviton-kinetic" mixing. Obviously, General Relativity provides specific forms for both of these expressions (which we will provide in Sect. 6.1.1), but we will point out now that the mixing term $\mathcal{P} = 0$ in GR, and, up to a total derivative,

$$\mathcal{Q}_{GR}^{\rho\alpha\beta\sigma\mu\nu} = g^{\mu[\nu} g^{\alpha]\beta} g^{\rho\sigma} + 2g^{\beta\sigma} g^{\alpha[\nu} g^{\rho]\mu} \tag{4.23}$$

To compute the field equations we perform a variation of the perturbed field variables, $\delta_{\mathrm{L}} g_{\mu\nu} \to \delta_{\mathrm{L}} g_{\mu\nu} + \hat{\delta}\delta_{\mathrm{L}} g_{\mu\nu}$. This induces a response $\mathcal{L}_{\{2\}} \to \mathcal{L}_{\{2\}} + \hat{\delta}\mathcal{L}_{\{2\}}$, where

$$4\hat{\delta}\mathcal{L}_{\{2\}} = \Big[A^{\alpha\beta} + B^{\rho\alpha\beta}\nabla_\rho\Big]\hat{\delta}\delta_{\mathrm{L}} g_{\alpha\beta}, \tag{4.24}$$

with

$$A^{\mu\nu} \equiv \Big[\mathcal{W}^{\mu\nu\alpha\beta} + \mathcal{P}^{\mu\nu\rho\alpha\beta}\nabla_\rho\Big]\delta_{\mathrm{L}} g_{\alpha\beta}, \tag{4.25a}$$

$$B^{\rho\alpha\beta} \equiv \Big[\mathcal{P}^{\mu\nu\rho\alpha\beta} + \mathcal{Q}^{\sigma\mu\nu\rho\alpha\beta}\nabla_\sigma\Big]\delta_{\mathrm{L}} g_{\mu\nu}. \tag{4.25b}$$

It is important to realize that (4.24) contains a total derivative, which can be isolated by rewriting the derivative term. Doing so (without removing any terms or integrating by parts) yields

$$4\hat{\delta}\mathcal{L}_{\{2\}} = \Big[A^{\alpha\beta} - \nabla_\rho B^{\rho\alpha\beta}\Big]\hat{\delta}\delta_{\mathrm{L}}g_{\alpha\beta} + \nabla_\rho\Big[B^{\rho\alpha\beta}\hat{\delta}\delta_{\mathrm{L}}g_{\alpha\beta}\Big]. \tag{4.26}$$

The second term is a total derivative and can be used to identify the surface current,

$$\vartheta^\rho = \hat{\delta}\delta_{\mathrm{L}}g_{\alpha\beta}(\mathcal{P}^{\mu\nu\rho\alpha\beta}\delta_{\mathrm{L}}g_{\mu\nu} + 2\mathcal{Q}^{\sigma\mu\nu\rho\alpha\beta}\nabla_\sigma\delta_{\mathrm{L}}g_{\mu\nu}), \tag{4.27}$$

where the term in the brackets can be interpreted as the canonical momentum. We are now in a position to state that the variations $\hat{\delta}\delta_{\mathrm{L}}g_{\mu\nu}$ must vanish on the boundary for a well posed variational principle to be employed. The functional derivative of $\mathcal{L}_{\{2\}}$ with respect to the perturbed field variable is given by

$$4\frac{\hat{\delta}}{\hat{\delta}\delta_{\mathrm{L}}g_{\alpha\beta}}\mathcal{L}_{\{2\}} = A^{\alpha\beta} - \nabla_\rho B^{\rho\alpha\beta}. \tag{4.28}$$

This can be used to construct the Lagrangian perturbed energy-momentum tensor, where we obtain

$$\begin{aligned}\delta_{\mathrm{L}}U^{\alpha\beta} = {}& \frac{1}{2}\Big[\mathcal{Q}^{\sigma\mu\nu\rho\alpha\beta}\Big]\nabla_\rho\nabla_\sigma\delta_{\mathrm{L}}g_{\mu\nu} - \frac{1}{2}\Big[\mathcal{P}^{\alpha\beta\sigma\mu\nu} - \mathcal{P}^{\mu\nu\sigma\alpha\beta} - \nabla_\rho\mathcal{Q}^{\sigma\mu\nu\rho\alpha\beta}\Big]\nabla_\sigma\delta_{\mathrm{L}}g_{\mu\nu} \\ & - \frac{1}{2}\Big[\mathcal{W}^{\mu\nu\alpha\beta} + U^{\alpha\beta}g^{\mu\nu} - \nabla_\rho\mathcal{P}^{\mu\nu\rho\alpha\beta}\Big]\delta_{\mathrm{L}}g_{\mu\nu}.\end{aligned} \tag{4.29}$$

Equation (4.29) can be written as an operator expansion:

$$\delta_{\mathrm{L}}U^{\alpha\beta} = \hat{\mathbb{W}}^{\alpha\beta\mu\nu}\delta_{\mathrm{L}}g_{\mu\nu}, \tag{4.30a}$$

where $\hat{\mathbb{W}}^{\alpha\beta\mu\nu}$ is expanded up to second derivatives;

$$\hat{\mathbb{W}}^{\alpha\beta\mu\nu} = \mathbb{E}^{\alpha\beta\mu\nu} + \mathbb{F}^{\sigma\alpha\beta\mu\nu}\nabla_\sigma + \mathbb{G}^{\rho\sigma\alpha\beta\mu\nu}\nabla_\rho\nabla_\sigma, \tag{4.30b}$$

where we identify

$$\mathbb{E}^{\alpha\beta\mu\nu} \equiv -\frac{1}{2}\Big[\mathcal{W}^{\mu\nu\alpha\beta} + U^{\alpha\beta}g^{\mu\nu} - \nabla_\rho\mathcal{P}^{\mu\nu\rho\alpha\beta}\Big], \tag{4.31a}$$

$$\mathbb{F}^{\sigma\alpha\beta\mu\nu} \equiv -\frac{1}{2}\Big[\mathcal{P}^{\alpha\beta\sigma\mu\nu} - \mathcal{P}^{\mu\nu\sigma\alpha\beta} - \nabla_\rho\mathcal{Q}^{\sigma\mu\nu\rho\alpha\beta}\Big], \tag{4.31b}$$

$$\mathbb{G}^{\rho\sigma\alpha\beta\mu\nu} \equiv \frac{1}{2}\mathcal{Q}^{\sigma\mu\nu\rho\alpha\beta}. \tag{4.31c}$$

We now observe that (4.31) provides us with an understanding as to how the coupling coefficients $\mathcal{Q}$, $\mathcal{P}$, $\mathcal{W}$ in the Lagrangian for perturbations $\mathcal{L}_{\{2\}}$ combine to construct the generalized modifications to the gravitational field equations.

The Eulerian perturbed dark energy momentum tensor is given by

$$\delta_{\mathrm{E}} U^{\mu\nu} = \hat{\mathbb{W}}^{\mu\nu\alpha\beta}\left(\delta_{\mathrm{E}} g_{\alpha\beta} + 2\nabla_{(\alpha}\xi_{\beta)}\right) - \xi^{\alpha}\nabla_{\alpha}U^{\mu\nu} + 2U^{\alpha(\mu}\nabla_{\alpha}\xi^{\nu)}. \tag{4.32}$$

In terms of the coefficients in the operator expansion, the Eulerian perturbed conservation equation is given by

$$\begin{aligned}
&\Big[\mathbb{G}^{\rho\sigma\mu\nu\alpha\beta}\Big]\nabla_{\mu}\nabla_{\rho}\nabla_{\sigma}\left(\delta_{\mathrm{E}} g_{\alpha\beta} + 2\nabla_{(\alpha}\xi_{\beta)}\right)\\
&\quad + \Big[\nabla_{\mu}\mathbb{G}^{\rho\sigma\mu\nu\alpha\beta} + \mathbb{F}^{\sigma\rho\nu\alpha\beta}\Big]\nabla_{\rho}\nabla_{\sigma}\left(\delta_{\mathrm{E}} g_{\alpha\beta} + 2\nabla_{(\alpha}\xi_{\beta)}\right)\\
&\quad + 2\Big[\nabla_{\rho}\mathbb{F}^{\mu\rho\nu\alpha\beta} + \mathbb{E}^{\mu\nu\alpha\beta} + U^{\alpha(\mu}g^{\nu)\beta}\Big]\nabla_{\mu}\nabla_{\alpha}\xi_{\beta} - \Big[\nabla_{\mu}\nabla_{\alpha}U^{\mu\nu}\Big]\xi^{\alpha}\\
&\quad + \Big[\nabla_{\rho}\mathbb{F}^{\mu\rho\nu\alpha\beta} + \mathbb{E}^{\mu\nu\alpha\beta} + \frac{1}{2}\left(U^{\mu\nu}g^{\alpha\beta} - U^{\alpha\beta}g^{\mu\nu} + 2g^{\nu\beta}U^{\alpha\mu}\right)\Big]\nabla_{\mu}\delta_{\mathrm{E}} g_{\alpha\beta}\\
&\quad + \Big[\nabla_{\mu}\mathbb{E}^{\mu\nu\alpha\beta}\Big]\left(\delta_{\mathrm{E}} g_{\alpha\beta} + 2\nabla_{(\alpha}\xi_{\beta)}\right) = 0.
\end{aligned} \tag{4.33}$$

This theory has been useful in showing how the formalism for the Lagrangian for perturbations can be extended to include derivatives of the metric; we obtained covariant forms of all the important dynamical quantities. This entire theory, however, will be encompassed by the second order scalar field theory we are about to discuss.

4.3 High-Order Scalar Field Theories

We will now turn our attention to field theories containing a scalar field ϕ, its first and second derivatives. The motivation for studying perturbations of high-order scalar field theories is rather obvious when one inspects high-order scalar field theories such as the Horndeski or Galileon theory. In those theories, large complicated combinations of first and second derivatives of a scalar field ϕ are combined to produce second order field equations. Perturbing these theories directly is complicated and model dependent. As we show below, one cannot write down a general set of Lorentz invariant building blocks for a general high-order scalar field theory. The problem becomes more acute when we realize that we do not know, *a priori*, what the underlying Lagrangian should look like; indeed, if the field equations need to be at most second order at all! However, one can write down the Lagrangian which governs perturbations. It has become commonplace to study these rather complicated theories, and so it is now rather important that we begin to construct an effective field theory to classify perturbations of these theories.

We will study the perturbations of "high order" scalar field theories. The field content we will consider is of the form

$$\mathcal{L} = \mathcal{L}(g_{\mu\nu}, \phi, \nabla_\mu \phi, \nabla_\mu \nabla_\nu \phi, \ldots). \tag{4.34}$$

A general second order scalar field theory has a Lagrangian given by

$$\mathcal{L} = \mathcal{L}(g_{\mu\nu}, \phi, \nabla_\mu \phi, \nabla_\mu \nabla_\nu \phi) \tag{4.35}$$

The Lagrangian is a scalar quantity, and so (for a Lorentz invariant theory) the field content must combine in such a way as to leave the Lagrangian being a scalar. That is, we will never have the Lagrangian being an explicit function of $\nabla_\mu \phi$ or $\nabla_\mu \nabla_\nu \phi$ (this is of course a trivial point). To see what impact this has on our ability to write down a general field content in terms of scalar quantities consider for a moment the Lagrangian of a first-order theory, $\mathcal{L} = \mathcal{L}(g_{\mu\nu}, \phi, \nabla_\mu \phi)$. There are only two Lorentz invariant scalars which can be built from the field content: ϕ and $g^{\mu\nu} \nabla_\mu \phi \nabla_\nu \phi$. So, to write $\mathcal{L} = \mathcal{L}(\phi, g^{\mu\nu} \nabla_\mu \phi \nabla_\nu \phi)$ is *not a loss of generality* for a Lorentz invariant first order scalar field theory. One is not able to write a similar catch-all expression for a 2-and-higher order scalar field theory. The Lagrangian for a second order field theory will be constructed from scalar terms such as

$$\phi, \quad (\nabla_\mu \phi)(\nabla^\mu \phi), \quad (\nabla^\mu \phi)(\nabla^\nu \phi)(\nabla_\mu \nabla_\nu \phi), \quad (\nabla^\mu \nabla^\nu \phi)(\nabla_\mu \nabla_\nu \phi), \tag{4.36a}$$

$$(\nabla_\mu \phi)(\nabla^\mu \nabla^\nu \phi)(\nabla_\nu \nabla_\alpha \phi)(\nabla^\alpha \nabla^\beta \phi)(\nabla_\beta \phi), \quad \ldots \tag{4.36b}$$

which is not an exhaustive list. In fact, an exhaustive list cannot be written down and thus we are not able to explicitly write down the building blocks of a general second order scalar field theory (again, realize that this is fundamentally different from the case of a first order scalar field theory); this issue will obviously become more acute with higher-order field theories. However, and this is perhaps where our formalism "comes into its own", we can write down a Lagrangian which completely describes the perturbative sector.

To illustrate how our effective action formalism can be applied to high order scalar field theories to obtain useable results, we will concentrate on a second order scalar field theory from hereon.

4.3.1 The Lagrangian for Perturbations: $\mathcal{L}_{\{2\}}$

We will study the set of theories given by

$$\mathcal{L} = \mathcal{L}(g_{\mu\nu}, \partial_\alpha g_{\mu\nu}, \phi, \partial_\alpha \phi, \partial_\alpha \partial_\beta \phi). \tag{4.37}$$

One may be concerned that this field content only contains partial derivatives, and not covariant derivatives. The latter set of theories would be given by

$$\mathcal{L} = \mathcal{L}(g_{\mu\nu}, \phi, \nabla_\mu \phi, \nabla_\mu \nabla_\nu \phi), \tag{4.38}$$

which is encompassed by (4.37). It is important to realize, however, that (4.38) has imposed a theoretical restriction upon the theory: derivatives of the metric are forced to only appear inside Christoffel symbols. We will not impose this restriction from the outset, but we will show how to impose it later on. The field content (4.38) will encompass, for example, the Galileon theory in flat spacetime [3] but not the covariant Galileon theory [4, 5], because that would require curvature tensors in the field content. Nevertheless, this field theory allows for a very rich landscape of possibilities.

The field content of the Lagrangian for perturbations from the theory (4.37) is given by

$$\mathcal{L}_{\{2\}} = \mathcal{L}_{\{2\}}(\delta_{\mathrm{L}} g_{\mu\nu}, \nabla_\alpha \delta_{\mathrm{L}} g_{\mu\nu}, \delta\phi, \nabla_\mu \delta\phi, \nabla_\mu \nabla_\nu \delta\phi). \tag{4.39}$$

Constructing the Lagrangian for perturbations by writing down all possible quadratic terms of these perturbed field variables yields

$$\begin{aligned}
\mathcal{L}_{\{2\}} &= \mathcal{A}\delta\phi^2 + \mathcal{B}^{\mu}\delta\phi\nabla_\mu\delta\phi + \frac{1}{2}\mathcal{C}^{\mu\nu}\nabla_\mu\delta\phi\nabla_\nu\delta\phi + \mathcal{D}^{\mu\nu}\delta\phi\nabla_\mu\nabla_\nu\delta\phi \\
&+ \mathcal{E}^{\mu\alpha\beta}\nabla_\mu\delta\phi\nabla_\alpha\nabla_\beta\delta\phi + \frac{1}{2}\mathcal{F}^{\mu\nu\alpha\beta}\nabla_\mu\nabla_\nu\delta\phi\nabla_\alpha\nabla_\beta\delta\phi \\
&+ \mathcal{I}^{\rho\mu\nu}\nabla_\rho\delta_{\mathrm{L}} g_{\mu\nu}\delta\phi + \mathcal{J}^{\rho\mu\nu\alpha}\nabla_\rho\delta_{\mathrm{L}} g_{\mu\nu}\nabla_\alpha\delta\phi \\
&+ \mathcal{N}^{\rho\mu\nu\alpha\beta}\nabla_\rho\delta_{\mathrm{L}} g_{\mu\nu}\nabla_\alpha\nabla_\beta\delta\phi + \frac{1}{2}\mathcal{M}^{\rho\mu\nu\sigma\alpha\beta}\nabla_\rho\delta_{\mathrm{L}} g_{\mu\nu}\nabla_\sigma\delta_{\mathrm{L}} g_{\alpha\beta} \\
&+ \frac{1}{4}\bigg[\mathcal{V}^{\mu\nu}\delta\phi\delta_{\mathrm{L}} g_{\mu\nu} + \mathcal{Y}^{\alpha\mu\nu}\delta_{\mathrm{L}} g_{\mu\nu}\nabla_\alpha\delta\phi + \mathcal{Z}^{\mu\nu\alpha\beta}\delta_{\mathrm{L}} g_{\alpha\beta}\nabla_\mu\nabla_\nu\delta\phi \\
&+ \frac{1}{2}\mathcal{W}^{\mu\nu\alpha\beta}\delta_{\mathrm{L}} g_{\mu\nu}\delta_{\mathrm{L}} g_{\alpha\beta} + \mathcal{U}^{\rho\mu\nu\alpha\beta}\nabla_\rho\delta_{\mathrm{L}} g_{\mu\nu}\delta_{\mathrm{L}} g_{\alpha\beta}\bigg].
\end{aligned} \tag{4.40}$$

We had to introduce 15 coupling coefficients, each of which is only a function of the (background) unperturbed field variables. As should be apparent, this theory is much richer and computationally complex than the theories we discussed in Chap. 3.

We now obtain equations of motion and the perturbed energy-momentum tensor from this Lagrangian. It will turn out that the quantities which arise in the energy-momentum tensor are easier to do "general" calculations with than the coefficients that appear in the Lagrangian. However, the fundamental origin of the terms in the energy-momentum tensor is mysterious *without* reference to the Lagrangian.

We induce variations in the perturbed field variables

$$\delta\phi \to \delta\phi + \hat{\delta}\delta\phi, \quad \delta_{\rm L}g_{\mu\nu} \to \delta_{\rm L}g_{\mu\nu} + \hat{\delta}\delta_{\rm L}g_{\mu\nu}. \tag{4.41}$$

The Lagrangian will respond accordingly, $\mathcal{L}_{\{2\}} \to \mathcal{L}_{\{2\}} + \hat{\delta}\mathcal{L}_{\{2\}}$, where the response $\hat{\delta}\mathcal{L}_{\{2\}}$ can be compactly written as

$$\hat{\delta}\mathcal{L}_{\{2\}} = A\hat{\delta}\delta\phi \, + \, B^{\mu}\nabla_{\mu}\hat{\delta}\delta\phi \, + \, C^{\mu\nu}\nabla_{\mu}\nabla_{\nu}\hat{\delta}\delta\phi \, + \, D^{\mu\nu}\hat{\delta}\delta_{\rm L}g_{\mu\nu} \, + \, E^{\alpha\mu\nu}\nabla_{\alpha}\hat{\delta}\delta_{\rm L}g_{\mu\nu}, \tag{4.42}$$

where we defined

$$A \equiv \left[2\mathcal{A} + \mathcal{B}^{\mu}\nabla_{\mu} + \mathcal{D}^{\mu\nu}\nabla_{\mu}\nabla_{\nu}\right]\delta\phi + \left[\frac{1}{4}\mathcal{V}^{\mu\nu} + \mathcal{I}^{\rho\mu\nu}\nabla_{\rho}\right]\delta_{\rm L}g_{\mu\nu}, \tag{4.43a}$$

$$B^{\mu} \equiv \left[\mathcal{B}^{\mu} + \mathcal{C}^{\mu\nu}\nabla_{\nu} + \mathcal{E}^{\mu\alpha\beta}\nabla_{\alpha}\nabla_{\beta}\right]\delta\phi + \left[\frac{1}{4}\mathcal{Y}^{\mu\alpha\beta} + \mathcal{J}^{\rho\alpha\beta\mu}\nabla_{\rho}\right]\delta_{\rm L}g_{\alpha\beta}, \tag{4.43b}$$

$$C^{\mu\nu} \equiv \left[\mathcal{D}^{\mu\nu} + \mathcal{E}^{\alpha\mu\nu}\nabla_{\alpha} + \mathcal{F}^{\mu\nu\alpha\beta}\nabla_{\alpha}\nabla_{\beta}\right]\delta\phi + \left[\frac{1}{4}\mathcal{Z}^{\mu\nu\alpha\beta} + \mathcal{N}^{\rho\alpha\beta\mu\nu}\nabla_{\rho}\right]\delta_{\rm L}g_{\alpha\beta}, \tag{4.43c}$$

$$4D^{\mu\nu} \equiv \left[\mathcal{V}^{\mu\nu} + \mathcal{Y}^{\alpha\mu\nu}\nabla_{\alpha} + \mathcal{Z}^{\alpha\beta\mu\nu}\nabla_{\alpha}\nabla_{\beta}\right]\delta\phi + \left[\mathcal{W}^{\mu\nu\alpha\beta} + \mathcal{U}^{\rho\alpha\beta\mu\nu}\nabla_{\rho}\right]\delta_{\rm L}g_{\alpha\beta}, \tag{4.43d}$$

$$E^{\rho\mu\nu} \equiv \left[\mathcal{I}^{\rho\mu\nu} + \mathcal{J}^{\rho\mu\nu\alpha}\nabla_{\alpha} + \mathcal{N}^{\rho\mu\nu\alpha\beta}\nabla_{\alpha}\nabla_{\beta}\right]\delta\phi + \left[\frac{1}{4}\mathcal{U}^{\rho\mu\nu\alpha\beta} + \mathcal{M}^{\rho\mu\nu\sigma\alpha\beta}\nabla_{\sigma}\right]\delta_{\rm L}g_{\alpha\beta}. \tag{4.43e}$$

Isolating all total derivatives, (4.42) becomes

$$\begin{aligned}\hat{\delta}\mathcal{L}_{\{2\}} = &\left[A - \nabla_{\mu}B^{\mu} + \nabla_{\mu}\nabla_{\nu}C^{\mu\nu}\right]\hat{\delta}\delta\phi + \left[D^{\mu\nu} - \nabla_{\alpha}E^{\alpha\mu\nu}\right]\hat{\delta}\delta_{\rm L}g_{\mu\nu} \\ &+ \nabla_{\rho}\left[\hat{\delta}\delta\phi B^{\rho} - \hat{\delta}\delta\phi\nabla_{\nu}C^{\rho\nu} + C^{\rho\nu}\nabla_{\nu}\hat{\delta}\delta\phi + E^{\rho\mu\nu}\hat{\delta}\delta_{\rm L}g_{\mu\nu}\right].\end{aligned} \tag{4.44}$$

Therefore, the functional derivatives of the Lagrangian for perturbations are

$$\frac{\hat{\delta}}{\hat{\delta}\delta\phi}\mathcal{L}_{\{2\}} = A - \nabla_{\mu}B^{\mu} + \nabla_{\mu}\nabla_{\nu}C^{\mu\nu}, \tag{4.45a}$$

$$\frac{\hat{\delta}}{\hat{\delta}\delta_{\mathrm{L}} g_{\mu\nu}} \mathcal{L}_{\{2\}} = D^{\mu\nu} - \nabla_\alpha E^{\alpha\mu\nu}, \tag{4.45b}$$

and the surface current is

$$\vartheta^\rho = (B^\rho - \nabla_\nu C^{\rho\nu})\hat{\delta}\delta\phi + C^{\rho\nu}\nabla_\nu \hat{\delta}\delta\phi + E^{\rho\mu\nu}\hat{\delta}\delta_{\mathrm{L}} g_{\mu\nu}. \tag{4.45c}$$

Equation (4.45a) will provide the equation of motion of the $\delta\phi$-field, Eq. (4.45b) provides the contribution to the perturbed gravitational field equations and (4.45c) provides us with the conditions that $\hat{\delta}\delta\phi$, $\hat{\delta}\delta_{\mathrm{L}} g_{\mu\nu}$ and the derivative $\nabla_\mu \hat{\delta}\delta\phi$ must vanish on the boundary for the variational principle to be applicable.

4.3.1.1 Perturbed Euler–Lagrange Equation

By requiring (4.45a) to vanish, we obtain the equation of motion of the $\delta\phi$-field,

$$\nabla_\mu \nabla_\nu C^{\mu\nu} - \nabla_\mu B^\mu + A = 0, \tag{4.46}$$

where A, B, C are defined in (4.43). After explicitly expanding all terms, we obtain the following expression for the equation of motion of the $\delta\phi$-field:

$$\begin{aligned}
&\Big[\mathcal{F}^{\mu\nu\alpha\beta}\Big]\nabla_\mu\nabla_\nu\nabla_\alpha\nabla_\beta\delta\phi + \Big[\mathcal{E}^{\alpha\mu\beta} - \mathcal{E}^{\mu\alpha\beta}\Big]\nabla_\mu\nabla_\alpha\nabla_\beta\delta\phi \\
&\quad + \Big[2\mathcal{D}^{\mu\nu} - \mathcal{C}^{\mu\nu} - \nabla_\alpha\mathcal{E}^{\alpha\mu\nu} + \nabla_\alpha\nabla_\beta\mathcal{F}^{\alpha\beta\mu\nu}\Big]\nabla_\mu\nabla_\nu\delta\phi \\
&\quad + \Big[\nabla_\alpha\nabla_\beta\mathcal{E}^{\nu\alpha\beta} - \nabla_\mu\mathcal{C}^{\mu\nu}\Big]\nabla_\nu\delta\phi + \Big[2\mathcal{A} - \nabla_\mu\mathcal{B}^\mu + \nabla_\mu\nabla_\nu\mathcal{D}^{\mu\nu}\Big]\delta\phi = \delta\mathcal{S},
\end{aligned} \tag{4.47}$$

where the scalar source term is given by

$$\begin{aligned}
\delta\mathcal{S} \equiv\ & \frac{1}{4}\Big[\nabla_\mu\mathcal{Y}^{\mu\alpha\beta} - \mathcal{V}^{\alpha\beta} - \nabla_\mu\nabla_\nu\mathcal{Z}^{\mu\nu\alpha\beta}\Big]\delta_{\mathrm{L}} g_{\alpha\beta} \\
& + \Big[\nabla_\mu\mathcal{J}^{\rho\alpha\beta\mu} - \mathcal{I}^{\rho\alpha\beta} + \frac{1}{4}\mathcal{Y}^{\rho\alpha\beta} - \nabla_\mu\nabla_\nu\mathcal{N}^{\rho\alpha\beta\mu\nu}\Big]\nabla_\rho\delta_{\mathrm{L}} g_{\alpha\beta} \\
& + \Big[\mathcal{J}^{\nu\alpha\beta\mu} - \frac{1}{4}\mathcal{Z}^{\mu\nu\alpha\beta}\Big]\nabla_\mu\nabla_\nu\delta_{\mathrm{L}} g_{\alpha\beta} - \Big[\mathcal{N}^{\rho\alpha\beta\mu\nu}\Big]\nabla_\mu\nabla_\nu\nabla_\rho\delta_{\mathrm{L}} g_{\alpha\beta}.
\end{aligned} \tag{4.48}$$

We can use the Ricci identity

$$(\nabla_\alpha \nabla_\beta - \nabla_\beta \nabla_\alpha) T_\mu = -R^\nu{}_{\mu\alpha\beta} T_\nu, \tag{4.49}$$

to eliminate third derivatives in the equation of motion by rewriting the second term of (4.47) as

$$\begin{aligned}(\mathcal{E}^{\alpha\mu\beta} - \mathcal{E}^{\mu\alpha\beta}) \nabla_\mu \nabla_\alpha \nabla_\beta \delta\phi &= \mathcal{E}^{\alpha\mu\beta} (\nabla_\mu \nabla_\alpha - \nabla_\alpha \nabla_\mu) \nabla_\beta \delta\phi \\ &= -\mathcal{E}^{\alpha\mu\beta} R^\nu{}_{\beta\mu\alpha} \nabla_\nu \delta\phi,\end{aligned} \tag{4.50}$$

where $R^\nu{}_{\beta\mu\alpha}$ is the Riemann tensor of the background. Hence, the equation of motion (4.47) becomes

$$\begin{aligned}&\left[\mathcal{F}^{\mu\nu\alpha\beta}\right] \nabla_\mu \nabla_\nu \nabla_\alpha \nabla_\beta \delta\phi + \left[2\mathcal{D}^{\mu\nu} - \mathcal{C}^{\mu\nu} - \nabla_\alpha \mathcal{E}^{\alpha\mu\nu} + \nabla_\alpha \nabla_\beta \mathcal{F}^{\alpha\beta\mu\nu}\right] \nabla_\mu \nabla_\nu \delta\phi \\ &+ \left[\nabla_\alpha \nabla_\beta \mathcal{E}^{\nu\alpha\beta} - \nabla_\mu \mathcal{C}^{\mu\nu} - \mathcal{E}^{\alpha\mu\beta} R^\nu{}_{\beta\mu\alpha}\right] \nabla_\nu \delta\phi \\ &+ \left[2\mathcal{A} - \nabla_\mu \mathcal{B}^\mu + \nabla_\mu \nabla_\nu \mathcal{D}^{\mu\nu}\right] \delta\phi = \delta\mathcal{S}.\end{aligned} \tag{4.51}$$

The Euler–Lagrange equation of motion for the $\delta\phi$-field is *a priori* (i.e. when $\mathcal{F} \neq 0$) of fourth order in derivatives of $\delta\phi$, and is sourced by up-to third derivatives of the Lagrangian perturbed metric. One can observe from (4.51) that the simpler subset of theories with $\mathcal{F}^{\mu\nu\alpha\beta} = 0$ the equation of motion becomes only second order in derivatives of $\delta\phi$. This reduction of the order of the equations of motion by including a background curvature tensor was also observed in, for example, kinetic gravity braiding [6].

If, for some reason, $\mathcal{F} = 0$, then the equation of motion (4.51) becomes second order, and can be written schematically as

$$G^{\mu\nu}_{\text{eff}} \nabla_\mu \nabla_\nu \delta\phi + \left[\cdots\right]^\nu \nabla_\nu \delta\phi + \left[\cdots\right] \delta\phi = \delta\mathcal{S}, \tag{4.52}$$

where the important quantity we want to highlight is the effective metric, $G^{\mu\nu}_{\text{eff}}$, which is given by

$$G^{\mu\nu}_{\text{eff}} \equiv 2\mathcal{D}^{\mu\nu} - \mathcal{C}^{\mu\nu} - \nabla_\alpha \mathcal{E}^{\alpha\mu\nu}. \tag{4.53}$$

Scalar field perturbations will be confined by the light-cone of this metric, rather than that of the spacetime metric.

4.3.1.2 Perturbed Dark Energy Momentum Tensor

The perturbed dark energy momentum tensor is computed from (4.45b) and yields

$$\delta_{\text{L}} U^{\mu\nu} = -\frac{1}{2}\left[4D^{\mu\nu} - 4\nabla_\alpha E^{\alpha\mu\nu} + U^{\mu\nu} g^{\alpha\beta} \delta_{\text{L}} g_{\alpha\beta}\right], \tag{4.54}$$

where D, E are defined in (4.43). Inserting D, E from (4.43), the Lagrangian perturbed energy-momentum tensor (4.54) is given by the rather compact expression

$$\delta_L U^{\mu\nu} = \hat{\mathbb{Y}}^{\mu\nu}\delta\phi + \hat{\mathbb{W}}^{\mu\nu\alpha\beta}\delta_L g_{\alpha\beta}, \tag{4.55a}$$

where $\hat{\mathbb{Y}}^{\mu\nu}$, $\hat{\mathbb{W}}^{\mu\nu\alpha\beta}$ are two operators,

$$\hat{\mathbb{Y}}^{\mu\nu} \equiv \mathbb{A}^{\mu\nu} + \mathbb{B}^{\alpha\mu\nu}\nabla_\alpha + \mathbb{C}^{\alpha\beta\mu\nu}\nabla_\alpha\nabla_\beta + \mathbb{D}^{\rho\alpha\beta\mu\nu}\nabla_\rho\nabla_\alpha\nabla_\beta, \tag{4.55b}$$

$$\hat{\mathbb{W}}^{\mu\nu\alpha\beta} \equiv \mathbb{E}^{\mu\nu\alpha\beta} + \mathbb{F}^{\rho\mu\nu\alpha\beta}\nabla_\rho + \mathbb{G}^{\rho\sigma\mu\nu\alpha\beta}\nabla_\rho\nabla_\sigma, \tag{4.55c}$$

where the coefficients which appear in these operators are defined by

$$\mathbb{A}^{\mu\nu} \equiv -\frac{1}{2}\left[\mathcal{V}^{\mu\nu} - 4\nabla_\rho\mathcal{I}^{\rho\mu\nu}\right], \tag{4.56a}$$

$$\mathbb{B}^{\alpha\mu\nu} \equiv -\frac{1}{2}\left[\mathcal{Y}^{\alpha\mu\nu} - 4(\mathcal{I}^{\alpha\mu\nu} + \nabla_\rho\mathcal{J}^{\rho\mu\nu\alpha})\right], \tag{4.56b}$$

$$\mathbb{C}^{\alpha\beta\mu\nu} \equiv -\frac{1}{2}\left[\mathcal{Z}^{\alpha\beta\mu\nu} - 4(\mathcal{J}^{\beta\mu\nu\alpha} + \nabla_\rho\mathcal{N}^{\rho\mu\nu\alpha\beta})\right], \tag{4.56c}$$

$$\mathbb{D}^{\rho\alpha\beta\mu\nu} \equiv 2\mathcal{N}^{\rho\mu\nu\alpha\beta}, \tag{4.56d}$$

$$\mathbb{E}^{\mu\nu\alpha\beta} \equiv -\frac{1}{2}\left[\mathcal{W}^{\mu\nu\alpha\beta} + U^{\mu\nu}g^{\alpha\beta} - \nabla_\rho\mathcal{U}^{\rho\mu\nu\alpha\beta}\right], \tag{4.56e}$$

$$\mathbb{F}^{\rho\mu\nu\alpha\beta} \equiv -\frac{1}{2}\left[\mathcal{U}^{\rho\alpha\beta\mu\nu} - \mathcal{U}^{\rho\mu\nu\alpha\beta} - 4\nabla_\epsilon\mathcal{M}^{\epsilon\mu\nu\rho\alpha\beta}\right], \tag{4.56f}$$

$$\mathbb{G}^{\rho\sigma\mu\nu\alpha\beta} \equiv 2\mathcal{M}^{\rho\mu\nu\sigma\alpha\beta}. \tag{4.56g}$$

It is useful to regard the set of tensors $\mathbb{A}, \mathbb{B}, \mathbb{C}, \mathbb{D}$ as controlling the perturbations in the scalar-field-sector and the set $\mathbb{E}, \mathbb{F}, \mathbb{G}$ as controlling the perturbations in the metric-sector. The bold-face tensors (4.56) contain a large amount of useful information. Their structure dictates how the coupling coefficients combine to construct the perturbed dark energy momentum tensor. We are able to observe that the coupling coefficient $\mathcal{J}$ affects the first and second derivative of the perturbed scalar field. Similarly, $\mathcal{U}$ affects the undifferentiated and first differential of the perturbed metric. These expressions are very useful windows into the structure of all second order scalar field theories.

4.3.1.3 Perturbed Conservation Equation

The Eulerian perturbed conservation equation $\delta_{\mathrm{E}}(\nabla_{\mu}U^{\mu\nu}) = 0$ using (4.55) yields

$$\begin{aligned}
&\mathbb{D}^{\rho\alpha\beta\mu\nu}\nabla_{\mu}\nabla_{\rho}\nabla_{\alpha}\nabla_{\beta}\delta\phi + (\nabla_{\mu}\mathbb{D}^{\rho\alpha\beta\mu\nu} + \mathbb{C}^{\alpha\beta\rho\nu})\nabla_{\rho}\nabla_{\alpha}\nabla_{\beta}\delta\phi \\
&+ (\nabla_{\mu}\mathbb{C}^{\alpha\beta\mu\nu} + \mathbb{B}^{\beta\alpha\nu})\nabla_{\alpha}\nabla_{\beta}\delta\phi + (\nabla_{\mu}\mathbb{B}^{\alpha\mu\nu} + \mathbb{A}^{\alpha\nu})\nabla_{\alpha}\delta\phi \\
&+ (\nabla_{\mu}\mathbb{A}^{\mu\nu})\delta\phi = \delta_{\mathrm{E}}J^{\nu} + 2E^{\nu},
\end{aligned} \tag{4.57a}$$

where the sources are written as a contribution from the Eulerian perturbed metric and a contribution from the ξ^{μ}-field,

$$\begin{aligned}
\delta_{\mathrm{E}}J^{\nu} \equiv &- \nabla_{\mu}\mathbb{E}^{\mu\nu\alpha\beta}\delta_{\mathrm{E}}g_{\alpha\beta} \\
&- (\nabla_{\mu}\mathbb{F}^{\rho\mu\nu\alpha\beta} + \mathbb{E}^{\rho\nu\alpha\beta} + \frac{1}{2}U^{\rho\nu}g^{\alpha\beta} - \frac{1}{2}U^{\alpha\beta}g^{\rho\nu} + g^{\nu\beta}U^{\alpha\rho})\nabla_{\rho}\delta_{\mathrm{E}}g_{\alpha\beta} \\
&- (\nabla_{\mu}\mathbb{G}^{\rho\sigma\mu\nu\alpha\beta} + \mathbb{F}^{\rho\sigma\nu\alpha\beta})\nabla_{\rho}\nabla_{\sigma}\delta_{\mathrm{E}}g_{\alpha\beta} \\
&- \mathbb{G}^{\rho\sigma\mu\nu\alpha\beta}\nabla_{\mu}\nabla_{\rho}\nabla_{\sigma}\delta_{\mathrm{E}}g_{\alpha\beta},
\end{aligned} \tag{4.57b}$$

$$\begin{aligned}
E^{\nu} \equiv &- \nabla_{\mu}\mathbb{E}^{\mu\nu\alpha\beta}\nabla_{(\alpha}\xi_{\beta)} \\
&- (\nabla_{\mu}\mathbb{F}^{\rho\mu\nu\alpha\beta} + \mathbb{E}^{\rho\nu\alpha\beta} + \frac{1}{2}U^{\rho\nu}g^{\alpha\beta} - \frac{1}{2}U^{\alpha\beta}g^{\rho\nu} + g^{\nu\beta}U^{\alpha\rho})\nabla_{\rho}\nabla_{(\alpha}\xi_{\beta)} \\
&- (\nabla_{\mu}\mathbb{G}^{\rho\sigma\mu\nu\alpha\beta} + \mathbb{F}^{\rho\sigma\nu\alpha\beta})\nabla_{\rho}\nabla_{\sigma}\nabla_{(\alpha}\xi_{\beta)} \\
&- \mathbb{G}^{\rho\sigma\mu\nu\alpha\beta}\nabla_{\mu}\nabla_{\rho}\nabla_{\sigma}\nabla_{(\alpha}\xi_{\beta)}.
\end{aligned} \tag{4.57c}$$

The perturbed conservation equation is *a priori* high order in derivatives of all fields in the theory. When $\mathbb{C} = \mathbb{D} = \mathbb{F} = \mathbb{G} = 0$, the conservation equation (4.57a) becomes at most second order in derivatives of $\delta\phi$ and the sources (4.57b) and (4.57c) become at most second order in derivatives of $\delta_{\mathrm{E}}g_{\mu\nu}$, ξ^{μ}. These are, however, very heavy-handed conditions to place upon the theory.

4.3.1.4 (3 + 1) Decomposition of the Coupling Coefficients

We will use the (3 + 1) decomposition to obtain the absolute maximal freedom in the components $\delta_{\mathrm{E}}U^{\mu}{}_{\nu}$. To begin, we note down the symmetries of the tensors in $\mathcal{L}_{\{2\}}$ that appear in $\delta_{\mathrm{L}}U^{\mu\nu}$. The symmetries are

$$\mathcal{V}^{\mu\nu} = \mathcal{V}^{(\mu\nu)}, \quad \mathcal{Y}^{\alpha\mu\nu} = \mathcal{Y}^{\alpha(\mu\nu)}, \quad \mathcal{I}^{\alpha\mu\nu} = \mathcal{I}^{\alpha(\mu\nu)}, \tag{4.58a}$$

$$\mathcal{J}^{\rho\mu\nu\alpha} = \mathcal{J}^{\rho(\mu\nu)\alpha}, \quad \mathcal{Z}^{\alpha\beta\mu\nu} = \mathcal{Z}^{(\alpha\beta)(\mu\nu)}, \tag{4.58b}$$

$$\mathcal{W}^{\mu\nu\alpha\beta} = \mathcal{W}^{(\mu\nu)(\alpha\beta)} = \mathcal{W}^{\alpha\beta\mu\nu}, \quad \mathcal{N}^{\rho\mu\nu\alpha\beta} = \mathcal{N}^{\rho(\mu\nu)(\alpha\beta)}, \tag{4.58c}$$

$$\mathcal{U}^{\rho\mu\nu\alpha\beta} = \mathcal{U}^{\rho(\mu\nu)(\alpha\beta)}, \quad \mathcal{M}^{\rho\mu\nu\sigma\alpha\beta} = \mathcal{M}^{\rho(\mu\nu)\sigma(\alpha\beta)} = \mathcal{M}^{\sigma\alpha\beta\rho\mu\nu}. \tag{4.58d}$$

Guided by these symmetries, the (3+1) decomposition of the tensors appearing in (4.55) is given by

$$\mathcal{V}^{\mu\nu} = A_{\mathcal{V}} u^\mu u^\nu + B_{\mathcal{V}} \gamma^{\mu\nu}, \tag{4.59a}$$

$$\mathcal{I}^{\rho\mu\nu} = A_{\mathcal{I}} u^\rho u^\mu u^\nu + B_{\mathcal{I}} u^\rho \gamma^{\mu\nu} + 2C_{\mathcal{I}} \gamma^{\rho(\mu} u^{\nu)}, \tag{4.59b}$$

$$\mathcal{Y}^{\rho\mu\nu} = A_{\mathcal{Y}} u^\rho u^\mu u^\nu + B_{\mathcal{Y}} u^\rho \gamma^{\mu\nu} + 2C_{\mathcal{Y}} \gamma^{\rho(\mu} u^{\nu)}, \tag{4.59c}$$

$$\begin{aligned} \mathcal{J}^{\rho\mu\nu\alpha} = {} & A_{\mathcal{J}} u^\rho u^\mu u^\nu u^\alpha + B_{\mathcal{J}} \gamma^{\alpha\rho} u^\mu u^\nu + C_{\mathcal{J}} u^\rho \gamma^{\mu\nu} u^\alpha + 2D_{\mathcal{J}} \gamma^{\rho(\mu} u^{\nu)} u^\alpha \\ & + 2E_{\mathcal{J}} \gamma^{\alpha(\mu} u^{\nu)} u^\rho + F_{\mathcal{J}} \gamma^{\alpha\rho} \gamma^{\mu\nu} + 2G_{\mathcal{J}} \gamma^{\alpha(\mu} \gamma^{\nu)\rho}, \end{aligned} \tag{4.59d}$$

$$\begin{aligned} \mathcal{Z}^{\alpha\beta\mu\nu} = {} & A_{\mathcal{Z}} u^\alpha u^\beta u^\mu u^\nu + B_{\mathcal{Z}} u^\alpha u^\beta \gamma^{\mu\nu} + C_{\mathcal{Z}} u^\mu u^\nu \gamma^{\alpha\beta} \\ & + 2D_{\mathcal{Z}} \left(u^\alpha u^{(\mu} \gamma^{\nu)\beta} + u^\beta u^{(\mu} \gamma^{\nu)\alpha} \right) + E_{\mathcal{Z}} \gamma^{\alpha\beta} \gamma^{\mu\nu} + 2F_{\mathcal{Z}} \gamma^{\alpha(\mu} \gamma^{\nu)\beta}, \end{aligned} \tag{4.59e}$$

$$\begin{aligned} \mathcal{W}^{\mu\nu\alpha\beta} = {} & A_{\mathcal{W}} u^\mu u^\nu u^\alpha u^\beta + B_{\mathcal{W}} \left(\gamma^{\mu\nu} u^\alpha u^\beta + \gamma^{\alpha\beta} u^\mu u^\nu \right) \\ & + 2C_{\mathcal{W}} \left(\gamma^{\mu(\alpha} u^{\beta)} u^\nu + \gamma^{\nu(\alpha} u^{\beta)} u^\mu \right) + D_{\mathcal{W}} \gamma^{\mu\nu} \gamma^{\alpha\beta} + 2E_{\mathcal{W}} \gamma^{\mu(\alpha} \gamma^{\beta)\nu}, \end{aligned} \tag{4.59f}$$

$$\begin{aligned} \mathcal{N}^{\rho\mu\nu\alpha\beta} = {} & A_{\mathcal{N}} u^\rho u^\mu u^\nu u^\alpha u^\beta + B_{\mathcal{N}} u^\rho u^\alpha u^\beta \gamma^{\mu\nu} + C_{\mathcal{N}} u^\rho u^\mu u^\nu \gamma^{\alpha\beta} \\ & + 2D_{\mathcal{N}} \left(u^\rho u^\alpha u^{(\mu} \gamma^{\nu)\beta} + u^\rho u^\beta u^{(\mu} \gamma^{\nu)\alpha} \right) + 2E_{\mathcal{N}} \gamma^{\rho(\alpha} u^{\beta)} u^\mu u^\nu \\ & + F_{\mathcal{N}} u^\rho \gamma^{\alpha\beta} \gamma^{\mu\nu} + 2G_{\mathcal{N}} u^\rho \gamma^{\mu(\alpha} \gamma^{\beta)\nu} + 2H_{\mathcal{N}} \gamma^{\rho(\mu} u^{\nu)} \gamma^{\alpha\beta} + 2I_{\mathcal{N}} \gamma^{\rho(\alpha} u^{\beta)} \gamma^{\mu\nu} \\ & + 2J_{\mathcal{N}} \left(\gamma^{\rho(\alpha} u^{\nu)} \gamma^{\mu\beta} + \gamma^{\rho(\alpha} u^{\mu)} \gamma^{\nu\beta} + \gamma^{\rho(\beta} u^{\nu)} \gamma^{\alpha\mu} + \gamma^{\rho(\beta} u^{\mu)} \gamma^{\alpha\nu} \right) \\ & + 2K_{\mathcal{N}} \gamma^{\rho(\mu} u^{\nu)} u^\alpha u^\beta, \end{aligned} \tag{4.59g}$$

$$\begin{aligned} \mathcal{U}^{\rho\mu\nu\alpha\beta} = {} & A_{\mathcal{U}} u^\rho u^\mu u^\nu u^\alpha u^\beta + B_{\mathcal{U}} u^\rho u^\alpha u^\beta \gamma^{\mu\nu} + C_{\mathcal{U}} u^\rho u^\mu u^\nu \gamma^{\alpha\beta} \\ & + 2D_{\mathcal{U}} \left(u^\rho u^\alpha u^{(\mu} \gamma^{\nu)\beta} + u^\rho u^\beta u^{(\mu} \gamma^{\nu)\alpha} \right) + 2E_{\mathcal{U}} \gamma^{\rho(\alpha} u^{\beta)} u^\mu u^\nu \\ & + F_{\mathcal{U}} u^\rho \gamma^{\alpha\beta} \gamma^{\mu\nu} + 2G_{\mathcal{N}} u^\rho \gamma^{\mu(\alpha} \gamma^{\beta)\nu} + 2H_{\mathcal{U}} \gamma^{\rho(\mu} u^{\nu)} \gamma^{\alpha\beta} + 2I_{\mathcal{U}} \gamma^{\rho(\alpha} u^{\beta)} \gamma^{\mu\nu} \\ & + 2J_{\mathcal{U}} \left(\gamma^{\rho(\alpha} u^{\nu)} \gamma^{\mu\beta} + \gamma^{\rho(\alpha} u^{\mu)} \gamma^{\nu\beta} + \gamma^{\rho(\beta} u^{\nu)} \gamma^{\alpha\mu} + \gamma^{\rho(\beta} u^{\mu)} \gamma^{\alpha\nu} \right) \\ & + 2K_{\mathcal{U}} \gamma^{\rho(\mu} u^{\nu)} u^\alpha u^\beta, \end{aligned} \tag{4.59h}$$

$$\begin{aligned}
\mathcal{M}^{\rho\mu\nu\sigma\alpha\beta} &= A_{\mathcal{M}} u^\rho u^\mu u^\nu u^\sigma u^\alpha u^\beta + B_{\mathcal{M}} u^\rho \gamma^{\mu\nu} u^\sigma u^\alpha u^\beta + C_{\mathcal{M}} u^\rho u^\mu u^\nu u^\sigma \gamma^{\alpha\beta} \\
&+ 2D_{\mathcal{M}} (\gamma^{\rho(\mu} u^{\nu)} u^\sigma u^\alpha u^\beta + \gamma^{\sigma(\mu} u^{\nu)} u^\rho u^\alpha u^\beta) \\
&+ 2E_{\mathcal{M}} (\gamma^{\rho(\alpha} u^{\beta)} u^\sigma u^\mu u^\nu + \gamma^{\sigma(\alpha} u^{\beta)} u^\rho u^\mu u^\nu) \\
&+ 2F_{\mathcal{M}} (\gamma^{\sigma(\mu} u^{\nu)} u^\rho \gamma^{\alpha\beta} + \gamma^{\rho(\mu} u^{\nu)} u^\sigma \gamma^{\alpha\beta} + \gamma^{\rho(\alpha} u^{\beta)} u^\sigma \gamma^{\mu\nu} + \gamma^{\sigma(\alpha} u^{\beta)} u^\rho \gamma^{\mu\nu}) \\
&+ G_{\mathcal{M}} (\gamma^{\rho\sigma} \gamma^{\mu\nu} u^\alpha u^\beta + \gamma^{\rho\sigma} \gamma^{\alpha\beta} u^\mu u^\nu) \\
&+ 2H_{\mathcal{M}} (\gamma^{\rho(\mu} \gamma^{\nu)\sigma} u^\alpha u^\beta + \gamma^{\rho(\alpha} \gamma^{\beta)\sigma} u^\mu u^\nu) \\
&+ 2I_{\mathcal{M}} (\gamma^{\rho\sigma} \gamma^{\alpha(\mu} u^{\nu)} u^\beta + \gamma^{\rho\sigma} \gamma^{\beta(\mu} u^{\nu)} u^\alpha) \\
&+ 2J_{\mathcal{M}} (\gamma^{\rho(\mu} \gamma^{\nu)\sigma} \gamma^{\alpha\beta} + \gamma^{\rho(\alpha} \gamma^{\beta)\sigma} \gamma^{\mu\nu}) \\
&+ 2K_{\mathcal{M}} \gamma^{\rho\sigma} \gamma^{\alpha(\mu} \gamma^{\nu)\beta} + L_{\mathcal{M}} \gamma^{\rho\sigma} \gamma^{\mu\nu} \gamma^{\alpha\beta}.
\end{aligned} \tag{4.59i}$$

Before we impose any theoretical structure whatsoever we see that there are 60 "free" functions of time in (4.59). As soon as we impose a small piece of theoretical structure, such as covariant derivatives of the scalar field at the background which we will discuss in Sect. 4.3.3.3 and leads up to Eqs. (4.74), the number of free functions reduces from 60 to 26.

4.3.2 Operator Expansion of $\delta_{\mathrm{L}} U^{\mu\nu}$

If ones aim was to provide the general modifications to the perturbed gravitational field equations, then all that needs to be provided is an expression for the "new" gravitational source, $\delta_{\mathrm{L}} U^{\mu\nu}$. This can be written down without reference to the underlying Lagrangian, which is the approach taken by [7–9], and was succesfully used for Galileon theories in [10]. We will show how to write down a "general" expression for modifications to the field equations, in the form of an operator expansion, for a theory which will encompass the second order scalar field theory we have been discussing. A possible weakness in using this strategy alone is that one is somewhat ignorant as to the "physical" meaning of the terms in the operator expansion: one is not able to (easily) determine whether or not a given term was due to some interactions in the field theory. Obviously, as soon as the expansion is motivated from a Lagrangian, the origin of all terms in the operator expansion becomes manifest.

The operator expansion of $\delta_{\mathrm{L}} U^{\mu\nu}$ which would encompass a generic second order scalar field theory is

$$\delta_{\mathrm{L}} U^{\mu\nu} = \hat{\mathbb{Y}}^{\mu\nu} \delta\phi + \hat{\mathbb{W}}^{\mu\nu\alpha\beta} \delta_{\mathrm{L}} g_{\alpha\beta}, \tag{4.60}$$

where the two differential operators, $\hat{\mathbb{Y}}^{\mu\nu}$, $\hat{\mathbb{W}}^{\mu\nu\alpha\beta}$, are expanded as

$$\hat{\mathbb{Y}}^{\mu\nu} = \mathbb{A}^{\mu\nu} + \mathbb{B}^{\rho\mu\nu} \nabla_\rho + \mathbb{C}^{\rho\sigma\mu\nu} \nabla_\rho \nabla_\sigma + \mathbb{D}^{\rho\sigma\pi\mu\nu} \nabla_\rho \nabla_\sigma \nabla_\pi, \tag{4.61a}$$

$$\hat{\mathbb{W}}^{\mu\nu\alpha\beta} = \mathbb{E}^{\mu\nu\alpha\beta} + \mathbb{F}^{\rho\mu\nu\alpha\beta} \nabla_\rho + \mathbb{G}^{\rho\sigma\mu\nu\alpha\beta} \nabla_\rho \nabla_\sigma. \tag{4.61b}$$

In principle, $\mathbb{A}, \ldots \mathbb{G}$ are the free tensors (up to being subject to the perturbed conservation equation); but in (4.56) we identified the tensors $\{\mathbb{A} \ldots \mathbb{G}\}$ in terms of the coupling coefficients in the Lagrangian. We are able to observe that the coupling coefficients in the Lagrangian for perturbations (i.e. the $\mathcal{V}, \mathcal{Y}, \mathcal{I}, \mathcal{W}$ etc), combine in a rather non-trivial way to construct the generalized modifications to the field equations (again, we note that if one was unaware of the Lagrangian for perturbations, the origin and structure of these terms would be somewhat of a mystery).

4.3.3 Imposing Theoretical Restrictions

So far we have been working with an incredibly general theory, where we found that there are 60 free parameters in the (3 + 1) decomposition. This is a rather unmanageable amount of freedom to have in a theory. The theory does not necessarily satisfy reparameterization invariance, have only covariant derivatives, nor contain second order field equations. These three properties are, however, highly desirable for fundamental theories. We will show how to impose these properties upon a (still general) theory.

4.3.3.1 Linking Conditions

We obtain the linking conditions by enforcing the time-like component of the conservation equation to be identical to the Euler–Lagrange equation. By contracting the perturbed Euler–Lagrange equation (4.47) with a time-like vector τ_μ and comparing with the conservation equation (4.57a) we obtain

$$\tau_\nu \mathbb{D}^{\alpha\beta\gamma\mu\nu} = \mathcal{F}^{\mu\alpha\beta\gamma}, \tag{4.62a}$$

$$\tau_\nu(\mathbb{C}^{\alpha\beta\mu\nu} + \nabla_\rho \mathbb{D}^{\mu\alpha\beta\rho\nu}) = 2\mathcal{E}^{[\alpha\mu]\beta}, \tag{4.62b}$$

$$\tau_\nu(\mathbb{B}^{\alpha\mu\nu} + \nabla_\rho \mathbb{C}^{\mu\alpha\rho\nu}) = 2\mathcal{D}^{\mu\alpha} - \mathcal{C}^{\mu\nu} - \nabla_\rho \mathcal{E}^{\rho\mu\alpha} + \nabla_\rho \nabla_\beta \mathcal{F}^{\rho\beta\mu\alpha}, \tag{4.62c}$$

$$\tau_\nu(\mathbb{A}^{\rho\nu} + \nabla_\mu \mathbb{B}^{\rho\mu\nu}) = \nabla_\alpha \nabla_\beta \mathcal{E}^{\rho\alpha\beta} - \nabla_\mu \mathcal{C}^{\mu\rho}, \tag{4.62d}$$

$$\tau_\nu \nabla_\mu \mathbb{A}^{\mu\nu} = 2\mathcal{A} - \nabla_\mu \mathcal{B}^\mu + \nabla_\mu \nabla_\nu \mathcal{D}^{\mu\nu}, \tag{4.62e}$$

$$\tau_\mu(\delta_{\mathrm{E}} J^\mu + 2E^\nu) = \mathcal{S}. \tag{4.62f}$$

By explicitly inserting the definitions (4.56) the linking conditions (4.62) read

$$\tau_\nu \mathcal{N}^{\alpha\mu\nu\beta\gamma} = \frac{1}{2}\mathcal{F}^{\mu\alpha\beta\gamma}, \tag{4.63a}$$

$$\tau_\nu\big[-\mathcal{Z}^{\alpha\beta\mu\nu}+4\mathcal{J}^{\beta\mu\nu\alpha}+4\nabla_\rho\mathcal{N}^{\rho\mu\nu\alpha\beta}+4\nabla_\rho\mathcal{N}^{\mu\rho\nu\alpha\beta}\big]=4\mathcal{E}^{[\alpha\mu]\beta}, \tag{4.63b}$$

$$\begin{aligned}\tau_\nu\big[-\mathcal{Y}^{\alpha\mu\nu}+4\mathcal{I}^{\alpha\mu\nu}+4\nabla_\rho\mathcal{J}^{\rho\mu\nu\alpha}-\nabla_\rho\mathcal{Z}^{\mu\alpha\rho\nu}+4\nabla_\rho\mathcal{J}^{\alpha\rho\nu\mu}+4\nabla_\rho\nabla_\beta\mathcal{N}^{\beta\rho\nu\mu\alpha}\big]\\ =4\mathcal{D}^{\mu\alpha}-2\mathcal{C}^{\mu\nu}-2\nabla_\rho\mathcal{E}^{\rho\mu\alpha}+2\nabla_\rho\nabla_\beta\mathcal{F}^{\rho\beta\mu\alpha},\end{aligned} \tag{4.63c}$$

$$\begin{aligned}\tau_\nu\big[-\mathcal{V}^{\rho\nu}+4\nabla_\mu\mathcal{I}^{\mu\rho\nu}-\nabla_\mu\mathcal{Y}^{\rho\mu\nu}+4\nabla_\mu\mathcal{I}^{\rho\mu\nu}+4\nabla_\mu\nabla_\beta\mathcal{J}^{\beta\mu\nu\rho}\big]\\ =2\nabla_\alpha\nabla_\beta\mathcal{E}^{\rho\alpha\beta}-2\nabla_\mu\mathcal{C}^{\mu\rho},\end{aligned} \tag{4.63d}$$

$$\tau_\nu\big[-\nabla_\mu\mathcal{V}^{\mu\nu}+4\nabla_\mu\nabla_\rho\mathcal{I}^{\rho\mu\nu}\big]=4\mathcal{A}-2\nabla_\mu\mathcal{B}^{\mu}+2\nabla_\mu\nabla_\nu\mathcal{D}^{\mu\nu}. \tag{4.63e}$$

4.3.3.2 Reparameterization Invariance

The conditions under which the theory is reparameterization invariant (i.e. when the vector field decouples from the scalar field) are called the *decoupling conditions*. To obtain these conditions, we first note that the interaction between $\delta\phi$ and ξ^μ in the Lagrangian is given by $\mathcal{L}^{\rm int}_{\{2\}}\supset\mathcal{L}_{\{2\}}$, where

$$\begin{aligned}\mathcal{L}^{\rm int}_{\{2\}}&=2\big[\mathcal{I}^{\rho\mu\nu}\delta\phi+\mathcal{J}^{\rho\mu\nu\alpha}\nabla_\alpha\delta\phi+\mathcal{N}^{\rho\mu\nu\alpha\beta}\nabla_\alpha\nabla_\beta\delta\phi\big]\nabla_\rho\nabla_{(\mu}\xi_{\nu)}\\ &\quad+\frac{1}{2}\big[\mathcal{V}^{\mu\nu}\delta\phi+\mathcal{Y}^{\alpha\mu\nu}\nabla_\alpha\delta\phi+\mathcal{Z}^{\alpha\beta\mu\nu}\nabla_\alpha\nabla_\beta\delta\phi\big]\nabla_{(\mu}\xi_{\nu)}\\ &=\xi_\nu\Big[\mathbb{D}^{\rho\alpha\beta\mu\nu}\nabla_\mu\nabla_\rho\nabla_\alpha\nabla_\beta\delta\phi+(\nabla_\mu\mathbb{D}^{\rho\alpha\beta\mu\nu}+\mathbb{C}^{\alpha\beta\rho\nu})\nabla_\rho\nabla_\alpha\nabla_\beta\delta\phi\\ &\quad+(\nabla_\mu\mathbb{C}^{\alpha\beta\mu\nu}+\mathbb{B}^{\beta\alpha\nu})\nabla_\alpha\nabla_\beta\delta\phi+(\nabla_\mu\mathbb{B}^{\alpha\mu\nu}+\mathbb{A}^{\alpha\nu})\nabla_\alpha\delta\phi+(\nabla_\mu\mathbb{A}^{\mu\nu})\delta\phi\Big],\end{aligned} \tag{4.64}$$

where to write the second equality we removed a total derivative and used the bold-face tensors "$\mathbb{X}$" as defined in (4.56). The Eulerian perturbed conservation equation (4.57a) can be written schematically as

$$F^\nu=\delta_{\rm E}J^\nu+2E^\nu, \tag{4.65}$$

where F^ν is the evolution equation for the perturbed scalar field and is given by the left-hand-side of (4.57a), which is also the term multiplying ξ^μ in the interaction Lagrangian (4.64), $\delta_{\rm E}J^\nu$ is the perturbed source, given by (4.57b) and E^ν is the evolution equation for the ξ^μ field, given by (4.57c). Contracting (4.65) with ξ^μ yields

$$\xi_\nu F^\nu=\xi_\nu\delta_{\rm E}J^\nu+2\xi_\nu E^\nu. \tag{4.66}$$

The perturbed scalar field decouples from ξ^μ in $\mathcal{L}_{\{2\}}$ when (4.64) vanishes; i.e. when

$$\xi_\nu F^\nu = 0. \tag{4.67}$$

Imposing (4.67) upon (4.64) provides us with the covariant conditions

$$\xi_\nu \mathbb{D}^{\rho\alpha\beta\mu\nu} = 0, \quad \xi_\nu(\nabla_\mu \mathbb{D}^{\rho\alpha\beta\mu\nu} + \mathbb{C}^{\alpha\beta\rho\nu}) = 0, \tag{4.68a}$$

$$\xi_\nu(\nabla_\mu \mathbb{C}^{\alpha\beta\mu\nu} + \mathbb{B}^{\beta\alpha\nu}) = 0, \quad \xi_\nu(\nabla_\mu \mathbb{B}^{\alpha\mu\nu} + \mathbb{A}^{\alpha\nu}) = 0, \quad \xi_\nu \nabla_\mu \mathbb{A}^{\mu\nu} = 0. \tag{4.68b}$$

After imposing (4.67), (4.66) becomes $\xi_\nu \delta_{\mathrm{E}} J^\nu + 2\xi_\nu E^\nu = 0$. For arbitrary values of $\delta_{\mathrm{E}} g_{\mu\nu}$, this is satisfied when their coefficients vanish in $\xi_\nu \delta_{\mathrm{E}} J^\nu$, providing us with the covariant conditions

$$\xi_\nu \nabla_\mu \mathbb{E}^{\mu\nu\alpha\beta} = 0, \tag{4.68c}$$

$$\xi_\nu \left(\nabla_\mu \mathbb{F}^{\rho\mu\nu\alpha\beta} + \mathbb{E}^{\rho\nu\alpha\beta} + \frac{1}{2} U^{\rho\nu} g^{\alpha\beta} - \frac{1}{2} U^{\alpha\beta} g^{\rho\nu} + g^{\nu\beta} U^{\alpha\rho} \right) = 0, \tag{4.68d}$$

$$\xi_\nu(\nabla_\mu \mathbb{G}^{\rho\sigma\mu\nu\alpha\beta} + \mathbb{F}^{\rho\sigma\nu\alpha\beta}) = 0, \tag{4.68e}$$

$$\xi_\nu \mathbb{G}^{\rho\sigma\mu\nu\alpha\beta} = 0. \tag{4.68f}$$

To reinforce the point, (4.68a,b) are the conditions that $\delta\phi$ and ξ^μ decouple in $\mathcal{L}_{\{2\}}$ and (4.68c–f) are the conditions that ξ^μ is not sourced by the Eulerian perturbed metric. These are the covariant forms of the decoupling conditions.

We will now write down these decoupling conditions in terms of the coupling coefficients that appear in $\mathcal{L}_{\{2\}}$. Inserting (4.56) into (4.68) we obtain

$$\xi_\nu \big[\nabla_\mu \mathcal{V}^{\mu\nu} - 4\nabla_\mu \nabla_\rho \mathcal{I}^{\rho\mu\nu} \big] = 0, \tag{4.69a}$$

$$\xi_\nu \big[4\nabla_\mu \nabla_\rho \mathcal{J}^{\rho\mu\nu\alpha} + 8\nabla_\rho \mathcal{I}^{(\alpha\rho)\nu} - \nabla_\mu \mathcal{Y}^{\alpha\mu\nu} - \mathcal{V}^{\alpha\nu} \big] = 0, \tag{4.69b}$$

$$\begin{aligned} \xi_\nu \big[& 4\nabla_\mu \nabla_\rho \mathcal{N}^{\rho\mu\nu\alpha\beta} + 4\nabla_\rho \mathcal{J}^{\beta\rho\nu\alpha} + 4\nabla_\rho \mathcal{J}^{\rho\alpha\nu\beta} \\ & - \nabla_\mu \mathcal{Z}^{\alpha\beta\mu\nu} - \mathcal{Y}^{\beta\alpha\nu} + 4\mathcal{I}^{\beta\alpha\nu} \big] = 0, \end{aligned} \tag{4.69c}$$

$$\xi_\nu \big[4\nabla_\sigma \mathcal{N}^{\rho\sigma\nu\alpha\beta} + 4\nabla_\sigma \mathcal{N}^{\sigma\rho\nu\alpha\beta} - \mathcal{Z}^{\alpha\beta\rho\nu} + 4\mathcal{J}^{\beta\rho\nu\alpha} \big] = 0, \tag{4.69d}$$

$$\xi_\nu \mathcal{N}^{\rho\mu\nu\alpha\beta} = 0, \tag{4.69e}$$

$$\xi_\nu \big[\nabla_\mu \mathcal{W}^{\mu\nu\alpha\beta} - \nabla_\mu \nabla_\rho \mathcal{U}^{\rho\mu\nu\alpha\beta} \big] = 0, \tag{4.69f}$$

$$\xi_\nu\big[4\nabla_\mu\nabla_\epsilon\mathcal{M}^{\epsilon\mu\nu\rho\alpha\beta} + 2\nabla_\mu\mathcal{U}^{(\rho\mu)\nu\alpha\beta} - \nabla_\mu\mathcal{U}^{\rho\alpha\beta\mu\nu} - \mathcal{W}^{\rho\nu\alpha\beta} - U^{\alpha\beta}g^{\rho\nu} + 2g^{\nu\beta}U^{\alpha\rho}\big] = 0, \tag{4.69g}$$

$$\xi_\nu\big[4\nabla_\mu\mathcal{M}^{\rho\mu\nu\sigma\alpha\beta} + 4\nabla_\mu\mathcal{M}^{\mu\sigma\nu\rho\alpha\beta} + \mathcal{U}^{\rho\sigma\nu\alpha\beta} - \mathcal{U}^{\rho\alpha\beta\sigma\nu}\big] = 0, \tag{4.69h}$$

$$\xi_\nu\mathcal{M}^{\rho\mu\nu\sigma\alpha\beta} = 0. \tag{4.69i}$$

4.3.3.3 Partial to Covariant Derivatives

Here we will show how to restrict our attention of the theory with partial derivatives (4.37) to the theory containing only covariant derivatives (4.38). To do this, we recall that the perturbed Christoffel symbol $\delta_{\rm L}\Gamma^\alpha{}_{\mu\nu}$ arises when we explicitly compute the variation of the second covariant derivative of a scalar field:

$$\delta_{\rm L}(\nabla_\mu\nabla_\nu\phi) = \nabla_\mu\nabla_\nu\delta\phi - \nabla_\alpha\phi\delta_{\rm L}\Gamma^\alpha{}_{\mu\nu}. \tag{4.70}$$

The perturbed Christoffel symbol $\delta_{\rm L}\Gamma^\alpha{}_{\mu\nu}$ is related to the derivative of the perturbed metric $\nabla_\alpha\delta_{\rm L}g_{\mu\nu}$ by multiplication of combinations of the background metric:

$$\delta_{\rm L}\Gamma^\rho{}_{\mu\nu} = \frac{1}{2}g^{\rho\beta}\Big[\nabla_\mu\delta_{\rm L}g_{\nu\beta} + \nabla_\nu\delta_{\rm L}g_{\mu\beta} - \nabla_\beta\delta_{\rm L}g_{\mu\nu}\Big] = -\mathrm{S}^{\sigma\epsilon\theta\rho}{}_{\mu\nu}\nabla_\sigma\delta_{\rm L}g_{\epsilon\theta}, \tag{4.71}$$

where we defined the *structure tensor*,

$$\mathrm{S}^{\pi\epsilon\gamma\rho}{}_{\mu\nu} \equiv \frac{1}{2}g^{\pi\rho}\delta^\epsilon{}_\mu\delta^\gamma{}_\nu - g^{\gamma\rho}\delta^\pi{}_{(\mu}\delta^\epsilon{}_{\nu)}. \tag{4.72}$$

The consequence of this is that not all of the 15 tensors in $\mathcal{L}_{\{2\}}$ (4.40) are independent: the way in which the derivative of the perturbed metric enters the theory is not free, it enters via the perturbed Christoffel symbol, which is always accompanied by the second covariant derivative of the perturbed scalar field variable. From (4.38) and (4.71) this implies we can impose the piece of structure upon $\mathcal{L}_{\{2\}}$:

$$\frac{\hat{\delta}^2}{\hat{\delta}X\hat{\delta}\nabla_\pi\delta_{\rm L}g_{\epsilon\gamma}}\mathcal{L}_{\{2\}} = \left(\mathrm{S}^{\pi\epsilon\gamma\lambda}{}_{\mu\nu}\nabla_\lambda\phi\right)\frac{\hat{\delta}^2}{\hat{\delta}X\hat{\delta}(\nabla_\mu\nabla_\nu\delta\phi)}\mathcal{L}_{\{2\}}, \tag{4.73}$$

where $X \in \{\delta\phi, \nabla_\mu\delta\phi, \nabla_\mu\nabla_\nu\delta\phi\}$. This structure completely determines $\mathcal{I}$ from $\mathcal{D}$, $\mathcal{J}$ from $\mathcal{E}$, $\mathcal{N}$ and $\mathcal{M}$ from $\mathcal{F}$, and $\mathcal{U}$ from $\mathcal{Z}$ via background field variables. Hence, after imposing this structure, the number of coupling coefficients in the second order Lagrangian reduces from 15 to 10. The explicit form of this structure is

$$\mathcal{I}^{\rho\mu\nu} = \mathrm{S}^{\rho\mu\nu\lambda}{}_{\alpha\beta}\mathcal{D}^{\alpha\beta}\nabla_\lambda\phi, \tag{4.74a}$$

$$\mathcal{J}^{\epsilon\rho\sigma\alpha} = \mathbb{S}^{\epsilon\rho\sigma\lambda}{}_{\mu\nu}\mathcal{E}^{\alpha\mu\nu}\nabla_\lambda\phi, \tag{4.74b}$$

$$\mathcal{N}^{\epsilon\rho\sigma\alpha\beta} = \mathbb{S}^{\epsilon\rho\sigma\lambda}{}_{\mu\nu}\mathcal{F}^{\mu\nu\alpha\beta}\nabla_\lambda\phi, \tag{4.74c}$$

$$\mathcal{U}^{\rho\mu\nu\alpha\beta} = \mathbb{S}^{\rho\mu\nu\lambda}{}_{\epsilon\gamma}\mathcal{Z}^{\epsilon\gamma\alpha\beta}\nabla_\lambda\phi, \tag{4.74d}$$

$$\mathcal{M}^{\rho\mu\nu\sigma\alpha\beta} = \mathbb{S}^{\rho\mu\nu\lambda}{}_{\epsilon\gamma}\mathbb{S}^{\sigma\alpha\beta\xi}{}_{\omega\pi}\mathcal{F}^{\epsilon\gamma\omega\pi}\nabla_\lambda\phi\nabla_\xi\phi. \tag{4.74e}$$

By imposing the structure (4.74) the freedom in the theory substantially reduces: one can see that the number of free parameters in the (3+1) decomposition of the dark energy momentum tensor reduces from 60 to 26.

We now use the structure (4.74) to rewrite the coefficients $\{\mathbb{A}\cdots\mathbb{G}\}$ which we defined in (4.56) and which are used to write perturbed dark energy momentum tensor; we will use an over-head-check symbol to distinguish the coefficients obtained after imposing the structure. The scalar-field-sector tensors become

$$\check{\mathbb{A}}^{\mu\nu} = -\frac{1}{2}\Big[\mathcal{V}^{\mu\nu} - 4\mathbb{S}^{\rho\mu\nu\lambda}{}_{\alpha\beta}\nabla_\rho(\mathcal{D}^{\alpha\beta}\nabla_\lambda\phi)\Big], \tag{4.75a}$$

$$\check{\mathbb{B}}^{\alpha\mu\nu} = -\frac{1}{2}\Big[\mathcal{Y}^{\alpha\mu\nu} - 4\mathbb{S}^{\alpha\mu\nu\lambda}{}_{\pi\omega}\mathcal{D}^{\pi\omega}\nabla_\lambda\phi - 4\mathbb{S}^{\rho\mu\nu\lambda}{}_{\pi\omega}\nabla_\rho(\mathcal{E}^{\alpha\pi\omega}\nabla_\lambda\phi)\Big], \tag{4.75b}$$

$$\check{\mathbb{C}}^{\alpha\beta\mu\nu} = -\frac{1}{2}\Big[\mathcal{Z}^{\alpha\beta\mu\nu} - 4\mathbb{S}^{\beta\mu\nu\lambda}{}_{\pi\omega}\mathcal{E}^{\alpha\pi\omega}\nabla_\lambda\phi - 4\mathbb{S}^{\rho\mu\nu\lambda}{}_{\pi\omega}\nabla_\rho(\mathcal{F}^{\pi\omega\alpha\beta}\nabla_\lambda\phi)\Big], \tag{4.75c}$$

$$\check{\mathbb{D}}^{\rho\alpha\beta\mu\nu} = 2\mathbb{S}^{\rho\mu\nu\lambda}{}_{\pi\omega}\mathcal{F}^{\pi\omega\alpha\beta}\nabla_\lambda\phi, \tag{4.75d}$$

and the metric-sector tensors become

$$\check{\mathbb{E}}^{\mu\nu\alpha\beta} = -\frac{1}{2}\Big[\mathcal{W}^{\mu\nu\alpha\beta} + U^{\mu\nu}g^{\alpha\beta} - \mathbb{S}^{\rho\mu\nu\lambda}{}_{\pi\omega}\nabla_\rho(\mathcal{Z}^{\pi\omega\alpha\beta}\nabla_\lambda\phi)\Big], \tag{4.75e}$$

$$\begin{aligned}\check{\mathbb{F}}^{\rho\mu\nu\alpha\beta} = &-\frac{1}{2}\Big[\mathbb{S}^{\rho\alpha\beta\lambda}{}_{\pi\omega}\mathcal{Z}^{\pi\omega\mu\nu}\nabla_\lambda\phi - \mathbb{S}^{\rho\mu\nu\lambda}{}_{\pi\omega}\mathcal{Z}^{\pi\omega\alpha\beta}\nabla_\lambda\phi \\ &- 4\mathbb{S}^{\epsilon\mu\nu\lambda}{}_{\gamma\xi}\mathbb{S}^{\rho\alpha\beta\zeta}{}_{\omega\pi}\nabla_\epsilon(\mathcal{F}^{\gamma\xi\omega\pi}\nabla_\lambda\phi\nabla_\zeta\phi)\Big],\end{aligned} \tag{4.75f}$$

$$\check{\mathbb{G}}^{\rho\sigma\mu\nu\alpha\beta} = 2\mathbb{S}^{\rho\mu\nu\lambda}{}_{\epsilon\gamma}\mathbb{S}^{\sigma\alpha\beta\zeta}{}_{\omega\pi}\mathcal{F}^{\epsilon\gamma\omega\pi}\nabla_\lambda\phi\nabla_\zeta\phi. \tag{4.75g}$$

We now observe that $\check{\mathbb{F}}$, $\check{\mathbb{G}}$ do not contain any tensors that are not already present in the scalar-field sector tensors. This will be the observable "link" between metric and scalar-field perturbations that is due to the structure. We also see from (4.75) that $\mathcal{D}, \mathcal{E}, \mathcal{F}$ only appear contracted with S.

4.3.3.4 Second Order Field Equations

We will now show how to impose second order field equations upon the theory. In (4.51) we presented the perturbed Euler–Lagrange equation, from which we can read off that the condition for second order field equations is

$$\mathcal{F}^{\mu\nu\alpha\beta} = 0. \tag{4.76}$$

We should note that this condition is not the only condition which yields second order field equations. However, the condition (4.76) certainly yields second order field equations.

Using (4.76), the equation of motion (4.51) becomes

$$\begin{aligned}&\left[2\mathcal{D}^{\mu\nu} - \mathcal{C}^{\mu\nu} - \nabla_\alpha \mathcal{E}^{\alpha\mu\nu}\right]\nabla_\mu\nabla_\nu\delta\phi\\ &+\left[\nabla_\alpha\nabla_\beta\mathcal{E}^{\nu\alpha\beta} - \nabla_\mu\mathcal{C}^{\mu\nu} - \mathcal{E}^{\alpha\mu\beta}R^\nu{}_{\beta\mu\alpha}\right]\nabla_\nu\delta\phi\\ &+\left[2\mathcal{A} - \nabla_\mu\mathcal{B}^\mu + \nabla_\mu\nabla_\nu\mathcal{D}^{\mu\nu}\right]\delta\phi = \delta\mathcal{S}.\end{aligned} \tag{4.77}$$

When the condition (4.76) is imposed upon the structure (4.74) we also find that

$$\mathcal{N}^{\rho\mu\nu\alpha\beta} = \mathcal{M}^{\rho\mu\nu\sigma\alpha\beta} = 0. \tag{4.78}$$

This condition can be used to remove terms in (4.56). After imposing (4.76) and the structure (4.75), only the tensors $\{\mathcal{V}, \mathcal{D}, \mathcal{Y}, \mathcal{E}, \mathcal{Z}, \mathcal{W}\}$ are free, and they collectively have 21 free functions in their (3+1) decomposition. The tensors (4.75) become

$$\check{\mathbb{A}}^{\mu\nu} = -\frac{1}{2}\left[\mathcal{V}^{\mu\nu} - 4\mathrm{S}^{\rho\mu\nu\lambda}{}_{\alpha\beta}\nabla_\rho(\mathcal{D}^{\alpha\beta}\nabla_\lambda\phi)\right], \tag{4.79a}$$

$$\check{\mathbb{B}}^{\alpha\mu\nu} = -\frac{1}{2}\left[\mathcal{Y}^{\alpha\mu\nu} - 4\mathrm{S}^{\alpha\mu\nu\lambda}{}_{\pi\omega}\mathcal{D}^{\pi\omega}\nabla_\lambda\phi - 4\mathrm{S}^{\rho\mu\nu\lambda}{}_{\pi\omega}\nabla_\rho(\mathcal{E}^{\alpha\pi\omega}\nabla_\lambda\phi)\right], \tag{4.79b}$$

$$\check{\mathbb{C}}^{\alpha\beta\mu\nu} = -\frac{1}{2}\left[\mathcal{Z}^{\alpha\beta\mu\nu} - 4\mathrm{S}^{\beta\mu\nu\lambda}{}_{\pi\omega}\mathcal{E}^{\alpha\pi\omega}\nabla_\lambda\phi\right], \tag{4.79c}$$

$$\check{\mathbb{D}}^{\rho\alpha\beta\mu\nu} = 0, \tag{4.79d}$$

$$\check{\mathbb{E}}^{\mu\nu\alpha\beta} = -\frac{1}{2}\Big[\mathcal{W}^{\mu\nu\alpha\beta} + U^{\mu\nu}g^{\alpha\beta} - \mathrm{S}^{\rho\mu\nu\lambda}{}_{\pi\omega}\nabla_{\rho}(\mathcal{Z}^{\pi\omega\alpha\beta}\nabla_{\lambda}\phi)\Big], \tag{4.79e}$$

$$\check{\mathbb{F}}^{\rho\mu\nu\alpha\beta} = -\frac{1}{2}\Big[\mathrm{S}^{\rho\alpha\beta\lambda}{}_{\pi\omega}\mathcal{Z}^{\pi\omega\mu\nu}\nabla_{\lambda}\phi - \mathrm{S}^{\rho\mu\nu\lambda}{}_{\pi\omega}\mathcal{Z}^{\pi\omega\alpha\beta}\nabla_{\lambda}\phi\Big], \tag{4.79f}$$

$$\check{\mathbb{G}}^{\rho\sigma\mu\nu\alpha\beta} = 0. \tag{4.79g}$$

Hence, we conclude that the perturbed dark energy momentum tensor which gives at most second order field equations is given by

$$\delta_{\mathrm{E}}U^{\mu}{}_{\nu} = \hat{\mathbb{Y}}^{\mu}{}_{\nu}\delta\phi + \hat{\mathbb{W}}^{\mu}{}_{\nu}{}^{\alpha\beta}\delta_{\mathrm{L}}g_{\alpha\beta} - g_{\nu\alpha}\pounds_{\xi}U^{\mu\alpha} + U^{\alpha\mu}\delta_{\mathrm{E}}g_{\nu\alpha}, \tag{4.80a}$$

where

$$\hat{\mathbb{Y}}^{\mu}{}_{\nu}\delta\phi = \mathbb{A}^{\mu}{}_{\nu}\delta\phi + \mathbb{B}^{\alpha\mu}{}_{\nu}\nabla_{\alpha}\delta\phi + \mathbb{C}^{\alpha\beta\mu}{}_{\nu}\nabla_{\alpha}\nabla_{\beta}\delta\phi, \tag{4.80b}$$

$$\hat{\mathbb{W}}^{\mu}{}_{\nu}{}^{\alpha\beta}\delta_{\mathrm{L}}g_{\alpha\beta} = \mathbb{E}^{\mu}{}_{\nu}{}^{\alpha\beta}\delta_{\mathrm{L}}g_{\alpha\beta} + \mathbb{F}^{\rho\mu}{}_{\nu}{}^{\alpha\beta}\nabla_{\rho}\delta_{\mathrm{L}}g_{\alpha\beta}. \tag{4.80c}$$

Not all components of the tensors $\mathbb{C}$, $\mathbb{F}$ are allowed to be non-zero.

4.3.4 Perturbed Fluid Variables for Second Order Field Equations

We will now present the perturbed fluid variables for this general second order scalar field theory imposed with the condition that the field equations should be second order. Later on in the calculation we will impose reparameterization invariance upon the theory. We will immediately build upon the results presented in Sect. 4.3.4.

The $(3+1)$ decomposition of the tensors which appear in the perturbed energy momentum tensor (4.80) is

$$\mathbb{A}^{\mu\nu} = A_{\mathbb{A}}u^{\mu}u^{\nu} + B_{\mathbb{A}}\gamma^{\mu\nu}, \tag{4.81a}$$

$$\mathbb{B}^{\alpha\mu\nu} = A_{\mathbb{B}}u^{\alpha}u^{\mu}u^{\nu} + B_{\mathbb{B}}u^{\alpha}\gamma^{\mu\nu} + 2C_{\mathbb{B}}\gamma^{\alpha(\mu}u^{\nu)}, \tag{4.81b}$$

$$\begin{aligned}\mathbb{C}^{\alpha\beta\mu\nu} = {} & A_{\mathbb{C}}u^{\alpha}u^{\beta}u^{\mu}u^{\nu} + B_{\mathbb{C}}u^{\alpha}u^{\beta}\gamma^{\mu\nu} + C_{\mathbb{C}}u^{\mu}u^{\nu}\gamma^{\alpha\beta} \\ & + 2D_{\mathbb{C}}\big(u^{\alpha}u^{(\mu}\gamma^{\nu)\beta} + u^{\beta}u^{(\mu}\gamma^{\nu)\alpha}\big) \\ & + E_{\mathbb{C}}\gamma^{\alpha\beta}\gamma^{\mu\nu} + 2F_{\mathbb{C}}\gamma^{\alpha(\mu}\gamma^{\nu)\beta},\end{aligned} \tag{4.81c}$$

$$\mathbb{E}^{\alpha\beta\mu\nu} = A_{\mathbb{E}} u^\alpha u^\beta u^\mu u^\nu + B_{\mathbb{E}} u^\alpha u^\beta \gamma^{\mu\nu} + C_{\mathbb{E}} u^\mu u^\nu \gamma^{\alpha\beta} + 2D_{\mathbb{E}} \left(u^\alpha u^{(\mu}\gamma^{\nu)\beta} + u^\beta u^{(\mu}\gamma^{\nu)\alpha}\right) + E_{\mathbb{E}} \gamma^{\alpha\beta}\gamma^{\mu\nu} + 2F_{\mathbb{E}}\gamma^{\alpha(\mu}\gamma^{\nu)\beta}, \tag{4.81d}$$

$$\begin{aligned}\mathbb{F}^{\rho\mu\nu\alpha\beta} = {} & A_{\mathbb{F}} u^\rho u^\mu u^\nu u^\alpha u^\beta + B_{\mathbb{F}} u^\rho u^\alpha u^\beta \gamma^{\mu\nu} + C_{\mathbb{F}} u^\rho u^\mu u^\nu \gamma^{\alpha\beta} \\ & + 2D_{\mathbb{F}}\left(u^\rho u^\alpha u^{(\mu}\gamma^{\nu)\beta} + u^\rho u^\beta u^{(\mu}\gamma^{\nu)\alpha}\right) + 2E_{\mathbb{F}}\gamma^{\rho(\alpha}u^{\beta)}u^\mu u^\nu \\ & + F_{\mathbb{F}} u^\rho \gamma^{\alpha\beta}\gamma^{\mu\nu} + 2G_{\mathbb{F}} u^\rho \gamma^{\mu(\alpha}\gamma^{\beta)\nu} + 2H_{\mathbb{F}}\gamma^{\rho(\mu}u^{\nu)}\gamma^{\alpha\beta} + 2I_{\mathbb{F}}\gamma^{\rho(\alpha}u^{\beta)}\gamma^{\mu\nu} \\ & + 2J_{\mathbb{F}}\left(\gamma^{\rho(\alpha}u^{\nu)}\gamma^{\mu\beta} + \gamma^{\rho(\alpha}u^{\mu)}\gamma^{\nu\beta} + \gamma^{\rho(\beta}u^{\nu)}\gamma^{\alpha\mu} + \gamma^{\rho(\beta}u^{\mu)}\gamma^{\alpha\nu}\right) \\ & + 2K_{\mathbb{F}}\gamma^{\rho(\mu}u^{\nu)}u^\alpha u^\beta. \end{aligned} \tag{4.81e}$$

Equation (4.79f) allows us to realise that $\mathbb{F}$ has no free components. The tensors $\{\mathbb{A}, \mathbb{B}, \mathbb{C}, \mathbb{E}\}$ have a maximum of $\{2, 3, 6, 6\} = 17$ free functions in their $(3 + 1)$ decomposition. We write the components of the perturbed metric in the synchronous gauge as

$$\delta_{\mathrm{E}} g_{\mu\nu} = \frac{1}{3} h\gamma_{\mu\nu} + h^{(\Pi)}_{\mu\nu}. \tag{4.82}$$

Substituting (4.81) and (4.82) into (4.80) allows the computation of the perturbed fluid variables by comparison with

$$\delta_{\mathrm{E}} U^\mu{}_\nu = \delta\rho u^\mu u_\nu + \delta P \gamma^\mu{}_\nu + 2(\rho + P)v^{(\mu}u_{\nu)} + P\Pi^\mu{}_\nu, \tag{4.83}$$

where we find

$$\begin{aligned}\delta\rho = {} & A_{\mathbb{A}}\delta\phi + A_{\mathbb{B}}\dot{\delta\phi} + A_{\mathbb{C}}\ddot{\delta\phi} + C_{\mathbb{C}}\nabla^2\delta\phi + \frac{1}{3}\left[3B_{\mathbb{E}} - 2E_{\mathbb{F}}K\right]h + C_{\mathbb{F}}\dot{h} \\ & + \left[(3B_{\mathbb{E}} - 2E_{\mathbb{F}}K)\frac{2}{3} - 3C_{\mathbb{F}}\frac{2}{9}K - 2E_{\mathbb{F}}\frac{1}{3}K\right]\nabla_\mu\xi^\mu + \left[2E_{\mathbb{F}} + 2C_{\mathbb{F}}\right]\nabla_\mu\dot{\xi}^\mu, \end{aligned} \tag{4.84a}$$

$$\begin{aligned}(\rho + P)V^\pi = {} & \gamma^{\pi\beta}\nabla_\beta\left[(C_{\mathbb{B}} - D_{\mathbb{C}}\frac{1}{3}K)\delta\phi + D_{\mathbb{C}}\dot{\delta\phi} + \frac{1}{3}(3H_{\mathbb{F}} + 2J_{\mathbb{F}})h\right] + 2J_{\mathbb{F}}\nabla_\rho h^{(\Pi)\rho\pi} \\ & + 2D_{\mathbb{F}}\ddot{\xi}^\pi + \left[K_{\mathbb{F}}\frac{2}{3}K - 2D_{\mathbb{E}} + \frac{2}{3}H_{\mathbb{F}}K + J_{\mathbb{F}}\frac{8}{3}K - \rho\right]\frac{1}{3}K\xi^\pi \\ & + (3H_{\mathbb{F}} + 2J_{\mathbb{F}})\frac{2}{3}\gamma^{\pi\beta}\nabla_\beta\nabla_\mu\xi^\mu \\ & + \left[\rho + 2D_{\mathbb{E}} - \frac{2}{3}K(D_{\mathbb{F}} + H_{\mathbb{F}} + K_{\mathbb{F}} + 4J_{\mathbb{F}})\right]\dot{\xi}^\pi \\ & + 4J_{\mathbb{F}}(\gamma^{\rho\alpha}\gamma^{\beta\pi} - \frac{1}{3}\gamma^{\rho\pi}\gamma^{\alpha\beta})\nabla_\rho\nabla_{(\alpha}\xi_{\beta)} + \frac{4}{3}J_{\mathbb{F}}K(\dot{\xi}^\pi - \frac{1}{3}K\xi^\pi), \end{aligned} \tag{4.84b}$$

$$\begin{aligned}\delta P =& B_{\mathbb{A}}\delta\phi + B_{\mathbb{B}}\dot{\delta\phi} + B_{\mathbb{C}}\ddot{\delta\phi} + \left(E_{\mathbb{C}} + \frac{2}{3}F_{\mathbb{C}}\right)\nabla^2\delta\phi \\ &+ \frac{1}{3}\left[3E_{\mathbb{E}} + 2F_{\mathbb{E}} - 2K(I_{\mathbb{F}} + \frac{2}{3}J_{\mathbb{F}}) + P\right]h + \left[F_{\mathbb{F}} + \frac{2}{3}G_{\mathbb{F}}\right]\dot{h} \\ &+ \left[\left(3E_{\mathbb{E}} + 2F_{\mathbb{E}} - 2K(I_{\mathbb{F}} + \frac{2}{3}J_{\mathbb{F}})\right)\frac{2}{3} + \frac{2}{3}P - 2(I_{\mathbb{F}} + \frac{2}{3}J_{\mathbb{F}})\frac{1}{3}K \right.\\ &\left. - 3(F_{\mathbb{F}} + \frac{2}{3}G_{\mathbb{F}})\frac{2}{9}K\right]\nabla_\alpha\xi^\alpha + \left[2(I_{\mathbb{F}} + \frac{2}{3}J_{\mathbb{F}}) + 3(F_{\mathbb{F}} + \frac{2}{3}G_{\mathbb{F}})\frac{2}{3}\right]\nabla_\mu\dot{\xi}^\mu,\end{aligned} \tag{4.84c}$$

$$\begin{aligned}P\Pi^{\rho\sigma} =& 2F_{\mathbb{C}}\left(\gamma^{\mu(\rho}\gamma^{\sigma)\nu} - \frac{1}{3}\gamma^{\rho\sigma}\gamma^{\mu\nu}\right)\nabla_\mu\nabla_\nu\delta\phi + (2F_{\mathbb{E}} - \frac{4}{3}J_{\mathbb{F}} + P)h^{(\Pi)\rho\sigma} \\ &+ 2(2F_{\mathbb{E}} - \frac{4}{3}J_{\mathbb{F}} + P)(\gamma^{\rho\alpha}\gamma^{\sigma\beta} - \frac{1}{3}\gamma^{\rho\sigma}\gamma^{\alpha\beta})\nabla_{(\alpha}\xi_{\beta)} \\ &+ 2G_{\mathbb{F}}\left[\dot{h}^{(\Pi)\rho\sigma} + (\gamma^{\rho\alpha}\gamma^{\sigma\beta} - \frac{1}{3}\gamma^{\rho\sigma}\gamma^{\alpha\beta})u^\mu\nabla_\mu\nabla_{(\alpha}\xi_{\beta)}\right] \\ &- 4J_{\mathbb{F}}\left[(\gamma^{\pi(\rho}\gamma^{\sigma)\alpha} - \frac{1}{3}\gamma^{\rho\sigma}\gamma^{\pi\alpha})(\nabla_\pi n_\alpha + \frac{1}{3}K\nabla_\pi\xi_\alpha + \frac{1}{3}\xi_\alpha\nabla_\pi K - \nabla_\pi\dot{\xi}_\alpha)\right.\\ &\left. + \frac{1}{3}K(n^{(\rho} + \frac{1}{3}K\xi^{(\rho} - \dot{\xi}^{(\rho})u^{\sigma)}\right].\end{aligned} \tag{4.84d}$$

This theory has non-trivial contributions to the anisotropic stress tensor, and has vector and scalar sources.

The number of free functions becomes substantially reduced when the coefficients also satisfy the decoupling conditions (4.68) (i.e. when the theory is reparameterization invariant), which, with the second order field equation constraint, read

$$\xi_\nu\mathbb{C}^{\alpha\beta\rho\nu} = 0, \tag{4.85a}$$

$$\xi_\nu(\nabla_\mu\mathbb{C}^{\alpha\beta\mu\nu} + \mathbb{B}^{\beta\alpha\nu}) = 0, \quad \xi_\nu(\nabla_\mu\mathbb{B}^{\alpha\mu\nu} + \mathbb{A}^{\alpha\nu}) = 0, \quad \xi_\nu\nabla_\mu\mathbb{A}^{\mu\nu} = 0, \tag{4.85b}$$

$$\xi_\nu\nabla_\mu\mathbb{E}^{\mu\nu\alpha\beta} = 0, \tag{4.85c}$$

$$\xi_\nu(\nabla_\mu\mathbb{F}^{\rho\mu\nu\alpha\beta} + \mathbb{E}^{\rho\nu\alpha\beta} + \frac{1}{2}U^{\rho\nu}g^{\alpha\beta} - \frac{1}{2}U^{\alpha\beta}g^{\rho\nu} + g^{\nu\beta}U^{\alpha\rho}) = 0, \tag{4.85d}$$

$$\xi_\nu\mathbb{F}^{\rho\sigma\nu\alpha\beta} = 0. \tag{4.85e}$$

Inserting the $(3+1)$ decomposition (4.81) into (4.85) reveals that reparameterization invariance is enforced by the following parameter choices:

$$B_{\mathbb{B}} = -C_{\mathbb{B}}, \quad \dot{C}_{\mathbb{B}} + KC_{\mathbb{B}} + B_{\mathbb{A}} = 0, \quad B_{\mathbb{C}} = D_{\mathbb{C}} = E_{\mathbb{C}} = F_{\mathbb{C}} = 0, \tag{4.86a}$$

$$C_{\mathbb{E}} + F_{\mathbb{E}} + D_{\mathbb{E}} = 0, \qquad E_{\mathbb{E}} = 0, \qquad \dot{D}_{\mathbb{E}} + K(D_{\mathbb{E}} + F_{\mathbb{E}}) = 0, \tag{4.86b}$$

$$C_{\mathbb{E}} = \frac{1}{2}(\rho + P), \qquad D_{\mathbb{E}} = -\frac{1}{2}\rho, \qquad F_{\mathbb{E}} = -\frac{1}{2}P, \tag{4.86c}$$

$$B_{\mathbb{F}} = D_{\mathbb{F}} = F_{\mathbb{F}} = G_{\mathbb{F}} = H_{\mathbb{F}} = I_{\mathbb{F}} = J_{\mathbb{F}} = K_{\mathbb{F}} = 0. \tag{4.86d}$$

After imposing (4.86) and $C_{\mathbb{F}} = -E_{\mathbb{F}} = -3B_{\mathbb{E}}/2K$, $A_{\mathbb{C}} = 0$ (these last conditions are found by manually enforcing decoupling and second order field equations), the perturbed fluid variables (4.84) become

$$\delta\rho = (A_{\mathbb{A}} - C_{\mathbb{C}}k^2)\delta\phi + A_{\mathbb{B}}\dot{\delta\phi} - \frac{3}{2K}B_{\mathbb{E}}\dot{h}, \tag{4.87a}$$

$$\theta = \frac{C_{\mathbb{B}}}{\rho + P}\delta\phi, \tag{4.87b}$$

$$\delta P = -(\dot{C}_{\mathbb{B}} + KC_{\mathbb{B}})\delta\phi - C_{\mathbb{B}}\dot{\delta\phi}, \tag{4.87c}$$

$$P\Pi^{\rho\sigma} = 0. \tag{4.87d}$$

We wrote $V^{\mu} = \gamma^{\mu\nu}\nabla_{\nu}\theta$ to isolate the velocity divergence field. There are only four free functions in (4.87). By rearranging the variables in (4.87) the perturbed pressure is given by

$$\delta P = -\frac{C_{\mathbb{B}}}{A_{\mathbb{B}}}\delta\rho + (\rho + P)\left[-\frac{\dot{C}_{\mathbb{B}} + KC_{\mathbb{B}}}{C_{\mathbb{B}}} + \frac{A_{\mathbb{A}}}{A_{\mathbb{B}}}\left(1 - \frac{C_{\mathbb{C}}}{A_{\mathbb{A}}}k^2\right)\right]\theta - \frac{3}{2K}\frac{C_{\mathbb{B}}B_{\mathbb{E}}}{A_{\mathbb{B}}}\dot{h}. \tag{4.88}$$

It is useful to note that $C_{\mathbb{C}} = B_{\mathbb{E}} = 0$ in a reparameterization invariant first order scalar field theory. We now have a very useful result: we have obtained a closed form of the perturbed pressure for a general second order scalar field theory imposed with (i) second order field equations and (ii) reparameterization invariance.

4.3.5 Summary

In this section we developed the formalism for the general second order scalar field theory. Although we started from a Lagrangian (which was specified by the coupling coefficients $(\mathcal{A}, \ldots, \mathcal{U})$, it is apparent that calculations are easier to perform with the tensors that appear in the energy-momentum tensor, $\mathbb{A}, \ldots, \mathbb{G}$.

We then showed how to impose theoretical restrictions upon the general theory, and explicitly showed what constraints this puts upon the "free" tensors in the Lagrangian. The route we fully developed took the general theory, imposed second order field equations and covariant derivatives (reducing the free functions from 60 to 17) then imposed the decoupling conditions, which left only 4 free functions. Equation (4.88) is our main result.

4.4 Scalar and High-Order Metric Theories

We will briefly study a class of theories which will encompass what are usually referred to as "scalar-tensor theories". The field content is a scalar field, its first derivative and the first and second derivatives of the metric:

$$\mathcal{L} = \mathcal{L}(g_{\mu\nu}, \partial_\alpha g_{\mu\nu}, \partial_\alpha\partial_\beta g_{\mu\nu}, \phi, \partial_\alpha\phi). \tag{4.89}$$

To write down the Lagrangian for perturbations requires 15 coupling coefficients:

$$\begin{aligned}\mathcal{L}_{\{2\}} &= \mathcal{A}(\delta\phi)^2 + \mathcal{B}^\mu\delta\phi\nabla_\mu\delta\phi + \frac{1}{2}\mathcal{C}^{\mu\nu}\nabla_\mu\delta\phi\nabla_\nu\delta\phi\\ &+ \frac{1}{4}\mathcal{V}^{\mu\nu}\delta\phi\delta_{\mathrm{L}}g_{\mu\nu} + \frac{1}{4}\mathcal{Y}^{\alpha\mu\nu}\nabla_\alpha\delta\phi\delta_{\mathrm{L}}g_{\mu\nu} + \frac{1}{4}\mathcal{H}^{\mu\nu\rho\sigma\epsilon\pi}\delta_{\mathrm{L}}g_{\mu\nu}\nabla_\rho\nabla_\sigma\delta_{\mathrm{L}}g_{\epsilon\pi}\\ &+ \frac{1}{8}\mathcal{W}^{\mu\nu\alpha\beta}\delta_{\mathrm{L}}g_{\mu\nu}\delta_{\mathrm{L}}g_{\alpha\beta} + \mathcal{K}^{\alpha\beta\mu\nu}\delta\phi\nabla_\alpha\nabla_\beta\delta_{\mathrm{L}}g_{\mu\nu} + \mathcal{S}^{\alpha\beta\mu\nu\rho}\nabla_\rho\delta\phi\nabla_\alpha\nabla_\beta\delta_{\mathrm{L}}g_{\mu\nu}\\ &+ \mathcal{I}^{\rho\mu\nu}\delta\phi\nabla_\rho\delta_{\mathrm{L}}g_{\mu\nu} + \frac{1}{2}\mathcal{M}^{\rho\mu\nu\sigma\alpha\beta}\nabla_\rho\delta_{\mathrm{L}}g_{\mu\nu}\nabla_\sigma\delta_{\mathrm{L}}g_{\alpha\beta}\\ &+ \frac{1}{2}\mathcal{R}^{\rho\sigma\mu\nu\epsilon\pi\alpha\beta}\nabla_\rho\nabla_\sigma\delta_{\mathrm{L}}g_{\mu\nu}\nabla_\epsilon\nabla_\pi\delta_{\mathrm{L}}g_{\alpha\beta} + \frac{1}{4}\mathcal{U}^{\rho\mu\nu\alpha\beta}\delta_{\mathrm{L}}g_{\alpha\beta}\nabla_\rho\delta_{\mathrm{L}}g_{\mu\nu}\\ &+ \mathcal{J}^{\rho\mu\nu\alpha}\nabla_\rho\delta_{\mathrm{L}}g_{\mu\nu}\nabla_\alpha\delta\phi + \mathcal{L}^{\rho\sigma\alpha\beta\xi\mu\nu}\nabla_\rho\nabla_\sigma\delta_{\mathrm{L}}g_{\alpha\beta}\nabla_\xi\delta_{\mathrm{L}}g_{\mu\nu}.\end{aligned} \tag{4.90}$$

To compute the field equation of $\delta\phi$ and the perturbed dark energy-momentum tensor we induce virtual variations

$$\delta\phi \rightarrow \delta\phi + \hat{\delta}\delta\phi, \quad \delta_{\mathrm{L}}g_{\mu\nu} \rightarrow \delta_{\mathrm{L}}g_{\mu\nu} + \hat{\delta}\delta_{\mathrm{L}}g_{\mu\nu}, \tag{4.91}$$

where the response in the Lagrangian for perturbations is

$$\hat{\delta}\mathcal{L}_{\{2\}} = A\hat{\delta}\delta\phi + B^\rho\nabla_\rho\hat{\delta}\delta\phi + C^{\mu\nu}\hat{\delta}\delta_{\mathrm{L}}g_{\mu\nu} + D^{\sigma\alpha\beta}\nabla_\sigma\hat{\delta}\delta_{\mathrm{L}}g_{\alpha\beta} + H^{\alpha\beta\mu\nu}\nabla_\alpha\nabla_\beta\hat{\delta}\delta_{\mathrm{L}}g_{\mu\nu}, \tag{4.92}$$

where we defined

$$A \equiv \left[2\mathcal{A} + \mathcal{B}^\mu\nabla_\mu\right]\delta\phi + \left[\frac{1}{4}\mathcal{V}^{\mu\nu} + \mathcal{I}^{\rho\mu\nu}\nabla_\rho + \mathcal{K}^{\alpha\beta\mu\nu}\nabla_\alpha\nabla_\beta\right]\delta_{\mathrm{L}}g_{\mu\nu}, \tag{4.93a}$$

$$B^{\rho} \equiv \left[\mathcal{B}^{\rho} + \mathcal{C}^{\mu\rho}\nabla_{\mu}\right]\delta\phi + \left[\frac{1}{4}\mathcal{Y}^{\rho\mu\nu} + \mathcal{J}^{\alpha\mu\nu\rho}\nabla_{\alpha} + \mathcal{S}^{\alpha\beta\mu\nu\rho}\nabla_{\alpha}\nabla_{\beta}\right]\delta_{\mathrm{L}}g_{\mu\nu}, \tag{4.93b}$$

$$4C^{\mu\nu} \equiv \left[\mathcal{V}^{\mu\nu} + \mathcal{Y}^{\alpha\mu\nu}\nabla_{\alpha}\right]\delta\phi + \left[\mathcal{W}^{\mu\nu\alpha\beta} + \mathcal{U}^{\rho\alpha\beta\mu\nu}\nabla_{\rho} + \mathcal{H}^{\mu\nu\rho\sigma\alpha\beta}\nabla_{\rho}\nabla_{\sigma}\right]\delta_{\mathrm{L}}g_{\alpha\beta}, \tag{4.93c}$$

$$\begin{aligned} D^{\sigma\alpha\beta} \equiv & \left[\mathcal{I}^{\sigma\alpha\beta} + \mathcal{J}^{\sigma\alpha\beta\rho}\nabla_{\rho}\right]\delta\phi \\ & + \left[\frac{1}{4}\mathcal{U}^{\sigma\alpha\beta\mu\nu} + \mathcal{M}^{\rho\mu\nu\sigma\alpha\beta}\nabla_{\rho} + \mathcal{L}^{\xi\pi\mu\nu\sigma\alpha\beta}\nabla_{\xi}\nabla_{\pi}\right]\delta_{\mathrm{L}}g_{\mu\nu}, \end{aligned} \tag{4.93d}$$

$$\begin{aligned} H^{\alpha\beta\mu\nu} \equiv & \left[\mathcal{K}^{\alpha\beta\mu\nu} + \mathcal{S}^{\alpha\beta\mu\nu\rho}\nabla_{\rho}\right]\delta\phi \\ & + \left[\frac{1}{4}\mathcal{H}^{\epsilon\pi\alpha\beta\mu\nu} + \mathcal{L}^{\alpha\beta\mu\nu\xi\epsilon\pi}\nabla_{\xi} + \mathcal{R}^{\rho\sigma\epsilon\pi\alpha\beta\mu\nu}\nabla_{\rho}\nabla_{\sigma}\right]\delta_{\mathrm{L}}g_{\epsilon\pi}. \end{aligned} \tag{4.93e}$$

Rewriting (4.92) to isolate the total derivative yields

$$\begin{aligned} \hat{\delta}\mathcal{L}_{\{2\}} = & \left[A - \nabla_{\alpha}B^{\alpha}\right]\hat{\delta}\delta\phi + \left[C^{\alpha\beta} - \nabla_{\sigma}D^{\sigma\alpha\beta} + \nabla_{\mu}\nabla_{\nu}H^{\nu\mu\alpha\beta}\right]\hat{\delta}\delta_{\mathrm{L}}g_{\alpha\beta} \\ & + \nabla_{\beta}\left[B^{\beta}\hat{\delta}\delta\phi + (D^{\beta\mu\nu} - \nabla_{\alpha}H^{\alpha\beta\mu\nu} + H^{\beta\alpha\mu\nu}\nabla_{\alpha})\hat{\delta}\delta_{\mathrm{L}}g_{\mu\nu}\right]. \end{aligned} \tag{4.94}$$

Thus we obtain the functional derivatives

$$\frac{\hat{\delta}}{\hat{\delta}\delta\phi}\mathcal{L}_{\{2\}} = A - \nabla_{\alpha}B^{\alpha}, \tag{4.95a}$$

$$\frac{\hat{\delta}}{\hat{\delta}\delta_{\mathrm{L}}g_{\alpha\beta}} = C^{\alpha\beta} - \nabla_{\sigma}D^{\sigma\alpha\beta} + \nabla_{\mu}\nabla_{\nu}H^{\nu\mu\alpha\beta}, \tag{4.95b}$$

and the surface current,

$$\vartheta^{\beta} = B^{\beta}\hat{\delta}\delta\phi + (D^{\beta\mu\nu} - \nabla_{\alpha}H^{\alpha\beta\mu\nu} + H^{\beta\alpha\mu\nu}\nabla_{\alpha})\hat{\delta}\delta_{\mathrm{L}}g_{\mu\nu}. \tag{4.96}$$

This shows us that $\hat{\delta}\delta\phi$, $\hat{\delta}\delta_{\rm L}g_{\mu\nu}$ and $\nabla_\rho\hat{\delta}\delta_{\rm L}g_{\mu\nu}$ must vanish on the boundary for a well posed variational principle to be employed.

4.4.1 Perturbed Euler–Lagrange Equation

The equation of motion of the $\delta\phi$-field is found from (4.95a) and is given by

$$\nabla_\mu B^\mu = A, \tag{4.97}$$

and after explicitly inserting the definitions of A, B from (4.93) we obtain

$$\mathcal{C}^{\mu\nu}\nabla_\nu\nabla_\mu\delta\phi + (\nabla_\nu\mathcal{C}^{\mu\nu})\nabla_\mu\delta\phi + (\nabla_\mu\mathcal{B}^\mu - 2\mathcal{A})\delta\phi = \delta S, \tag{4.98a}$$

where the source term is given by

$$\begin{aligned}\delta S \equiv &\left[\frac{1}{4}\mathcal{V}^{\mu\nu} + \mathcal{I}^{\rho\mu\nu}\nabla_\rho - \frac{1}{4}\nabla_\rho\mathcal{Y}^{\rho\mu\nu}\right]\delta_{\rm L}g_{\mu\nu} - \left[\nabla_\rho\mathcal{J}^{\alpha\mu\nu\rho} + \frac{1}{4}\mathcal{Y}^{\alpha\mu\nu}\right]\nabla_\alpha\delta_{\rm L}g_{\mu\nu}\\ &- \left[\mathcal{J}^{\beta\mu\nu\alpha} + \mathcal{K}^{\alpha\beta\mu\nu} + \nabla_\rho\mathcal{S}^{\alpha\beta\mu\nu\rho}\right]\nabla_\alpha\nabla_\beta\delta_{\rm L}g_{\mu\nu}\\ &- \left[\mathcal{S}^{\alpha\beta\mu\nu\rho}\right]\nabla_\rho\nabla_\alpha\nabla_\beta\delta_{\rm L}g_{\mu\nu}.\end{aligned} \tag{4.98b}$$

4.4.2 Perturbed Dark Energy Momentum Tensor

The perturbed energy-momentum tensor found from (4.95a) and is given by

$$\delta_{\rm L}U^{\alpha\beta} = -\frac{1}{2}\left[4C^{\alpha\beta} - 4\nabla_\sigma D^{\sigma\alpha\beta} + 4\nabla_\mu\nabla_\nu H^{\nu\mu\alpha\beta} + U^{\alpha\beta}g^{\mu\nu}\delta_{\rm L}g_{\mu\nu}\right]. \tag{4.99}$$

After inserting the definitions of C, D, H from (4.93), the perturbed dark energy-momentum tensor can be compactly written as an operator expansion

$$\delta_{\rm L}U^{\mu\nu} = \hat{\mathbb{Y}}^{\mu\nu}\delta\phi + \hat{\mathbb{W}}^{\mu\nu\alpha\beta}\delta_{\rm L}g_{\alpha\beta}, \tag{4.100a}$$

where the operators are expanded as

$$\hat{\mathbb{Y}}^{\alpha\beta} = \mathbb{A}^{\alpha\beta} + \mathbb{B}^{\theta\alpha\beta}\nabla_\theta + \mathbb{C}^{\rho\theta\alpha\beta}\nabla_\rho\nabla_\theta + \mathbb{D}^{\sigma\rho\theta\alpha\beta}\nabla_\sigma\nabla_\rho\nabla_\theta, \tag{4.100b}$$

$$\hat{\mathbb{W}}^{\alpha\beta\epsilon\pi} = \mathbb{E}^{\alpha\beta\epsilon\pi} + \mathbb{F}^{\rho\alpha\beta\epsilon\pi}\nabla_\rho + \mathbb{G}^{\sigma\rho\alpha\beta\epsilon\pi}\nabla_\sigma\nabla_\rho$$
$$+ \mathbb{H}^{\rho\gamma\lambda\alpha\beta\epsilon\pi}\nabla_\rho\nabla_\gamma\nabla_\lambda + \mathbb{I}^{\sigma\rho\gamma\lambda\alpha\beta\epsilon\pi}\nabla_\sigma\nabla_\rho\nabla_\gamma\nabla_\lambda, \tag{4.100c}$$

where the coefficients are given by

$$\mathbb{A}^{\alpha\beta} = -\frac{1}{2}\left[\mathcal{V}^{\alpha\beta} + 4\nabla_\sigma\nabla_\rho\mathcal{K}^{\rho\sigma\alpha\beta} - 4\nabla_\sigma\mathcal{I}^{\sigma\alpha\beta}\right], \tag{4.101a}$$

$$\mathbb{B}^{\theta\alpha\beta} = -\frac{1}{2}\Big[\mathcal{Y}^{\theta\alpha\beta} + 4\nabla_\rho\mathcal{K}^{\theta\rho\alpha\beta} + 4\nabla_\rho\mathcal{K}^{\rho\theta\alpha\beta}$$
$$+ 4\nabla_\sigma\nabla_\rho\mathcal{S}^{\rho\sigma\alpha\beta\theta} - 4\mathcal{I}^{\theta\alpha\beta} - 4\nabla_\rho\mathcal{J}^{\rho\alpha\beta\theta}\Big], \tag{4.101b}$$

$$\mathbb{C}^{\rho\theta\alpha\beta} = -\frac{1}{2}\left[4\mathcal{K}^{\theta\rho\alpha\beta} + 4\nabla_\xi\mathcal{S}^{\xi\rho\alpha\beta\theta} + 4\nabla_\xi\mathcal{S}^{\rho\xi\alpha\beta\theta} - 4\mathcal{J}^{\rho\alpha\beta\theta}\right], \tag{4.101c}$$

$$\mathbb{D}^{\sigma\rho\theta\alpha\beta} = -2\mathcal{S}^{\rho\sigma\alpha\beta\theta}, \tag{4.101d}$$

$$\mathbb{E}^{\alpha\beta\epsilon\pi} = -\frac{1}{2}\left[\mathcal{W}^{\alpha\beta\epsilon\pi} + U^{\alpha\beta}g^{\epsilon\pi} + \nabla_\sigma\nabla_\rho\mathcal{H}^{\epsilon\pi\rho\sigma\alpha\beta} - \nabla_\sigma\mathcal{U}^{\sigma\alpha\beta\epsilon\pi}\right], \tag{4.101e}$$

$$\mathbb{F}^{\rho\alpha\beta\epsilon\pi} = -\frac{1}{2}\Big[\nabla_\theta\mathcal{H}^{\epsilon\pi\rho\theta\alpha\beta} + \nabla_\theta\mathcal{H}^{\epsilon\pi\theta\rho\alpha\beta} - 4\nabla_\theta\mathcal{M}^{\rho\epsilon\pi\theta\alpha\beta}$$
$$+ \mathcal{U}^{\rho\epsilon\pi\alpha\beta} - \mathcal{U}^{\rho\alpha\beta\epsilon\pi} + 4\nabla_\xi\nabla_\zeta\mathcal{L}^{\zeta\xi\alpha\beta\rho\epsilon\pi}\Big], \tag{4.101f}$$

$$\mathbb{G}^{\sigma\rho\alpha\beta\epsilon\pi} = -\frac{1}{2}\Big[\mathcal{H}^{\alpha\beta\sigma\rho\epsilon\pi} + \mathcal{H}^{\epsilon\pi\rho\sigma\alpha\beta} - 4\mathcal{M}^{\rho\epsilon\pi\sigma\alpha\beta}$$
$$+ 4\nabla_\theta\nabla_\xi\mathcal{R}^{\sigma\rho\epsilon\pi\xi\theta\alpha\beta} + 4\nabla_\lambda\mathcal{L}^{\sigma\rho\epsilon\pi\lambda\alpha\beta}\Big], \tag{4.101g}$$

$$\mathbb{H}^{\rho\gamma\lambda\alpha\beta\epsilon\pi} = -\frac{1}{2}\left[4\nabla_\theta\mathcal{R}^{\gamma\lambda\epsilon\pi\theta\rho\alpha\beta} + 4\nabla_\theta\mathcal{R}^{\gamma\lambda\epsilon\pi\rho\theta\alpha\beta} + 4\mathcal{L}^{\gamma\rho\alpha\beta\lambda\epsilon\pi}\right], \tag{4.101h}$$

$$\mathbb{I}^{\sigma\rho\gamma\lambda\alpha\beta\epsilon\pi} = 2\mathcal{R}^{\gamma\lambda\epsilon\pi\rho\sigma\alpha\beta}. \tag{4.101i}$$

The coefficients (4.101) enable us to observe that the coupling coefficients in the Lagrangian for perturbations have combined in a highly non-trivial way to construct the generalized modification to the gravitational field equations.

4.4.3 Theoretical Restrictions

We will now sketch how to impose theoretical restrictions upon the theory to reduce its complexity and isolate some interesting structure.

From (4.100) it is clear that the theory will have second order field equations when

$$\mathbb{C} = \mathbb{D} = \mathbb{F} = \mathbb{G} = \mathbb{H} = \mathbb{I} = 0. \tag{4.102}$$

By imposing the conditions (4.102) upon (4.101) we find the following constraints become imposed upon the coupling coefficients in the Lagrangian:

$$\mathcal{R} = \mathcal{L} = \mathcal{S} = 0, \quad \mathcal{H}^{\alpha\beta\sigma\rho\epsilon\pi} + \mathcal{H}^{\epsilon\pi\rho\sigma\alpha\beta} = 4\mathcal{M}^{\rho\epsilon\pi\sigma\alpha\beta}, \tag{4.103}$$

$$\mathcal{U}^{\rho\epsilon\pi\alpha\beta} = \mathcal{U}^{\rho\alpha\beta\epsilon\pi}, \quad \mathcal{K}^{\theta\rho\alpha\beta} = \mathcal{J}^{\rho\alpha\beta\theta}. \tag{4.104}$$

These conditions are very "strong", and we do not expect them to be respected by established theories; they are interesting nonetheless and are particularly useful for constructing toy models.

A well motivated subset of the general scalar-tensor theories has a field content given by

$$\mathcal{L} = \mathcal{L}(g_{\mu\nu}, R^{\alpha}{}_{\beta\mu\nu}, \phi, \nabla_{\mu}\phi), \tag{4.105}$$

and the Lagrangian for perturbations will be built from quadratic combinations of

$$\mathcal{L}_{\{2\}} = \mathcal{L}_{\{2\}}(\delta_{\mathrm{L}} g_{\mu\nu}, \delta_{\mathrm{L}} R^{\alpha}{}_{\beta\mu\nu}, \delta\phi, \nabla_{\mu}\delta\phi), \tag{4.106}$$

where one can compute

$$\delta_{\mathrm{L}} R^{\alpha}{}_{\beta\mu\nu} = 2\mathrm{S}^{\sigma\epsilon\theta\alpha}{}_{\beta[\mu}\nabla_{\nu]}\nabla_{\sigma}\delta_{\mathrm{L}} g_{\epsilon\theta} + 2\left(\Gamma^{\rho}{}_{\beta[\mu}\mathrm{S}^{\sigma\epsilon\theta\alpha}{}_{\nu]\rho} - \Gamma^{\alpha}{}_{\rho[\mu}\mathrm{S}^{\sigma\epsilon\theta\rho}{}_{\nu]\beta}\right)\nabla_{\sigma}\delta_{\mathrm{L}} g_{\epsilon\theta}, \tag{4.107}$$

where $\mathrm{S}^{\sigma\epsilon\theta\alpha}{}_{\beta\mu}$ is the structure tensor we defined in (4.72). Notice that this will constrain the way in which the derivatives of the metric combine to construct the Lagrangian for perturbations and the field equations. This is an equivalent of the structure we imposed in Sect. 4.3.3.3.

Another well motivated subset of theories are the scalar-tensor theories, whose field content is just the Ricci scalar, a scalar field ϕ and its kinetic term $\mathcal{X}$,

$$\mathcal{L} = \mathcal{L}(R, \mathcal{X}, \phi). \tag{4.108}$$

This field content is clearly only a subset of (4.89) but will completely encompass Brans-Dicke and all $F(R)$ theories. The subset (4.108) will illustrate how theories are contained within the generalized Lagrangian (4.90), and can be extracted by setting various coupling coefficients to zero. The Lagrangian for perturbations for this theory can be explicitly calculated via

$$\begin{aligned}\mathcal{L}_{\{2\}} &= \mathcal{L}_{,\phi\phi}(\delta\phi)^2 + \mathcal{L}_{,\mathcal{X}\mathcal{X}}(\delta\mathcal{X})^2 + \mathcal{L}_{,RR}(\delta R)^2 + 2\mathcal{L}_{,\phi\mathcal{X}}\delta\phi\delta\mathcal{X} + 2\mathcal{L}_{,\phi R}\delta\phi\delta R \\ &\quad + 2\mathcal{L}_{,\mathcal{X}R}\delta\mathcal{X}\delta R + \mathcal{L}_{,\mathcal{X}}\left(\delta^2\mathcal{X} + g^{\mu\nu}\delta\mathcal{X}\delta g_{\mu\nu}\right) + \mathcal{L}_{,R}\left(\delta^2 R + g^{\mu\nu}\delta R\delta g_{\mu\nu}\right) \\ &\quad + \mathcal{L}_{,\phi}g^{\mu\nu}\delta\phi\delta g_{\mu\nu} + \frac{1}{4}\mathcal{L}\left(g^{\mu\nu}g^{\alpha\beta} - 2g^{\mu(\alpha}g^{\beta)\nu}\right)\delta g_{\mu\nu}\delta g_{\alpha\beta}. \end{aligned} \tag{4.109}$$

The variations of the Ricci scalar can be written compactly as

$$\delta R = V^{\rho\sigma\mu\nu}\nabla_\rho\nabla_\sigma\delta g_{\mu\nu} - R^{\mu\nu}\delta g_{\mu\nu}, \tag{4.110a}$$

$$\begin{aligned}\delta^2 R &= Y^{\xi\gamma\alpha\beta\epsilon\pi}\delta g_{\xi\gamma}\nabla_\alpha\nabla_\beta\delta g_{\epsilon\pi} + Z^{\rho\alpha\beta\sigma\epsilon\pi}\nabla_\rho\delta g_{\alpha\beta}\nabla_\sigma\delta g_{\epsilon\pi} \\ &\quad + \left(R^{\alpha(\rho}g^{\sigma)\beta} + R^{\beta(\rho}g^{\sigma)\alpha}\right)\delta g_{\alpha\beta}\delta g_{\rho\sigma}, \end{aligned} \tag{4.110b}$$

where V, Y, Z are *known* functions of the metric, and are given by

$$V^{\rho\sigma\mu\nu} = g^{\rho\mu}g^{\sigma\nu} - g^{\rho\sigma}g^{\mu\nu}, \tag{4.111a}$$

$$\begin{aligned}Y^{\mu\nu\rho\sigma\epsilon\theta} &= g^{\rho(\mu}g^{\nu)\lambda}\mathrm{S}^{\sigma\epsilon\theta}{}_{\lambda}{}^{\alpha}{}_{\alpha} - g^{\beta(\mu}g^{\nu)\lambda}\mathrm{S}^{\sigma\epsilon\rho}{}_{\lambda}{}^{\theta}{}_{\beta} \\ &\quad + 2g^{\mu(\alpha}g^{\rho)\nu}\mathrm{S}^{\sigma\epsilon\theta\beta}{}_{\alpha\beta} - 2g^{\mu(\alpha}g^{\beta)\nu}\mathrm{S}^{\sigma\epsilon\theta\rho}{}_{\alpha\beta}, \end{aligned} \tag{4.111b}$$

$$\begin{aligned}Z^{\alpha\mu\nu\beta\gamma\pi} &= 2\mathrm{S}^{\alpha\mu\nu\sigma}{}_{\sigma\epsilon}\mathrm{S}^{\beta\gamma\pi\epsilon\lambda}{}_{\lambda} - 2\mathrm{S}^{\alpha\mu\nu\sigma}{}_{\lambda}{}^{\epsilon}\mathrm{S}^{\beta\gamma\pi\lambda}{}_{\sigma\epsilon} \\ &\quad + g^{\alpha(\mu}g^{\nu)\lambda}\mathrm{S}^{\beta\gamma\pi}{}_{\lambda}{}^{\rho}{}_{\rho} - g^{\rho(\mu}g^{\nu)\lambda}\mathrm{S}^{\beta\gamma\pi}{}_{\lambda}{}^{\alpha}{}_{\rho}, \end{aligned} \tag{4.111c}$$

where $\mathrm{S}^{\mu\nu\rho\alpha\beta\sigma}$ is the structure tensor, defined in (4.72). Similarly, the variations of the kinetic scalar can be written as

$$\delta\mathcal{X} = E^{\mu\nu}\delta g_{\mu\nu} + F^{\mu}\nabla_\mu\delta\phi, \tag{4.112}$$

$$\delta^2\mathcal{X} = I^{\mu\nu}\nabla_\mu\delta\phi\nabla_\nu\delta\phi + J^{\alpha\mu\nu}\nabla_\alpha\delta\phi\delta g_{\mu\nu} + K^{\mu\nu\alpha\beta}\delta g_{\alpha\beta}\delta g_{\mu\nu}, \tag{4.113}$$

where E, F, I, J, K are *known* functions of derivatives of the scalar field and the metric; they are given by

$$E^{\mu\nu} = \frac{1}{2}\nabla^\mu\phi\nabla^\nu\phi, \quad F^\mu = -\nabla^\mu\phi, \tag{4.114a}$$

$$I^{\mu\nu} = -g^{\mu\nu}, \qquad J^{\alpha\mu\nu} = 2g^{\alpha(\mu}\nabla^{\nu)}\phi, \qquad K^{\mu\nu\alpha\beta} = -\nabla^{(\mu}\phi g^{\nu)(\alpha}\nabla^{\beta)}\phi. \quad (4.114b)$$

Using these expressions, the Lagrangian for perturbations (4.109) can be written as

$$\begin{aligned}
\mathcal{L}_{\{2\}} &= \mathcal{A}(\delta\phi)^2 + \mathcal{B}^\mu \delta\phi \nabla_\mu \delta\phi + \frac{1}{2}\mathcal{C}^{\mu\nu}\nabla_\mu\delta\phi\nabla_\nu\delta\phi \\
&\quad + \frac{1}{4}\Big[\mathcal{V}^{\mu\nu}\delta\phi\delta_{\mathrm{L}}g_{\mu\nu} + \mathcal{Y}^{\alpha\mu\nu}\nabla_\alpha\delta\phi\delta_{\mathrm{L}}g_{\mu\nu} \\
&\qquad + \mathcal{H}^{\mu\nu\rho\sigma\epsilon\pi}\delta_{\mathrm{L}}g_{\mu\nu}\nabla_\rho\nabla_\sigma\delta_{\mathrm{L}}g_{\epsilon\pi} + \frac{1}{2}\mathcal{W}^{\mu\nu\alpha\beta}\delta_{\mathrm{L}}g_{\mu\nu}\delta_{\mathrm{L}}g_{\alpha\beta}\Big] \\
&\quad + \mathcal{K}^{\alpha\beta\mu\nu}\delta\phi\nabla_\alpha\nabla_\beta\delta_{\mathrm{L}}g_{\mu\nu} + \mathcal{S}^{\alpha\beta\mu\nu\rho}\nabla_\rho\delta\phi\nabla_\alpha\nabla_\beta\delta_{\mathrm{L}}g_{\mu\nu} \\
&\quad + \frac{1}{2}\mathcal{M}^{\rho\mu\nu\sigma\alpha\beta}\nabla_\rho\delta_{\mathrm{L}}g_{\mu\nu}\nabla_\sigma\delta_{\mathrm{L}}g_{\alpha\beta} + \frac{1}{2}\mathcal{R}^{\rho\sigma\mu\nu\epsilon\pi\alpha\beta}\nabla_\rho\nabla_\sigma\delta_{\mathrm{L}}g_{\mu\nu}\nabla_\epsilon\nabla_\pi\delta_{\mathrm{L}}g_{\alpha\beta}.
\end{aligned} \quad (4.115)$$

The 11 tensors in (4.115) are explicitly given by

$$\mathcal{A} = -\frac{1}{2}\mathcal{L}_{,\phi\phi}, \quad \mathcal{B}^\mu = -\mathcal{L}_{,\phi\mathcal{X}}F^\mu, \quad \mathcal{C}^{\mu\nu} = -\mathcal{L}_{,\mathcal{X}}I^{\mu\nu} - \mathcal{L}_{,\mathcal{X}\mathcal{X}}F^\mu F^\nu, \quad (4.116a)$$

$$\mathcal{V}^{\mu\nu} = -4\mathcal{L}_{,\phi\mathcal{X}}E^{\mu\nu} - 2\mathcal{L}_{,\phi}g^{\mu\nu} + 4\mathcal{L}_{,\phi R}R^{\mu\nu}, \quad (4.116b)$$

$$\mathcal{Y}^{\alpha\mu\nu} = 4\mathcal{L}_{,\mathcal{X}R}F^\alpha R^{\mu\nu} - 2\mathcal{L}_{,\mathcal{X}}J^{\alpha\mu\nu} - 2\mathcal{L}_{,\mathcal{X}}g^{\mu\nu}F^\alpha - 4\mathcal{L}_{,\mathcal{X}\mathcal{X}}E^{\mu\nu}F^\alpha, \quad (4.116c)$$

$$\begin{aligned}
\frac{1}{4}\mathcal{H}^{\epsilon\pi\rho\sigma\mu\nu} &= -\frac{1}{2}\mathcal{L}_{,R}\left(Y^{\epsilon\pi\rho\sigma\mu\nu} + g^{\epsilon\pi}V^{\rho\sigma\mu\nu}\right) \\
&\quad - \mathcal{L}_{,\mathcal{X}R}V^{\rho\sigma\mu\nu}E^{\epsilon\pi} + \mathcal{L}_{,RR}V^{\rho\sigma\mu\nu}R^{\epsilon\pi},
\end{aligned} \quad (4.116d)$$

$$\mathcal{M}^{\rho\alpha\beta\sigma\mu\nu} = -\mathcal{L}_{,R}Z^{\rho\alpha\beta\sigma\epsilon\pi}, \quad \mathcal{R}^{\rho\sigma\mu\nu\alpha\beta\epsilon\pi} = -\mathcal{L}_{,RR}V^{\rho\sigma\mu\nu}V^{\alpha\beta\epsilon\pi}, \quad (4.116e)$$

$$\mathcal{K}^{\rho\sigma\mu\nu} = -\mathcal{L}_{,R\phi}V^{\rho\sigma\mu\nu}, \qquad \mathcal{S}^{\rho\sigma\mu\nu\alpha} = \mathcal{L}_{,\mathcal{X}R}V^{\rho\sigma\mu\nu}F^\alpha, \quad (4.116f)$$

$$\begin{aligned}
\frac{1}{8}\mathcal{W}^{\mu\nu\alpha\beta} &= -\frac{1}{2}\mathcal{L}_{,R}\left(R^{\alpha(\mu}g^{\nu)\beta} + R^{\beta(\mu}g^{\nu)\alpha} - g^{\alpha\beta}R^{\mu\nu}\right) - \frac{1}{2}\mathcal{L}_{,\mathcal{X}\mathcal{X}}E^{\mu\nu}E^{\alpha\beta} \\
&\quad - \frac{1}{2}\mathcal{L}_{,RR}R^{\mu\nu}R^{\alpha\beta} + \mathcal{L}_{,\mathcal{X}R}R^{\mu\nu}E^{\alpha\beta} - \frac{1}{2}\mathcal{L}_{,\mathcal{X}}\left(K^{\mu\nu\alpha\beta} + g^{\alpha\beta}E^{\mu\nu}\right) \\
&\quad - \frac{1}{8}\mathcal{L}\left(g^{\mu\nu}g^{\alpha\beta} - 2g^{\mu(\alpha}g^{\beta)\nu}\right).
\end{aligned} \quad (4.116g)$$

Notice that (4.115) is obtained from the general Lagrangian (4.90) by setting

$$\mathcal{I} = \mathcal{U} = \mathcal{L} = \mathcal{J} = 0. \quad (4.117)$$

This is a direct consequence of our choosing the field content to be (4.108), rather than the larger and more general field content (4.89). Thus, our choice of including the Ricci scalar R in the field content had the effect of "switching off" various interaction terms in the larger (and more general) theory.

4.4.4 Summary

In this section we developed the formalism for a general scalar-tensor theory. We obtained the Lagrangian for perturbations, equation of motion and perturbed dark energy momentum tensor. We have uncovered various general results. For instance, by setting some of the coupling coefficients to zero in the general Lagrangian we would be able to probe theories of a particular type (e.g. those with $\mathcal{L}(\phi, \mathcal{X}, R)$ are found via the restrictions (4.117)).

Perhaps one of the more interesting results is the similarity between the energy momentum tensor for this general scalar-tensor theory (4.100) and the corresponding expression for the second order scalar theory (4.55). Imposing second order field equations upon the theory will enforce very similar constraints upon the tensors $\mathbb{A}, \ldots, \mathbb{I}$ that appear in $\delta_{\mathrm{L}} U^{\mu\nu}$, whether they were constructed from a scalar-tensor theory or high-derivative scalar theory. What this means, for example, is that the closed form for δP derived for the second order scalar field theory, (4.88) can be expected to hold for large portions of scalar-tensor theories (imposed with second order field equations).

4.5 Vector Field Theories

We now show how vector field theories fit into our formalism. There are a few theories in the literature which include vector fields as mediators of gravity. The æther theories [11–15] use a vector field, A^{μ}, which is constrained to be time-like, $A^{\mu} A_{\mu} = -k^2$. In generalized Einstein æther the action is

$$S = \int \mathrm{d}^4 x \sqrt{-g}\left[R + \mathcal{F}(\mathcal{K}) + \lambda(A^{\mu} A_{\mu} + 1)\right], \tag{4.118}$$

where λ is a Lagrange multiplier whose role is to enforce the time-like constraint on A^{μ} and $\mathcal{F}(\mathcal{K})$ is an arbitrary function of the kinetic term,

$$\mathcal{K} = K^{\mu\nu\alpha\beta} \nabla_{\mu} A_{\alpha} \nabla_{\nu} A_{\beta}. \tag{4.119}$$

The coefficient-tensor $K^{\mu\nu\alpha\beta}$ is given as

$$K^{\mu\nu\alpha\beta} = c_1 g^{\mu\nu} g^{\alpha\beta} + c_2 g^{\mu\alpha} g^{\nu\beta} + c_3 g^{\mu\beta} g^{\nu\alpha} + c_4 g^{\alpha\beta} A^{\mu} A^{\nu}. \tag{4.120}$$

The generalized TeVeS action in a single frame is given by [16, 17]

$$S = \int \mathrm{d}^4x\sqrt{-g}\Big[R - \mathbb{K}^{\mu\nu\alpha\beta}\nabla_\mu A_\nu \nabla_\alpha A_\beta + \frac{1}{A_\alpha A^\alpha}V(\mu)\Big], \tag{4.121}$$

where μ is a non-dynamical field and $\mathbb{K}^{\mu\nu\alpha\beta}$ is a tensor which is decomposed with combinations of the metric and vector field A_μ. The vector field in single-frame TeVeS is not constrained to be of unit norm.

The æther and single-frame TeVeS actions are both specific choices of the theory and will be encompassed by what we are going to write down.

4.5.1 Field Content

We will consider a generic first order field content (i.e. at most, first derivatives of the extra fields appear),

$$\mathcal{L} = \mathcal{L}(A^\mu, \nabla_\mu A^\nu, g_{\mu\nu}). \tag{4.122}$$

This has implicitly assumed that the field content contains the first derivative of the metric, because $\nabla_\mu A^\nu = \partial_\mu A^\nu + \Gamma^\nu{}_{\mu\alpha}A^\alpha$ and $\Gamma^\nu{}_{\mu\alpha} \sim \partial g$. Thus, the field content (4.122) is a subset of the more general field content

$$\mathcal{L} = \mathcal{L}(A^\mu, \partial_\mu A^\nu, g_{\mu\nu}, \partial_\alpha g_{\mu\nu}). \tag{4.123}$$

In a manner similar to second order scalar fields one can observe that there are an infinite number of Lorentz invariant scalar quantities that can be constructed from the field content (4.122). For example,

$$\nabla_\mu A^\mu, \quad A^\mu A^\nu \nabla_\mu A_\nu, \quad \nabla_\mu A_\nu \nabla^\mu A^\nu, \quad A^\mu \nabla_\mu A_\nu \nabla^\nu A^\alpha A_\alpha,$$
$$A^\mu \nabla_\mu A_\nu \nabla^\nu A^\alpha \nabla_\alpha A_\beta A^\beta, \quad \cdots \tag{4.124}$$

The point is that there is an infinite tower of possible terms and so we cannot write down a *generic* Lagrangian for a first order vector field theory. One useful way to proceed is to identify and collect all possible terms with a particular number of derivatives (i.e. those without derivatives, linear in derivatives, quadratic in derivatives etc). The only possible scalar that can be constructed without derivatives is

$$\mathcal{S}_0 = A^\mu A_\mu. \tag{4.125a}$$

The entire set of scalars that that can constructed that is linear in the derivatives is

$$\mathcal{S}_1 = \Big\{\Sigma, \quad A^\mu A^\nu \nabla_\mu A_\nu\Big\}. \tag{4.125b}$$

The entire set of scalars that can be constructed that is quadratic in the derivatives is

$$\mathcal{S}_2 = \Big\{ g^{\mu\alpha} g^{\nu\beta} \nabla_\mu A_\nu \nabla_\alpha A_\beta,\ g^{\mu\beta} g^{\nu\alpha} \nabla_\mu A_\nu \nabla_\alpha A_\beta,$$
$$A^\beta A^\mu g^{\alpha\nu} \nabla_\mu A_\nu \nabla_\alpha A_\beta,\ A^\nu A^\beta g^{\alpha\mu} \nabla_\mu A_\nu \nabla_\alpha A_\beta \Big\}. \qquad (4.125c)$$

We could then state that a generic first order vector field theory containing at most quadratic terms in the derivatives is a function of the Lorentz invariant scalars

$$\mathcal{L} = \mathcal{L}(\mathcal{S}_0, \mathcal{S}_1, \mathcal{S}_2). \qquad (4.126)$$

In æther theories only $\mathcal{S}_0$ and $\mathcal{S}_2$ are used—the derivative does not appear linearly in the Lagrangian in Einstein æther and generalized æther theories. We have an argument for this (it is important to recall that in all æther theories the vector field is constrained to be time-like, $A^\mu A_\mu = -k^2$). The terms in the set $\mathcal{S}_1$ are $\nabla^\mu A_\mu$, which is a total derivative and therefore does not contribute to the dynamics, and $A^\mu A^\nu \nabla_\mu A_\nu$. This term can be rewritten as

$$A^\mu A^\nu \nabla_\mu A_\nu = \nabla_\mu (A^\mu A^\nu A_\nu) - A^\mu \nabla_\mu (A^\nu A_\nu). \qquad (4.127)$$

In all æther theories the final term vanishes because $A^\nu A_\nu = -k^2$, a constant, and we are just left with a total derivative, which can be neglected. Therefore, if we do not impose A^μ to be time-like we are allowed to have a term of the form $A^\mu A^\nu \nabla_\mu A_\nu$ which will non-trivially contribute to the dynamics.

4.5.2 Lagrangian for Perturbations of a Vector–Tensor Theory

We will consider a general first order vector field theory. This will include Einstein æther theories and its generalizations. The field content we consider is

$$\mathcal{L} = \mathcal{L}(A^\mu, \nabla_\nu A^\mu, g_{\mu\nu}). \qquad (4.128)$$

Whilst we could not write down an exhaustive list of the possible scalars which the *background* Lagrangian could be built from, we can write down all terms in the *perturbed* Lagrangian. The Lagrangian for perturbations of a vector field theory containing at most first covariant derivatives of the vector field is given by

$$\mathcal{L}_{\{2\}} = \mathcal{L}_{\{2\}}(\delta g_{\mu\nu}, \delta A^\mu, \delta(\nabla_\mu A^\nu)) = \mathcal{L}_{\{2\}}(\delta g_{\mu\nu}, \delta A^\mu, \nabla_\mu \delta A^\nu, \delta \Gamma^\alpha{}_{\mu\nu}). \qquad (4.129)$$

As in the scalar field case, we will write the field content we will use to construct the Lagrangian for perturbations as

$$\mathcal{L}_{\{2\}} = \mathcal{L}_{\{2\}}(\delta_{\rm L} g_{\mu\nu}, \delta_{\rm L} A_\mu, \nabla_\mu \delta_{\rm L} A_\nu, \nabla_\alpha \delta_{\rm L} g_{\mu\nu}). \qquad (4.130)$$

One should be careful to realise that index raising and lowering does not commute with the perturbation operator, $\delta_{\mathrm{L}} A^{\beta} = g^{\mu\beta}(\delta_{\mathrm{L}} A_{\mu} - A^{\nu}\delta_{\mathrm{L}} g_{\mu\nu})$.

Using a now familiar strategy, the Lagrangian for perturbations is given by

$$\begin{aligned}\mathcal{L}_{\{2\}} = {} & \mathcal{G}^{\mu\nu}\delta_{\mathrm{L}} A_{\mu}\delta_{\mathrm{L}} A_{\nu} + \mathcal{O}^{\mu\alpha\beta}\delta_{\mathrm{L}} A_{\mu}\nabla_{\alpha}\delta_{\mathrm{L}} A_{\beta} + \frac{1}{2}\mathcal{H}^{\mu\nu\alpha\beta}\nabla_{\mu}\delta_{\mathrm{L}} A_{\nu}\nabla_{\alpha}\delta_{\mathrm{L}} A_{\beta} \\ & + \frac{1}{4}\Big[\mathcal{Q}^{\mu\alpha\beta}\delta_{\mathrm{L}} A_{\mu}\delta_{\mathrm{L}} g_{\alpha\beta} + \mathcal{T}^{\mu\nu\alpha\beta}\delta_{\mathrm{L}} g_{\mu\nu}\nabla_{\alpha}\delta_{\mathrm{L}} A_{\beta} \\ & + \frac{1}{2}\mathcal{W}^{\mu\nu\alpha\beta}\delta_{\mathrm{L}} g_{\mu\nu}\delta_{\mathrm{L}} g_{\alpha\beta} + \mathcal{U}^{\rho\mu\nu\alpha\beta}\nabla_{\rho}\delta_{\mathrm{L}} g_{\mu\nu}\delta_{\mathrm{L}} g_{\alpha\beta}\Big] \\ & + \mathcal{S}^{\mu\rho\alpha\beta}\delta_{\mathrm{L}} A_{\mu}\nabla_{\rho}\delta_{\mathrm{L}} g_{\alpha\beta} + \mathcal{P}^{\mu\nu\rho\alpha\beta}\nabla_{\mu}\delta_{\mathrm{L}} A_{\nu}\nabla_{\rho}\delta_{\mathrm{L}} g_{\alpha\beta} \\ & + \frac{1}{2}\mathcal{M}^{\rho\mu\nu\sigma\alpha\beta}\nabla_{\rho}\delta_{\mathrm{L}} g_{\mu\nu}\nabla_{\sigma}\delta_{\mathrm{L}} g_{\alpha\beta}. \end{aligned} \tag{4.131}$$

We introduced 10 coupling coefficients. To compute equations of motion we induce variations in the perturbed field variables,

$$\delta_{\mathrm{L}} A_{\mu} \to \delta_{\mathrm{L}} A_{\mu} + \hat{\delta}\delta_{\mathrm{L}} A_{\mu}, \quad \delta_{\mathrm{L}} g_{\mu\nu} \to \delta_{\mathrm{L}} g_{\mu\nu} + \hat{\delta}\delta_{\mathrm{L}} g_{\mu\nu}, \tag{4.132}$$

where the response in the Lagrangian for perturbations is

$$\hat{\delta}\mathcal{L}_{\{2\}} = E^{\mu}\hat{\delta}\delta_{\mathrm{L}} A_{\mu} + B^{\alpha\beta}\nabla_{\alpha}\hat{\delta}\delta_{\mathrm{L}} A_{\beta} + C^{\alpha\beta}\hat{\delta}\delta_{\mathrm{L}} g_{\alpha\beta} + D^{\rho\alpha\beta}\nabla_{\rho}\hat{\delta}\delta_{\mathrm{L}} g_{\alpha\beta}, \tag{4.133}$$

where

$$E^{\mu} \equiv 2\mathcal{G}^{\mu\nu}\delta_{\mathrm{L}} A_{\nu} + \mathcal{O}^{\mu\alpha\beta}\nabla_{\alpha}\delta_{\mathrm{L}} A_{\beta} + \frac{1}{4}\mathcal{Q}^{\mu\alpha\beta}\delta_{\mathrm{L}} g_{\alpha\beta} + \mathcal{S}^{\mu\rho\alpha\beta}\nabla_{\rho}\delta_{\mathrm{L}} g_{\alpha\beta}, \tag{4.134a}$$

$$B^{\alpha\beta} \equiv \mathcal{O}^{\mu\alpha\beta}\delta_{\mathrm{L}} A_{\mu} + \mathcal{H}^{\mu\nu\alpha\beta}\nabla_{\mu}\delta_{\mathrm{L}} A_{\nu} + \frac{1}{4}\mathcal{T}^{\mu\nu\alpha\beta}\delta_{\mathrm{L}} g_{\mu\nu} + \mathcal{P}^{\alpha\beta\rho\mu\nu}\nabla_{\rho}\delta_{\mathrm{L}} g_{\mu\nu}, \tag{4.134b}$$

$$4C^{\alpha\beta} \equiv \mathcal{Q}^{\mu\alpha\beta}\delta_{\mathrm{L}} A_{\mu} + \mathcal{T}^{\alpha\beta\mu\nu}\nabla_{\mu}\delta_{\mathrm{L}} A_{\nu} + \mathcal{W}^{\mu\nu\alpha\beta}\delta_{\mathrm{L}} g_{\mu\nu} + \mathcal{U}^{\rho\mu\nu\alpha\beta}\nabla_{\rho}\delta_{\mathrm{L}} g_{\mu\nu}, \tag{4.134c}$$

$$D^{\rho\alpha\beta} \equiv \mathcal{S}^{\mu\rho\alpha\beta}\delta_{\mathrm{L}} A_{\mu} + \mathcal{P}^{\mu\nu\rho\alpha\beta}\nabla_{\mu}\delta_{\mathrm{L}} A_{\nu} + \frac{1}{4}\mathcal{U}^{\rho\alpha\beta\mu\nu}\delta_{\mathrm{L}} g_{\mu\nu} + \mathcal{M}^{\sigma\mu\nu\rho\alpha\beta}\nabla_{\sigma}\delta_{\mathrm{L}} g_{\mu\nu}. \tag{4.134d}$$

Isolating the total derivative in $\delta\mathcal{L}_{\{2\}}$ yields

$$\begin{aligned}\hat{\delta}\mathcal{L}_{\{2\}} &= \Big[E^{\mu} - \nabla_{\nu}B^{\nu\mu}\Big]\hat{\delta}\delta_{\mathrm{L}}A_{\mu} + \Big[C^{\alpha\beta} - \nabla_{\rho}D^{\rho\alpha\beta}\Big]\hat{\delta}\delta_{\mathrm{L}}g_{\alpha\beta} \\ &\quad + \nabla_{\rho}(B^{\rho\beta}\hat{\delta}\delta_{\mathrm{L}}A_{\beta} + D^{\rho\alpha\beta}\hat{\delta}\delta_{\mathrm{L}}g_{\alpha\beta}).\end{aligned} \tag{4.135}$$

The functional derivatives and surface current are

$$\frac{\hat{\delta}}{\hat{\delta}\delta_{\mathrm{L}}A_{\mu}}\mathcal{L}_{\{2\}} = E^{\mu} - \nabla_{\nu}B^{\nu\mu}, \tag{4.136a}$$

$$\frac{\hat{\delta}}{\hat{\delta}\delta_{\mathrm{L}}g_{\alpha\beta}}\mathcal{L}_{\{2\}} = C^{\alpha\beta} - \nabla_{\rho}D^{\rho\alpha\beta}, \tag{4.136b}$$

$$\vartheta^{\mu} = B^{\mu\beta}\hat{\delta}\delta_{\mathrm{L}}A_{\beta} + D^{\mu\alpha\beta}\hat{\delta}\delta_{\mathrm{L}}g_{\alpha\beta}. \tag{4.136c}$$

The expression (4.136a) will provide the Euler–Lagrange equation for the δA^{μ}-field, (4.136b) will provide the contribution to the gravitational field equations at perturbed order and (4.136c) tells us that $\hat{\delta}\delta_{\mathrm{L}}A_{\mu}$ and $\hat{\delta}\delta_{\mathrm{L}}g_{\mu\nu}$ must vanish on the boundary for a well posed variational principle to be employed. These expressions could be compared with those we obtained for a second order scalar field theory (4.45).

4.5.2.1 Perturbed Euler–Lagrange Equation

The perturbed Euler–Lagrange equation of the perturbed vector field, $\delta_{\mathrm{L}}A_{\mu}$, is found by demanding that $\mathcal{L}_{\{2\}}$ is independent of variations in the perturbed field variable $\hat{\delta}\delta_{\mathrm{L}}A_{\mu}$. From (4.136a) this yields

$$\nabla_{\mu}B^{\mu\nu} = E^{\nu}. \tag{4.137}$$

Explicitly, this expands out to

$$\begin{aligned}&\mathcal{H}^{\alpha\beta\nu\mu}\nabla_{\nu}\nabla_{\alpha}\delta_{\mathrm{L}}A_{\beta} - (\mathcal{O}^{\mu\alpha\beta} - \mathcal{O}^{\beta\alpha\mu} - \nabla_{\nu}\mathcal{H}^{\alpha\beta\nu\mu})\nabla_{\alpha}\delta_{\mathrm{L}}A_{\beta} \\ &\quad - (2\mathcal{G}^{\mu\alpha} - \nabla_{\nu}\mathcal{O}^{\alpha\nu\mu})\delta_{\mathrm{L}}A_{\alpha} = \mathcal{J}^{\mu},\end{aligned} \tag{4.138}$$

where the source term is

$$\begin{aligned}\mathcal{J}^{\mu} &\equiv \frac{1}{4}\big(\mathcal{Q}^{\mu\alpha\beta} - \nabla_{\nu}\mathcal{T}^{\alpha\beta\nu\mu}\big)\delta_{\mathrm{L}}g_{\alpha\beta} + \big(\mathcal{S}^{\mu\rho\alpha\beta} - \frac{1}{4}\mathcal{T}^{\alpha\beta\rho\mu} - \nabla_{\nu}\mathcal{P}^{\nu\mu\rho\alpha\beta}\big)\nabla_{\rho}\delta_{\mathrm{L}}g_{\alpha\beta} \\ &\quad - \mathcal{P}^{\nu\mu\rho\alpha\beta}\nabla_{\nu}\nabla_{\rho}\delta_{\mathrm{L}}g_{\alpha\beta}.\end{aligned} \tag{4.139}$$

Notice that the equation of motion of the perturbed vector field is second order, and *a priori* is sourced by second derivatives of the perturbed metric.

4.5.2.2 Perturbed Energy-Momentum Tensor

The perturbed energy momentum tensor can be computed from (4.136b), yielding

$$-2\delta_{\mathrm{L}} U^{\alpha\beta} = 4C^{\alpha\beta} - 4\nabla_\rho D^{\rho\alpha\beta} + U^{\alpha\beta} g^{\mu\nu} \delta_{\mathrm{L}} g_{\mu\nu}. \tag{4.140}$$

This can be written as an operator expansion

$$\delta_{\mathrm{L}} U^{\mu\nu} = \hat{\mathbb{W}}^{\mu\nu\alpha\beta} \delta_{\mathrm{L}} g_{\alpha\beta} + \hat{\mathbb{Z}}^{\mu\nu\alpha} \delta_{\mathrm{L}} A_\alpha, \tag{4.141}$$

where the operators are expanded as

$$\hat{\mathbb{W}}^{\mu\nu\alpha\beta} = \mathbb{E}^{\mu\nu\alpha\beta} + \mathbb{F}^{\rho\mu\nu\alpha\beta} \nabla_\rho, \tag{4.142}$$

$$\hat{\mathbb{Z}}^{\mu\nu\alpha} = \mathbb{H}^{\mu\nu\alpha} + \mathbb{I}^{\rho\mu\nu\alpha} \nabla_\rho + \mathbb{J}^{\rho\sigma\mu\nu\alpha} \nabla_\rho \nabla_\sigma, \tag{4.143}$$

and where we identify

$$\mathbb{H}^{\alpha\beta\mu} = 4\nabla_\rho \mathcal{S}^{\mu\rho\alpha\beta} - \mathcal{Q}^{\mu\alpha\beta}, \tag{4.144a}$$

$$\mathbb{I}^{\mu\alpha\beta\nu} = \frac{1}{2}\left[4\nabla_\rho \mathcal{P}^{\mu\nu\rho\alpha\beta} - \mathcal{T}^{\alpha\beta\mu\nu} + 4\mathcal{S}^{\nu\mu\alpha\beta}\right], \tag{4.144b}$$

$$\mathbb{J}^{\rho\mu\alpha\beta\nu} = 2\mathcal{P}^{\mu\nu\rho\alpha\beta}, \tag{4.144c}$$

$$\mathbb{E}^{\alpha\beta\mu\nu} = \frac{1}{2}\left[\nabla_\rho \mathcal{U}^{\rho\alpha\beta\mu\nu} - \mathcal{W}^{\mu\nu\alpha\beta} - U^{\alpha\beta} g^{\mu\nu}\right], \tag{4.144d}$$

$$\mathbb{F}^{\rho\alpha\beta\mu\nu} = \frac{1}{2}\left[+\mathcal{U}^{\rho\alpha\beta\mu\nu} + 4\nabla_\xi \mathcal{M}^{\rho\mu\nu\xi\alpha\beta} - \mathcal{U}^{\rho\mu\nu\alpha\beta}\right]. \tag{4.144e}$$

Notice that $\delta_{\mathrm{L}} U^{\mu\nu}$ contains two derivatives of $\delta_{\mathrm{L}} A_\mu$, which means that the conservation equation will contain three derivatives of $\delta_{\mathrm{L}} A_\mu$. The tensor which will control these third derivatives is $\mathcal{P}$, which arose as the mixing between $\nabla_\mu \delta_{\mathrm{L}} A_\nu$ and $\nabla_\mu \delta_{\mathrm{L}} g_{\alpha\beta}$ in $\mathcal{L}_{\{2\}}$ (these will vanish for a vector field theory whose kinetic term is constructed from the gauge-invariant combination $F^{\mu\nu} F_{\mu\nu}$, because the derivatives of the vector field do not couple to the derivatives of the metric). What this means is that *a priori* the conservation equation and Euler–Lagrange equations will contain different information, and so we will not be able to identify the general form of the linking conditions.

4.6 More Examples

There are many more theories and field contents which one could write down and apply our formalism to. We will not develop these theories to anywhere near the extent as the theories we have already considered, but we will briefly outline how they can be studied in our framework.

4.6.1 Bimetric Theories

There are a number of theories in the literature which contain two tensor fields; these theories are called bimetric theories. Bimetric theories arise in massive gravity theories [18–21], TeVeS and conformal gravities; a simple bimetric theory [22] has an action given by

$$S = \int \mathrm{d}^4x\,\sqrt{-g}\,R + \int \mathrm{d}^4x\,\sqrt{-\tilde{g}}\,\tilde{R} + \int \mathrm{d}^4x\,\sqrt{-g}\,\tilde{g}^{-1}g, \tag{4.145}$$

where $g_{\mu\nu}, \tilde{g}_{\mu\nu}$ are two tensor fields (the "metrics"), and $R, \tilde{R}$ are the Ricci scalars computed from the two metrics. The final term above is just a particular choice of interaction between the two metrics. Schematically, bimetric theories can be written as

$$\mathcal{L} = \mathcal{L}(g_{\mu\nu}, f_{\mu\nu}, \mathcal{T}_g, \mathcal{T}_f), \tag{4.146}$$

where $g_{\mu\nu}, f_{\mu\nu}$ are the two metrics and $\mathcal{T}_g, \mathcal{T}_f$ are their kinetic terms. For this field content the Lagrangian for perturbations will be given by

$$\begin{aligned}
8\mathcal{L}_{\{2\}} = {} & \mathcal{Q}_{gg}^{\sigma\mu\nu\rho\alpha\beta}\nabla_\sigma\delta_{\mathrm{L}}g_{\mu\nu}\nabla_\rho\delta_{\mathrm{L}}g_{\alpha\beta} + \mathcal{Q}_{ff}^{\sigma\mu\nu\rho\alpha\beta}\nabla_\sigma\delta_{\mathrm{L}}f_{\mu\nu}\nabla_\rho\delta_{\mathrm{L}}f_{\alpha\beta}\\
& + \mathcal{Q}_{gf}^{\sigma\mu\nu\rho\alpha\beta}\nabla_\sigma\delta_{\mathrm{L}}f_{\mu\nu}\nabla_\rho\delta_{\mathrm{L}}g_{\alpha\beta}\\
& + 2\mathcal{P}_{gg}^{\mu\nu\rho\alpha\beta}\delta_{\mathrm{L}}g_{\mu\nu}\nabla_\rho\delta_{\mathrm{L}}g_{\alpha\beta} + 2\mathcal{P}_{ff}^{\mu\nu\rho\alpha\beta}\delta_{\mathrm{L}}f_{\mu\nu}\nabla_\rho\delta_{\mathrm{L}}f_{\alpha\beta}\\
& + 2\mathcal{P}_{gf}^{\mu\nu\rho\alpha\beta}\delta_{\mathrm{L}}f_{\mu\nu}\nabla_\rho\delta_{\mathrm{L}}g_{\alpha\beta}\\
& + \mathcal{W}_{gg}^{\mu\nu\alpha\beta}\delta_{\mathrm{L}}g_{\mu\nu}\delta_{\mathrm{L}}g_{\alpha\beta} + \mathcal{W}_{gf}^{\mu\nu\alpha\beta}\delta_{\mathrm{L}}g_{\mu\nu}\delta_{\mathrm{L}}f_{\mu\nu} + \mathcal{W}_{ff}^{\mu\nu\alpha\beta}\delta_{\mathrm{L}}f_{\mu\nu}\delta_{\mathrm{L}}f_{\alpha\beta}.
\end{aligned} \tag{4.147}$$

The tensor $\mathcal{Q}_{ij}$ represents a generalization of the kinetic term of the gravitons.

4.6.2 Scalar–Vector–Tensor

There has been considerable interest in theories where gravity is mediated by scalar, vector and tensor fields. The most popular theory is TeVeS [23–27]. The field content for TeVeS is

$$\mathcal{L} = \mathcal{L}(g_{\mu\nu}, \phi, \nabla_\mu \phi, A^\mu, \nabla_\nu A^\mu). \tag{4.148}$$

So, the field content of the second order Lagrangian which will encompass TeVeS will be

$$\begin{aligned}\mathcal{L}_{\{2\}} &= \mathcal{L}_{\{2\}}(\delta g_{\mu\nu}, \delta\phi, \delta(\nabla_\mu \phi), \delta A^\mu, \delta(\nabla_\nu A^\mu)) \\ &= \mathcal{L}_{\{2\}}(\delta_{\mathrm{L}} g_{\mu\nu}, \delta\phi, \delta_{\mathrm{L}} A_\mu, \nabla_\rho \delta_{\mathrm{L}} g_{\mu\nu}, \nabla_\mu \delta\phi, \nabla_\mu \delta_{\mathrm{L}} A_\nu).\end{aligned} \tag{4.149}$$

This will require 21 *a priori* independent coupling coefficients to write down the general Lagrangian for perturbations.

4.6.3 High Order Scalars and Curvature Tensors

One of the notable classes of theories our formalism thus far has failed to encompass are those with second order derivatives of scalar fields and curvature tensors. There are a few theories of this type which recently have received attention in the literature: the Galileon [3, 22, 28–31], Horndeski [22, 32–34] and *Fab Four* [35–38] theories (as well as the more recent *Fab 5* theory [39]). The field content of these theories can be written as

$$\mathcal{L} = \mathcal{L}(g_{\mu\nu}, R_{\mu\nu\alpha\beta}, \phi, \nabla_\mu \phi, \nabla_\mu \nabla_\nu \phi). \tag{4.150}$$

Clearly, both of these theories are encompassed by the field content

$$\mathcal{L} = \mathcal{L}(g_{\mu\nu}, \partial_\alpha g_{\mu\nu}, \partial_\alpha \partial_\beta g_{\mu\nu}, \phi, \partial_\alpha \phi, \partial_\alpha \partial_\beta \phi). \tag{4.151}$$

Theories with field content of this form can clearly be encompassed by our generalized formalism for perturbations, requiring 21 coupling coefficients in the Lagrangian for perturbations to enable all freedom in the theory to be identified.

4.7 Summary

At a very technical and formal level we have provided expressions for the Lagrangian for perturbations and modifications to the gravitational field equations for very general classes of theories. Starting from the Lagrangian enables us to obtain a clear interpretation and understanding of how field couplings combine to construct the gravitational field equations. The generalized gravitational field equations for a theory containing the metric, a vector field and a scalar field (and their derivatives) can be written as

$$\delta_{\rm E} G^{\mu\nu} = 8\pi G \delta_{\rm E} T^{\mu\nu} + \delta_{\rm E} U^{\mu\nu}, \tag{4.152}$$

where the tensor parameterizing the deviations from General Relativity at the level of the field equations is given by

$$\delta_{\rm E} U^{\mu\nu} = \hat{\mathbb{W}}^{\mu\nu\alpha\beta}\delta_{\rm L} g_{\alpha\beta} + \hat{\mathbb{Z}}^{\mu\nu\alpha}\delta_{\rm L} A_\alpha + \hat{\mathbb{Y}}^{\mu\nu}\delta\phi - \xi^\alpha \nabla_\alpha U^{\mu\nu} + 2U^{\alpha(\mu}\nabla_\alpha \xi^{\nu)}. \tag{4.153}$$

The operators $\hat{\mathbb{W}}^{\mu\nu\alpha\beta}$, $\hat{\mathbb{Z}}^{\mu\nu\alpha}$, $\hat{\mathbb{Y}}^{\mu\nu}$, are expanded in orders of derivatives. In (4.56) these are given for a second order scalar field theory, (4.101) for a scalar-tensor theory and (4.144) for a vector field theory. These operators also show how the coupling coefficients in the Lagrangian combine in a highly non-trivial way to construct the field equations. The projectors (4.5) allows the perturbed fluid variables to be computed from contributions from the scalar, metric and Lie derivative sectors:

$$\delta\rho = \delta\rho_{[\delta\phi]} + \delta\rho_{[\delta g]} + \delta\rho_{[\pounds_\xi U]}, \tag{4.154a}$$

$$\delta P = \delta P_{[\delta\phi]} + \delta P_{[\delta g]} + \delta P_{[\pounds_\xi U]}, \tag{4.154b}$$

$$V^\mu = V^\mu{}_{[\delta\phi]} + V^\mu{}_{[\delta g]} + V^\mu{}_{[\pounds_\xi U]}, \tag{4.154c}$$

$$\Pi^{\mu\nu} = \Pi^{\mu\nu}{}_{[\delta\phi]} + \Pi^{\mu\nu}{}_{[\delta g]} + \Pi^{\mu\nu}{}_{[\pounds_\xi U]}. \tag{4.154d}$$

For example, the pressure perturbation contribution due to the perturbed metric is

$$\begin{aligned} \delta P_{[\delta g]} &= \mathrm{P}_{(\delta P)}{}^{\mu\nu}\hat{\mathbb{W}}_{\mu\nu}{}^{\alpha\beta}\delta_{\rm L} g_{\alpha\beta} \\ &= \frac{1}{3}\gamma^{\mu\nu}\hat{\mathbb{W}}_{\mu\nu}{}^{\alpha\beta}\delta_{\rm L} g_{\alpha\beta}, \end{aligned} \tag{4.155a}$$

and the anisotropic stress contribution due to the perturbed scalar field is given by

$$\begin{aligned} \Pi^{\alpha\beta}_{[\delta\phi]} &= \mathrm{P}_{(\Pi)\mu\nu}{}^{\alpha\beta}\hat{\mathbb{Y}}^{\mu\nu}\delta\phi \\ &= \frac{1}{P}\big(\gamma^\alpha{}_\mu\gamma^\beta{}_\nu - \frac{1}{3}\gamma^{\alpha\beta}\gamma_{\mu\nu}\big)\hat{\mathbb{Y}}^{\mu\nu}\delta\phi. \end{aligned} \tag{4.155b}$$

We realize that $\Pi^{\alpha\beta}_{[\delta\phi]}$ will not generically vanish. What this means is that scalar fields will directly source anisotropic stresses, but the field theory *requires* high-order derivatives of the scalar field to be present in the Lagrangian.

Everywhere we have used tensors: whilst this has necessitated the introduction of a (phenomenally) large number of indices,we have been able to provide formulae which are valid for perturbations about an arbitrary background (i.e. one could, with relative ease, write down the generalized gravitational field equations for perturbations about de Sitter or Minkowski backgrounds). This is a significant generalization which we have "for free" and which would not have been easily possible if one started out by constructing the generalized perturbation equations about an FRW background. By using tensors, one also obtains "for free" the sources to scalar, vector and tensor fluctuations.

One of the possible criticisms of our approach is that we have given our models "too much" freedom: for instance, there are a fantastically large number of free functions in the $3+1$ decomposition (4.59). Infact, what we have actually done is identify *all the freedom*: one can impose theoretical structure (e.g. covariant derivatives, spatial isotropy, Lorentz invariant scalars) which will significantly reduce the number of free parameters. For instance, we have shown that imposing a scalar-tensor structure upon a general Lagrangian immediately removes certain coupling coefficients (see the discussion preceding (4.117)), and second order field equations drastically reduces the number of free functions in a second order scalar field theory.

The results and formulae presented in this chapter are useful for two (rather distinct) purposes. First, our studies have highlighted general structures in the field equations. Vast classes of theories will have perturbed field equations which are of identical functional form. This could not (easily) have been discovered if one were to only study explicit theories. Indeed, ones results would be constrained by whichever theoretical restrictions were implicitly assumed. We have also been able to identify, in generality, the form of boundary terms. The second thing our studies have shown is something which is of great use to cosmological phenomenology: we have obtained the sources due to the dark sector to the perturbed gravitational field equations for general second order scalar field theories with realistic theoretical restrictions imposed. This will enable us to employ techniques to obtain, for example, cosmologically observable spectra, gravitational wave sources and new predictions for the growth of structure.

References

1. V.A. Rubakov, P.G. Tinyakov, Infrared-modified gravities and massive gravitons. Phys. Usp. **51**, 759–792 (2008) [arXiv:0802.4379]
2. K. Hinterbichler, Theoretical aspects of massive gravity. Rev. Mod. Phys. **84**, 671–710 (2012) [arXiv:1105.3735]
3. A. Nicolis, R. Rattazzi, E. Trincherini, The Galileon as a local modification of gravity. Phys. Rev. **D79**, 064036 (2009) [arXiv:0811.2197]
4. C. Deffayet, G. Esposito-Farese, A. Vikman, Covariant Galileon. Phys. Rev. **D79**, 084003 (2009) [arXiv:0901.1314]
5. C. Deffayet, S. Deser, G. Esposito-Farese, Generalized Galileons: all scalar models whose curved background extensions maintain second-order field equations and stress-tensors. Phys. Rev. **D80**,064015 (2009) [arXiv:0906.1967]
6. C. Deffayet, O. Pujolas, I. Sawicki, A. Vikman, Imperfect dark energy from kinetic gravity braiding. JCAP **1010**, 026 (2010) [arXiv:1008.0048]
7. C. Skordis, Consistent cosmological modifications to the Einstein equations. Phys. Rev. **D79**, 123527 (2009) [arXiv:0806.1238]

8. T. Baker, P. G. Ferreira, C. Skordis, J. Zuntz, Towards a fully consistent parameterization of modified gravity. Phys. Rev. **D84**, 124018 (2011) [arXiv:1107.0491]
9. J. Zuntz, T. Baker, P. Ferreira, C. Skordis, Ambiguous tests of general relativity on cosmological scales. JCAP **1206**, 032 (2012) [arXiv:1110.3830]
10. A. Barreira, B. Li, C. Baugh, S. Pascoli, Linear perturbations in Galileon gravity models [arXiv:1208.0600]
11. C. Eling, T. Jacobson, D. Mattingly, Einstein-aether theory [gr-qc/0410001]
12. T. Jacobson, D. Mattingly, Einstein-aether waves. Phys. Rev. **D70**, 024003 (2004) [gr-qc/0402005]
13. A. Tartaglia, N. Radicella, Vector field theories in cosmology. Phys. Rev. **D76**, 083501 (2007) [arXiv:0708.0675]
14. T.G. Zlosnik, P.G. Ferreira, G.D. Starkman, Modifying gravity with the Aether: an alternative to dark matter. Phys. Rev. **D75**, 044017 (2007) [astro-ph/0607411]
15. C. Armendariz-Picon, A. Diez-Tejedor, Aether unleashed. JCAP **0912**, 018 (2009) [arXiv:0904.0809]
16. T. Zlosnik, P. Ferreira, G.D. Starkman, The vector-tensor nature of Bekenstein's relativistic theory of modified gravity. Phys. Rev. **D74**, 044037 (2006) [gr-qc/0606039]
17. C. Skordis, The tensor-vector-scalar theory and its cosmology. Class. Quant. Grav. **26**, 143001 (2009) [arXiv:0903.3602]
18. C. de Rham, G. Gabadadze, A.J. Tolley, Resummation of massive gravity. Phys. Rev. Lett. **106**, 231101 (2011) [arXiv:1011.1232]
19. S. Hassan, R.A. Rosen, A. Schmidt-May, Ghost-free massive gravity with a general reference metric. JHEP **1202**, 026 (2012) [arXiv:1109.3230]
20. S.F. Hassan, R.A. Rosen, Bimetric gravity from ghost-free massive gravity. JHEP **02**, 126 (2012) [arXiv:1109.3515]
21. M.F. Paulos, A.J. Tolley, Massive gravity theories and limits of ghost-free bigravity models [arXiv:1203.4268]
22. T. Clifton, P.G. Ferreira, A. Padilla, C. Skordis, Modified gravity and cosmology. Phys. Rept. **513**, 1–189 (2012) [arXiv:1106.2476]
23. J.D. Bekenstein, Relativistic gravitation theory for the MOND paradigm. Phys. Rev. **D70**, 083509 (2004) [astro-ph/0403694]
24. T.G. Zlosnik, P.G. Ferreira, G.D. Starkman, The vector-tensor nature of Bekenstein's relativistic theory of modified gravity. Phys. Rev. **D74**, 044037 (2006) [gr-qc/0606039]
25. C. Skordis, D.F. Mota, P.G. Ferreira, C. Boehm, Large scale structure in Bekenstein's theory of relativistic modified Newtonian dynamics. Phys. Rev. Lett. **96**, 011301 (2006) [astro-ph/0505519]
26. C. Skordis, TeVeS Cosmology: covariant formalism for the background evolution and linear perturbation theory. Phys. Rev. **D74** 103513 (2006) [astro-ph/0511591]
27. C. Skordis, Generalizing TeVeS cosmology. Phys. Rev. **D77**, 123502 (2008) [arXiv:0801.1985]
28. G. Goon, K. Hinterbichler, M. Trodden, Symmetries for Galileons and DBI scalars on curved space. JCAP **1107**, 017 (2011) [arXiv:1103.5745]
29. C. Deffayet, X. Gao, D.A. Steer, G. Zahariade, From k-essence to generalised Galileons. Phys. Rev. **D84**, 064039 (2011) [arXiv:1103.3260]
30. S. Appleby, E.V. Linder, The paths of gravity in Galileon cosmology. JCAP **1203**, 043 (2012) [arXiv:1112.1981]
31. S.A. Appleby, E.V. Linder, Galileons on trial [arXiv:1204.4314]
32. G.W. Horndeski, Second-order scalar-tensor field equations in a four-dimensional space. Int. J. Theor. Phys. **10**, 363–384 (1974)
33. A. De Felice, T. Kobayashi, S. Tsujikawa, Effective gravitational couplings for cosmological perturbations in the most general scalar-tensor theories with second-order field equations. Phys. Lett. **B706** 123–133 (2011) [arXiv:1108.4242]
34. A. De Felice, S. Tsujikawa, Conditions for the cosmological viability of the most general scalar-tensor theories and their applications to extended Galileon dark energy models. JCAP **1202**, 007 (2012) [arXiv:1110.3878]

35. C. Charmousis, E.J. Copeland, A. Padilla, P.M. Saffin, General second order scalar-tensor theory, self tuning, and the Fab Four. Phys. Rev. Lett. **108** 051101 (2012) [arXiv:1106.2000]
36. C. Charmousis, E.J. Copeland, A. Padilla, P.M. Saffin, Self-tuning and the derivation of a class of scalar-tensor theories. Phys. Rev. **D85**, 104040 (2012) [arXiv:1112.4866]
37. J.-P. Bruneton et al. Fab Four: When John and George play, gravitation and cosmology [arXiv:1203.4446]
38. E.J. Copeland, A. Padilla, P.M. Saffin, The cosmology of the Fab-Four [arXiv:1208.3373]
39. S.A. Appleby, A. De Felice, E.V. Linder, Fab 5: Noncanonical Kinetic Gravity, Self Tuning, and Cosmic Acceleration [arXiv:1208.4163]

Chapter 5
Explicit Theories

5.1 Introduction

In this chapter we will provide expressions for the coupling coefficients in the Lagrangian for perturbations for explicit theories, as well as the values of the functions which appear in the (3+1) decomposition of the coupling coefficients. This will show the existence of a mapping between generalized and explicit theories.

5.2 Kinetic Scalar Fields

Here we will give the explicit forms of the tensors that appear in the dark scalar field case where only first order derivatives appear in the Lagrangian density of the dark sector. We will study the explicit case where the metric $g_{\mu\nu}$ and the derivative of the scalar field $\nabla_\mu\phi$ only appear in the dark sector through the kinetic term,

$$\mathcal{X} = -\frac{1}{2}g^{\mu\nu}\nabla_\mu\phi\nabla_\nu\phi. \tag{5.1}$$

Hence, the field content of a first order scalar field theory with a kinetic term can be written as

$$\mathcal{L} = \mathcal{L}(\phi, \mathcal{X}), \tag{5.2}$$

and the generalized gravitational action is

$$S = \int \mathrm{d}^4x\sqrt{-g}\left[R + 16\pi G\mathcal{L}_{\mathrm{m}} + \mathcal{L}(\phi, \mathcal{X})\right]. \tag{5.3}$$

The generalized gravitational field equations are given by $G^{\mu\nu} = 8\pi G T^{\mu\nu} + U^{\mu\nu}$ where the dark energy-momentum tensor is given by

J. Pearson, *Generalized Perturbations in Modified Gravity and Dark Energy*,
Springer Theses, DOI: 10.1007/978-3-319-01210-0_5,

$$U^{\mu\nu} = \mathcal{L}_{,\mathcal{X}} \nabla^\mu\phi\nabla^\nu\phi + \mathcal{L}g^{\mu\nu}, \tag{5.4}$$

and the Euler-Lagrange equation of motion for ϕ is given by

$$(\mathcal{L}_{,\mathcal{X}}g^{\mu\nu} - \mathcal{L}_{,\mathcal{X}\mathcal{X}}\nabla^\mu\phi\nabla^\nu\phi)\nabla_\mu\phi\nabla_\nu\phi + \mathcal{L}_{,\mathcal{X}\phi}\nabla^\mu\phi\nabla_\nu\phi + \mathcal{L}_{,\phi} = 0. \tag{5.5}$$

By explicit calculation we can identify the quantities $\{\mathcal{A}, \mathcal{B}^\mu, \mathcal{C}^{\mu\nu}, \mathcal{V}^{\mu\nu}, \mathcal{Y}^{\alpha\mu\nu}, \mathcal{W}^{\mu\nu\alpha\beta}\}$ which we introduced in our Lagrangian for perturbations (3.18). To aid our calculation it is useful to realize that the variations of the Lagrangian are

$$\delta\mathcal{L} = \mathcal{L}_{,\phi}\delta\phi + \mathcal{L}_{,\mathcal{X}}\delta\mathcal{X}, \tag{5.6a}$$

$$\delta^2\mathcal{L} = \mathcal{L}_{,\phi\phi}(\delta\phi)^2 + \mathcal{L}_{,\mathcal{X}\mathcal{X}}(\delta\mathcal{X})^2 + 2\mathcal{L}_{,\phi\mathcal{X}}\delta\phi\delta\mathcal{X} + \mathcal{L}_{,\mathcal{X}}\delta^2\mathcal{X}, \tag{5.6b}$$

and those of the kinetic term are

$$\delta\mathcal{X} = \frac{1}{2}\delta g_{\mu\nu}\nabla^\mu\phi\nabla^\nu\phi - \nabla^\mu\phi\nabla_\mu\delta\phi, \tag{5.6c}$$

$$\begin{aligned}\delta^2\mathcal{X} =& - g^{\mu\nu}\nabla_\mu\delta\phi\nabla_\nu\delta\phi + 2g^{\alpha(\mu}\nabla^{\nu)}\phi\nabla_\alpha\delta\phi\delta g_{\mu\nu}\\ &- \frac{1}{2}\left[g^{\mu(\alpha}\nabla^{\beta)}\phi\nabla^\nu\phi + g^{\nu(\alpha}\nabla^{\beta)}\phi\nabla^\mu\phi\right]\delta g_{\mu\nu}\delta g_{\alpha\beta}.\end{aligned} \tag{5.6d}$$

We now combine these expressions to form $\Diamond^2\mathcal{L}$, as defined in (2.18b) and compare the result with (3.18). By appropriate identifications, one finds that

$$\mathcal{A} = -\frac{1}{2}\mathcal{L}_{,\phi\phi}, \qquad \mathcal{B}^\mu = \mathcal{L}_{,\phi\mathcal{X}}\nabla^\mu\phi, \qquad \mathcal{C}^{\mu\nu} = \mathcal{L}_{,\mathcal{X}}g^{\mu\nu} - \mathcal{L}_{,\mathcal{X}\mathcal{X}}\nabla^\mu\phi\nabla^\nu\phi, \tag{5.7}$$

$$\mathcal{V}^{\mu\nu} = -2\left[g^{\mu\nu}\mathcal{L}_{,\phi} + \mathcal{L}_{,\mathcal{X}\phi}\nabla^\mu\phi\nabla^\nu\phi\right], \tag{5.8a}$$

$$\mathcal{Y}^{\alpha\mu\nu} = 2\left[\mathcal{L}_{,\mathcal{X}\mathcal{X}}\nabla^\alpha\phi\nabla^\mu\phi\nabla^\nu\phi + \mathcal{L}_{,\mathcal{X}}\left(g^{\mu\nu}\nabla^\alpha\phi - 2g^{\alpha(\mu}\nabla^{\nu)}\phi\right)\right], \tag{5.8b}$$

$$\begin{aligned}\mathcal{W}^{\alpha\beta\mu\nu} =& -\mathcal{L}_{,\mathcal{X}\mathcal{X}}\nabla^\mu\phi\nabla^\nu\phi\nabla^\alpha\phi\nabla^\beta\phi - \mathcal{L}_{,\mathcal{X}}\left(g^{\mu\nu}\nabla^\alpha\phi\nabla^\beta\phi + g^{\alpha\beta}\nabla^\mu\phi\nabla^\nu\phi\right)\\ &+ 2\mathcal{L}_{,\mathcal{X}}\left(g^{\mu(\alpha}\nabla^{\beta)}\phi\nabla^\nu\phi + g^{\nu(\alpha}\nabla^{\beta)}\phi\nabla^\mu\phi\right) - \mathcal{L}\left(g^{\mu\nu}g^{\alpha\beta} - 2g^{\mu(\alpha}g^{\beta)\nu}\right).\end{aligned} \tag{5.8c}$$

The perturbed dark energy-momentum tensor can be written as

$$\delta_{\mathrm{L}} U^{\mu\nu} = -\frac{1}{2}\left\{\mathcal{V}^{\mu\nu}\delta\phi + \mathcal{Y}^{\alpha\mu\nu}\nabla_{\alpha}\delta\phi\right\} - \frac{1}{2}\left\{\mathcal{W}^{\alpha\beta\mu\nu} + g^{\alpha\beta}U^{\mu\nu}\right\}\delta_{\mathrm{L}} g_{\alpha\beta}. \quad (5.9)$$

One can use (5.7) to compute, for example, the Euler-Lagrange equation (3.26) which was computed from the second order Lagrangian. For a canonical theory it is simple to use (5.7) to obtain the well known formula $\Box\delta\phi + V''\delta\phi = \delta_{\mathrm{E}} S$. This vindicates our use of the second order Lagrangian as the Lagrangian for the perturbed scalar field.

We now decompose the tensors (5.8) with an isotropic (3+1)-split; we write

$$\nabla_{\mu}\phi = -\dot{\phi}u_{\mu}, \qquad g_{\mu\nu} = \gamma_{\mu\nu} - u_{\mu}u_{\nu}, \qquad u^{\mu}u_{\mu} = -1, \qquad u^{\mu}\gamma_{\mu\nu} = 0. \quad (5.10)$$

Notice that with our signature choice, the definition of the covariant derivative of the scalar field means that $\nabla_{\mu}\phi = -\dot{\phi}u_{\mu} \to \partial_t\phi = \dot{\phi}$. The choice of using the minus sign is only important for terms linear or cubic in derivatives of ϕ; i.e. for $\mathcal{Y}^{\alpha\mu\nu}$. The energy-momentum tensor (5.4) can now be written as

$$U^{\mu\nu} = \rho u^{\mu}u^{\nu} + P\gamma^{\mu\nu}, \quad (5.11)$$

where the energy density and pressure are given by

$$\rho = 2\mathcal{X}\mathcal{L}_{,\mathcal{X}} - \mathcal{L}, \qquad P = \mathcal{L}. \quad (5.12)$$

Inserting the (3+1) split into the perturbed energy momentum tensor (5.8) and comparing with (3.31), one finds that

$$A_{\mathcal{V}} = -2\big(2\mathcal{X}\mathcal{L}_{,\mathcal{X}\phi} - \mathcal{L}_{,\phi}\big), \quad B_{\mathcal{V}} = -2\mathcal{L}_{,\phi}, \quad (5.13a)$$

$$A_{\mathcal{Y}} = -2\sqrt{2\mathcal{X}}\Big[2\mathcal{X}\mathcal{L}_{,\mathcal{X}\mathcal{X}} + \mathcal{L}_{,\mathcal{X}}\Big], \quad B_{\mathcal{Y}} = -C_{\mathcal{Y}} = -2\mathcal{L}_{,\mathcal{X}}\sqrt{2\mathcal{X}}, \quad (5.13b)$$

$$A_{\mathcal{W}} = -\Big[4\mathcal{X}^2\mathcal{L}_{,\mathcal{X}\mathcal{X}} + 2\rho + P\Big], \quad B_{\mathcal{W}} = -C_{\mathcal{W}} = -\rho, \quad D_{\mathcal{W}} = -E_{\mathcal{W}} = -P \quad (5.13c)$$

The sets of coefficients $(A_{\mathcal{V}}, B_{\mathcal{V}})$, $(A_{\mathcal{Y}}, B_{\mathcal{Y}})$ are in general different, however they only are different in non-canonical theories where there is an explicit coupling between the scalar field and its kinetic term and where a non-trivial function of the kinetic term appears in the Lagrangian density. It is also interesting to note that coefficients $D_{\mathcal{W}}$ and $E_{\mathcal{W}}$ which appear in the general decomposition of $\mathcal{E}_{\mu\nu\alpha\beta}$ (3.11c) are found to be equal but opposite in this scalar field scenario (in fact, their values are set to the pressure). Similarly, $B_{\mathcal{Y}} = -C_{\mathcal{Y}}$. In Sect. 3.4.2 we used this theory to compute the cosmological perturbations in the synchronous gauge. There we saw that $A_{\mathcal{W}}$, $B_{\mathcal{V}}$ do not make an appearance in the equations. Hence, from this observation and (5.13) we find that there are only 3 three free functions.

In Sect. 3.3.1 we mentioned that $\mathcal{C}^{\mu\nu}$ plays the role of an effective metric for the scalar field perturbations. Using the 3+1 decomposition for $\mathcal{C}^{\mu\nu}$ in (5.7) yields

$$\mathcal{C}^{\mu\nu} = \mathcal{L}_{,\mathcal{X}}\left[\gamma^{\mu\nu} - \left(1 + 2\mathcal{X}\frac{\mathcal{L}_{,\mathcal{X}\mathcal{X}}}{\mathcal{L}_{,\mathcal{X}}}\right)u^{\mu}u^{\nu}\right]. \tag{5.14}$$

5.3 Second Order Scalar Field Theory

We will now provide an explicit calculation of $\mathcal{L}_{\{2\}}$ for a second order scalar field theory. In Sect. 4.3 we discussed why we cannot write down an action which contains all Lorentz invariant scalar combinations of the field content. We will make a particular choice, and pick the theories containing the kinetic scalar, $\mathcal{X}$, and the Laplacian, $\Box\phi$. The action of the second order scalar field theory we consider is given by

$$S = \int \mathrm{d}^4x\,\sqrt{-g}\mathcal{L}(\phi, \mathcal{X}, \Omega), \tag{5.15}$$

where the kinetic scalar and Laplacian are defined as

$$\mathcal{X} \equiv -\frac{1}{2}g^{\mu\nu}\nabla_{\mu}\phi\nabla_{\nu}\phi, \qquad \Omega \equiv \Box\phi. \tag{5.16}$$

We will show how these scalars combine to construct the Lagrangian for perturbations. The variations of the Lagrangian $\mathcal{L} = \mathcal{L}(\phi, \mathcal{X}, \Omega)$ are given by

$$\delta\mathcal{L} = \mathcal{L}_{,\phi}\delta\phi + \mathcal{L}_{,\mathcal{X}}\delta\mathcal{X} + \mathcal{L}_{,\Omega}\delta\Omega, \tag{5.17a}$$

$$\begin{aligned} \delta^2\mathcal{L} = {} & \mathcal{L}_{,\phi\phi}\delta\phi\delta\phi + 2\mathcal{L}_{,\phi\mathcal{X}}\delta\phi\delta\mathcal{X} + 2\mathcal{L}_{,\phi\Omega}\delta\phi\delta\Omega + 2\mathcal{L}_{,\mathcal{X}\Omega}\delta\mathcal{X}\delta\Omega \\ & + \mathcal{L}_{,\mathcal{X}\mathcal{X}}\delta\mathcal{X}\delta\mathcal{X} + \mathcal{L}_{,\Omega\Omega}\delta\Omega\delta\Omega + \mathcal{L}_{,\mathcal{X}}\delta^2\mathcal{X} + \mathcal{L}_{,\Omega}\delta^2\Omega. \end{aligned} \tag{5.17b}$$

The first variations of the scalar quantities $\mathcal{X}$, Ω are

$$\delta\mathcal{X} = \frac{1}{2}\delta g_{\mu\nu}\nabla^{\mu}\phi\nabla^{\nu}\phi - \nabla^{\mu}\phi\nabla_{\mu}\delta\phi, \tag{5.18a}$$

$$\delta\Omega = \delta g^{\mu\nu}\nabla_{\mu}\nabla_{\nu}\phi + \Box\delta\phi - g^{\mu\nu}\nabla_{\alpha}\phi\delta\Gamma^{\alpha}{}_{\mu\nu}, \tag{5.18b}$$

and the second variations are given by

$$\begin{aligned} \delta^2\mathcal{X} = {} & -g^{\mu\nu}\nabla_{\mu}\delta\phi\nabla_{\nu}\delta\phi + 2g^{\alpha(\mu}\nabla^{\nu)}\phi\nabla_{\alpha}\delta\phi\delta g_{\mu\nu} \\ & - \frac{1}{2}\left[g^{\mu(\alpha}\nabla^{\beta)}\phi\nabla^{\nu}\phi + g^{\nu(\alpha}\nabla^{\beta)}\phi\nabla^{\mu}\phi\right]\delta g_{\mu\nu}\delta g_{\alpha\beta}, \end{aligned} \tag{5.18c}$$

$$\begin{aligned} \delta^2\Omega = {} & \delta^2 g^{\mu\nu}\nabla_{\mu}\nabla_{\nu}\phi + 2\delta g^{\mu\nu}\nabla_{\mu}\nabla_{\nu}\delta\phi \\ & - 2\delta g^{\mu\nu}\nabla_{\alpha}\phi\delta\Gamma^{\alpha}{}_{\mu\nu} - 2g^{\mu\nu}\nabla_{\alpha}\delta\phi\delta\Gamma^{\alpha}{}_{\mu\nu} - g^{\mu\nu}\nabla_{\alpha}\phi\delta^2\Gamma^{\alpha}{}_{\mu\nu}. \end{aligned} \tag{5.18d}$$

The second variation $\delta^2 g^{\mu\nu}$ is given by

$$\delta^2 g^{\mu\nu} = \left[g^{\mu(\alpha} g^{\beta)(\rho} g^{\sigma)\nu} + g^{\nu(\alpha} g^{\beta)(\rho} g^{\sigma)\mu} \right] \delta g_{\rho\sigma} \delta g_{\alpha\beta}, \tag{5.19}$$

and hence

$$\begin{aligned}\delta^2 \Omega = & \left[2 g^{\mu(\alpha} g^{\beta)(\rho} g^{\sigma)\nu} \nabla_\mu \nabla_\nu \phi \right] \delta g_{\rho\sigma} \delta g_{\alpha\beta} + 2 \delta g^{\mu\nu} \nabla_\mu \nabla_\nu \delta\phi \\ & - 2\delta g^{\mu\nu} \nabla_\alpha \phi \delta \Gamma^\alpha{}_{\mu\nu} - 2 g^{\mu\nu} \nabla_\alpha \delta\phi \delta\Gamma^\alpha{}_{\mu\nu} - g^{\mu\nu} \nabla_\alpha \phi \delta^2 \Gamma^\alpha{}_{\mu\nu}. \end{aligned} \tag{5.20}$$

The variations of the Christoffel symbol can be written as

$$\delta\Gamma^\rho{}_{\mu\nu} = -\mathrm{S}^{\sigma\epsilon\theta\rho}{}_{\mu\nu} \nabla_\sigma \delta g_{\epsilon\theta}, \qquad \delta^2 \Gamma^\alpha{}_{\mu\nu} = g^{\alpha(\gamma} g^{\lambda)\beta} \mathrm{S}^{\sigma\epsilon\rho}{}_{\beta\mu\nu} \delta g_{\lambda\gamma} \nabla_\sigma \delta g_{\epsilon\rho}, \tag{5.21}$$

where we have defined S to be the same as in (4.72), and is given by

$$\mathrm{S}^{\sigma\epsilon\theta}{}_{\beta\mu\nu} \equiv \frac{1}{2} \left(\delta^\sigma{}_\beta \delta^\epsilon{}_\mu \delta^\theta{}_\nu - 2 \delta^{(\sigma}{}_\mu \delta^{\epsilon)}{}_\nu \delta^\theta{}_\beta \right). \tag{5.22}$$

This has the attractive property of having zero variation $\delta \mathrm{S}^{\sigma\epsilon\theta}{}_{\beta\mu\nu} = 0$ and vanishing covariant derivative.

To obtain the energy-momentum tensor for the theory we use the above relations to write down the first variation of the action,

$$\begin{aligned}\delta S = \int \mathrm{d}^4 x \sqrt{-g} \Big[& \mathcal{L}_{,\phi} \delta\phi + \frac{1}{2} \mathcal{L}_{,\mathcal{X}} \nabla^\mu \phi \nabla^\nu \phi \delta g_{\mu\nu} - \mathcal{L}_{,\mathcal{X}} \nabla^\mu \phi \nabla_\mu \delta\phi \\ & + \mathcal{L}_{,\Omega} \Box \delta\phi - \mathcal{L}_{,\Omega} \nabla^\mu \nabla^\nu \phi \delta g_{\mu\nu} + \mathcal{L}_{,\Omega} g^{\mu\nu} \nabla_\alpha \phi \mathrm{S}^{\rho\epsilon\lambda\alpha}{}_{\mu\nu} \nabla_\rho \delta g_{\epsilon\lambda} + \frac{1}{2} g^{\mu\nu} \mathcal{L} \delta g_{\mu\nu} \Big]. \end{aligned} \tag{5.23}$$

After integrating by parts, we obtain the energy-momentum tensor,

$$\begin{aligned} U^{\mu\nu} = & (\mathcal{L}_{,\mathcal{X}} + 2\mathcal{L}_{,\Omega\phi}) \nabla^\mu \phi \nabla^\nu \phi - 2 \mathcal{L}_{,\Omega\mathcal{X}} \nabla^\alpha \phi \nabla^{(\mu} \phi \nabla^{\nu)} \nabla_\alpha \phi \\ & + 2 \mathcal{L}_{,\Omega\Omega} \nabla^{(\mu} \phi \nabla^{\nu)} \nabla_\alpha \nabla^\alpha \phi + g^{\mu\nu} \Big(\mathcal{L} - \nabla^\alpha (\mathcal{L}_{,\Omega} \nabla_\alpha \phi) \Big). \end{aligned} \tag{5.24}$$

In a similar way, one can compute the background Euler-Lagrange equation:

$$\Box \mathcal{L}_{,\Omega} + \nabla_\mu (\mathcal{L}_{,\mathcal{X}} \nabla^\mu \phi) + \mathcal{L}_{,\phi} = 0. \tag{5.25}$$

We use these expressions to compute the tensors in the decomposition of the Lagrangian for perturbations for a scalar field theory with field content $\mathcal{L} = \mathcal{L}(\phi, \mathcal{X}, \Omega)$. By comparing with (4.40), we find

$$\mathcal{A} = -\frac{1}{2}\mathcal{L}_{,\phi\phi}, \quad \mathcal{B}^{\mu} = \mathcal{L}_{,\phi\mathcal{X}}\nabla^{\mu}\phi, \quad \mathcal{C}^{\mu\nu} = \mathcal{L}_{,\mathcal{X}}g^{\mu\nu} - \mathcal{L}_{,\mathcal{X}\mathcal{X}}\nabla^{\mu}\phi\nabla^{\nu}\phi, \tag{5.26a}$$

$$\mathcal{D}^{\mu\nu} = -\mathcal{L}_{,\phi\Omega}g^{\mu\nu}, \quad \mathcal{E}^{\mu\alpha\beta} = \mathcal{L}_{,\mathcal{X}\Omega}\nabla^{\mu}\phi g^{\alpha\beta}, \quad \mathcal{F}^{\mu\nu\alpha\beta} = -\mathcal{L}_{,\Omega\Omega}g^{\mu\nu}g^{\alpha\beta}, \tag{5.26b}$$

$$\mathcal{V}^{\mu\nu} = -2\Big[\mathcal{L}_{,\mathcal{X}\phi}\nabla^{\mu}\phi\nabla^{\nu}\phi + \mathcal{L}_{,\phi}g^{\mu\nu} - 2\mathcal{L}_{,\phi\Omega}\nabla^{\mu}\nabla^{\nu}\phi\Big], \tag{5.26c}$$

$$\mathcal{Y}^{\alpha\mu\nu} = -2\Big[2\mathcal{L}_{,\mathcal{X}\Omega}\nabla^{\alpha}\phi\nabla^{\mu}\nabla^{\nu}\phi - \mathcal{L}_{,\mathcal{X}\mathcal{X}}\nabla^{\mu}\phi\nabla^{\nu}\phi\nabla^{\alpha}\phi$$
$$-\mathcal{L}_{,\mathcal{X}}(g^{\mu\nu}\nabla^{\alpha}\phi - 2g^{\alpha(\mu}\nabla^{\nu)}\phi)\Big], \tag{5.26d}$$

$$\mathcal{Z}^{\mu\nu\alpha\beta} = -2\Big[\mathcal{L}_{,\mathcal{X}\Omega}g^{\mu\nu}\nabla^{\alpha}\phi\nabla^{\beta}\phi - 2\mathcal{L}_{,\Omega\Omega}g^{\mu\nu}\nabla^{\alpha}\nabla^{\beta}\phi$$
$$+\mathcal{L}_{,\Omega}(g^{\mu\nu}g^{\alpha\beta} - 2g^{\mu(\alpha}g^{\beta)\nu})\Big], \tag{5.26e}$$

$$\mathcal{W}^{\mu\nu\alpha\beta} = -\mathcal{L}_{,\mathcal{X}\mathcal{X}}\nabla^{\mu}\phi\nabla^{\nu}\phi\nabla^{\alpha}\phi\nabla^{\beta}\phi - 4\mathcal{L}_{,\Omega\Omega}\nabla^{\mu}\nabla^{\nu}\phi\nabla^{\alpha}\nabla^{\beta}\phi$$
$$-\mathcal{L}_{,\mathcal{X}}\Big(g^{\alpha\beta}\nabla^{\mu}\phi\nabla^{\nu}\phi + g^{\mu\nu}\nabla^{\alpha}\phi\nabla^{\beta}\phi\Big)$$
$$+2\mathcal{L}_{,\Omega}\Big(g^{\mu\nu}\nabla^{\alpha}\nabla^{\beta}\phi + g^{\alpha\beta}\nabla^{\mu}\nabla^{\nu}\phi\Big)$$
$$+2\mathcal{L}_{,\mathcal{X}}\Big(g^{\mu(\alpha}\nabla^{\beta)}\phi\nabla^{\nu}\phi + g^{\nu(\alpha}\nabla^{\beta)}\phi\nabla^{\mu}\phi\Big)$$
$$-4\mathcal{L}_{,\Omega}\Big(g^{\mu(\alpha}\nabla^{\beta)}\nabla^{\nu}\phi + g^{\nu(\alpha}\nabla^{\beta)}\nabla^{\mu}\phi\Big)$$
$$+2\mathcal{L}_{,\mathcal{X}\Omega}\Big(\nabla^{\mu}\phi\nabla^{\nu}\phi\nabla^{\alpha}\nabla^{\beta}\phi + \nabla^{\alpha}\phi\nabla^{\beta}\phi\nabla^{\mu}\nabla^{\nu}\phi\Big)$$
$$-\mathcal{L}\Big(g^{\mu\nu}g^{\alpha\beta} - 2g^{\mu(\alpha}g^{\beta)\nu}\Big), \tag{5.26f}$$

$$\mathcal{I}^{\epsilon\rho\sigma} = \nabla_{\lambda}\phi \mathrm{S}^{\epsilon\rho\sigma\lambda}{}_{\mu\nu}\mathcal{D}^{\mu\nu}, \qquad \mathcal{J}^{\epsilon\rho\sigma\alpha} = \nabla_{\lambda}\phi \mathrm{S}^{\epsilon\rho\sigma\lambda}{}_{\mu\nu}\mathcal{E}^{\alpha\mu\nu}, \tag{5.26g}$$

$$\mathcal{N}^{\epsilon\rho\sigma\alpha\beta} = \nabla_{\lambda}\phi \mathrm{S}^{\epsilon\rho\sigma\lambda}{}_{\mu\nu}\mathcal{F}^{\mu\nu\alpha\beta}, \qquad \mathcal{U}^{\rho\mu\nu\alpha\beta} = \nabla_{\lambda}\phi \mathrm{S}^{\rho\mu\nu\lambda}{}_{\epsilon\gamma}\mathcal{Z}^{\epsilon\gamma\alpha\beta}, \tag{5.26h}$$

$$\mathcal{M}^{\rho\mu\nu\sigma\alpha\beta} = \nabla_{\lambda}\phi\nabla_{\xi}\phi \mathrm{S}^{\rho\mu\nu\lambda}{}_{\epsilon\gamma}\mathrm{S}^{\sigma\alpha\beta\xi}{}_{\omega\pi}\mathcal{F}^{\epsilon\gamma\omega\pi}. \tag{5.26i}$$

The tensors (5.38f–5.38h) are exactly those which are found by imposing the structure we discussed in Sect. 4.3.3.3. If we arranged the theory so that $\mathcal{F} = 0$ (e.g. by setting

$\mathcal{L}_{,\Omega\Omega} = 0$) then the effective metric governing the scalar field perturbations (4.53) is given by

$$G^{\mu\nu}_{\text{eff}} = \mathcal{L}_{,\mathcal{X}\mathcal{X}} \nabla^\mu\phi\nabla^\nu\phi - g^{\mu\nu}\big[2\mathcal{L}_{,\phi\Omega} + \mathcal{L}_{,\mathcal{X}} + \nabla_\alpha(\mathcal{L}_{,\Omega\mathcal{X}}\nabla^\alpha\phi)\big]. \tag{5.27}$$

An example extension is to a theory constructed from the scalar quantities

$$\mathcal{X} \equiv -\frac{1}{2}\nabla^\mu\phi\nabla_\mu\phi, \qquad \Omega \equiv \Box\phi, \qquad \Sigma \equiv \nabla^\mu\phi\nabla^\nu\phi\nabla_\mu\nabla_\nu\phi, \tag{5.28}$$

then, as an example, one can obtain the tensor

$$\begin{aligned} -\mathcal{F}^{\mu\nu\alpha\beta} &= \mathcal{L}_{,\Omega\Omega} g^{\mu\nu} g^{\alpha\beta} + \mathcal{L}_{,\Sigma\Sigma}\nabla^\mu\phi\nabla^\nu\phi\nabla^\alpha\phi\nabla^\beta\phi \\ &\quad + \mathcal{L}_{,\Omega\Sigma}\left(g^{\alpha\beta}\nabla^\mu\phi\nabla^\nu\phi + g^{\mu\nu}\nabla^\alpha\phi\nabla^\beta\phi\right). \end{aligned} \tag{5.29}$$

We will not use this field content.

We now obtain the explicit forms of the coefficients (4.59) which appear in the (3+1) decomposition of the coupling coefficients in the Lagrangian. To use the (3+1) decomposition for this theory, it is useful to observe the following results:

$$\nabla_\mu\phi = -\dot{\phi}u_\mu, \qquad \nabla_\mu\nabla_\nu\phi = \ddot{\phi}u_\mu u_\nu - \dot{\phi}K_{\mu\nu}, \qquad K_{\mu\nu} = \frac{1}{3}K\gamma_{\mu\nu}, \tag{5.30a}$$

$$\Omega = \Box\phi = -(\ddot{\phi} + \dot{\phi}K), \tag{5.30b}$$

and we remind that $K = 3\mathcal{H}$ for an FRW universe. Inserting (5.30) into the energy-momentum tensor (5.24) we find expressions for the energy density and pressure,

$$\rho = \dot{\phi}^2(\mathcal{L}_{,\mathcal{X}} + 2\mathcal{L}_{,\Omega\phi} + 2\mathcal{L}_{,\Omega\mathcal{X}}) + 2\mathcal{L}_{,\Omega\Omega}\dot{\phi}\dot{\Omega} - P, \tag{5.31a}$$

$$P = \mathcal{L} - \nabla_\alpha(\mathcal{L}_{,\Omega}\nabla^\alpha\phi) \tag{5.31b}$$

For the functions in the (3+1) decompositions of the coupling coefficients (5.26) we find

$$A_{\mathcal{D}} = -B_{\mathcal{D}} = \mathcal{L}_{,\phi\Omega}, \tag{5.32a}$$

$$A_{\mathcal{V}} = -2\left(\mathcal{L}_{,\phi\mathcal{X}}\dot{\phi}^2 - \mathcal{L}_{,\phi} - 2\mathcal{L}_{,\phi\Omega}\ddot{\phi}\right), \qquad B_{\mathcal{V}} = -2\left(\mathcal{L}_{,\phi} + \frac{2}{3}\mathcal{L}_{,\phi\Omega}K\dot{\phi}\right), \tag{5.32b}$$

$$A_{\mathcal{E}} = -B_{\mathcal{E}} = \mathcal{L}_{,\mathcal{X}\Omega}\dot{\phi}, \qquad C_{\mathcal{E}} = 0, \tag{5.32c}$$

$$A_{\mathcal{F}} = -B_{\mathcal{F}} = D_{\mathcal{F}} = -\mathcal{L}_{,\Omega\Omega}, \qquad C_{\mathcal{F}} = E_{\mathcal{F}} = 0, \tag{5.32d}$$

$$A_{\mathcal{Y}} = -2\dot{\phi}\left(\mathcal{L}_{,\mathcal{X}\mathcal{X}}\dot{\phi}^2 - 2\mathcal{L}_{,\mathcal{X}\Omega}\ddot{\phi} + \mathcal{L}_{,\mathcal{X}}\right), \tag{5.32e}$$

$$B_{\mathcal{Y}} = -2\dot{\phi}\left(\mathcal{L}_{,\mathcal{X}} + \frac{2}{3}\mathcal{L}_{,\mathcal{X}\Omega}\dot{\phi}K\right), \qquad C_{\mathcal{Y}} = 2\mathcal{L}_{,\mathcal{X}}\dot{\phi}, \tag{5.32f}$$

$$A_{\mathcal{Z}} = 2\left(\mathcal{L}_{,\Omega} + \mathcal{L}_{,\Omega\mathcal{X}}\dot{\phi}^2 - 2\mathcal{L}_{,\Omega\Omega}\ddot{\phi}\right), \tag{5.32g}$$

$$B_{\mathcal{Z}} = 2\left(\mathcal{L}_{,\Omega} + 2\mathcal{L}_{,\Omega\Omega}\ddot{\phi}\right), \qquad C_{\mathcal{Z}} = 2\left(\mathcal{L}_{,\Omega} + \frac{2}{3}\mathcal{L}_{,\Omega\Omega}\dot{\phi}K - \mathcal{L}_{,\mathcal{X}\Omega}\dot{\phi}^2\right), \tag{5.32h}$$

$$D_{\mathcal{Z}} = F_{\mathcal{Z}} = -2\mathcal{L}_{,\Omega}, \qquad E_{\mathcal{Z}} = 2\mathcal{L}_{,\Omega\mathcal{X}}\dot{\phi}^2 - C_{\mathcal{Z}}, \tag{5.32i}$$

$$A_{\mathcal{W}} = -\mathcal{L}_{,\mathcal{X}\mathcal{X}}\dot{\phi}^4 - 4\mathcal{L}_{,\Omega\Omega}\ddot{\phi}^2 + 4\mathcal{L}_{,\Omega}\ddot{\phi} + 4\mathcal{L}_{,\mathcal{X}\Omega}\dot{\phi}^2\ddot{\phi} - 2\mathcal{L}_{,\mathcal{X}}\dot{\phi}^2 + \mathcal{L}, \tag{5.32j}$$

$$B_{\mathcal{W}} = \frac{4}{3}\mathcal{L}_{,\Omega\Omega}K\dot{\phi}\ddot{\phi} - \mathcal{L}_{,\mathcal{X}}\dot{\phi}^2 + 2\mathcal{L}_{,\Omega}\left(\ddot{\phi} + \frac{1}{3}K\dot{\phi}\right) - \frac{2}{3}\mathcal{L}_{,\mathcal{X}\Omega}K\dot{\phi}^3 + \mathcal{L}, \tag{5.32k}$$

$$C_{\mathcal{W}} = \mathcal{L}_{,\mathcal{X}}\dot{\phi}^2 - 2\mathcal{L}_{,\Omega}\left(\ddot{\phi} + \frac{1}{3}K\dot{\phi}\right) - \mathcal{L}, \tag{5.32l}$$

$$D_{\mathcal{W}} = -\frac{4}{9}\mathcal{L}_{,\Omega\Omega}K^2\dot{\phi}^2 - \frac{4}{3}\mathcal{L}_{,\Omega}K\dot{\phi} - \mathcal{L}, \quad E_{\mathcal{W}} = -\left(D_{\mathcal{W}} + \frac{4}{9}\mathcal{L}_{,\Omega\Omega}K^2\dot{\phi}^2\right). \tag{5.32m}$$

There is rather a lot of structure which links the relative values of these coefficients. One can immediately read off that 8 of the coefficients are set by other coefficients or are exactly zero (these are $B_{\mathcal{D}}$, $B_{\mathcal{E}}$, $C_{\mathcal{E}}$, $B_{\mathcal{F}}$, $C_{\mathcal{F}}$, $D_{\mathcal{F}}$, $E_{\mathcal{F}}$, $F_{\mathcal{Z}}$); this reduces the number of free functions $26 \rightarrow 18$. There are some subtle differences in these coefficients and those we obtained for a first order scalar field theory (5.13), namely the appearance of the extrinsic curvature scalar, $K = 3\mathcal{H}$ (where the equality holds for an FRW background).

Whilst we have not explicitly presented the generalized fluid variables for this second order scalar field theory, it is clear that these variables will contain the same contributions from the first order scalar field theory, in addition to new contributions which are unique to the second order theory. The point is this: the anisotropic stress (7.16d) will now be given by an expression of the form

$$\Pi^{\mathrm{S}} = 2\frac{P - E_{\mathcal{W}}}{P}\left(\xi^{\mathrm{S}} + \frac{1}{2}\eta\right) + \cdots. \tag{5.33}$$

One can verify that $E_{\mathcal{W}} \neq P$ when $\mathcal{L}_{,\Omega} \neq 0$, and therefore the scalar field directly contributes towards the anisotropic stress. This is a very different phenomenology to the first order scalar field theory.

For the theory $\mathcal{L} = F(\phi, \mathcal{X})\Omega$, (5.32) becomes

$$A_{\mathcal{D}} = -B_{\mathcal{D}} = F_{,\phi}, \tag{5.34a}$$

$$A_{\mathcal{V}} = -2\left(F_{,\phi\mathcal{X}}\Omega\dot{\phi}^2 - F_{,\phi}\Omega - 2F_{,\phi}\ddot{\phi}\right), \qquad B_{\mathcal{V}} = -2F_{,\phi}\left(\Omega + \frac{2}{3}K\dot{\phi}\right), \tag{5.34b}$$

$$A_{\mathcal{E}} = -B_{\mathcal{E}} = F_{,\mathcal{X}}\dot{\phi}, \qquad C_{\mathcal{E}} = A_{\mathcal{F}} = B_{\mathcal{F}} = D_{\mathcal{F}} = C_{\mathcal{F}} = E_{\mathcal{F}} = F_{\mathcal{F}} = 0, \tag{5.34c}$$

$$A_{\mathcal{Y}} = -2\dot{\phi}\left(F_{,\mathcal{X}\mathcal{X}}\Omega\dot{\phi}^2 - 2F_{,\mathcal{X}}\ddot{\phi} + F_{,\mathcal{X}}\Omega\right), \tag{5.34d}$$

$$B_{\mathcal{Y}} = -2\dot{\phi}F_{,\mathcal{X}}\left(\Omega + \frac{2}{3}\dot{\phi}K\right), \qquad C_{\mathcal{Y}} = 2F_{,\mathcal{X}}\Omega\dot{\phi}, \tag{5.34e}$$

$$A_{\mathcal{Z}} = 2\left(F + F_{,\mathcal{X}}\dot{\phi}^2\right), \qquad C_{\mathcal{Z}} = 2\left(F - F_{,\mathcal{X}}\dot{\phi}^2\right), \tag{5.34f}$$

$$B_{\mathcal{Z}} = -D_{\mathcal{Z}} = -E_{\mathcal{Z}} = -F_{\mathcal{Z}} = 2F, \tag{5.34g}$$

$$A_{\mathcal{W}} = -F_{,\mathcal{X}\mathcal{X}}\Omega\dot{\phi}^4 + 4F\ddot{\phi} + 4F_{,\mathcal{X}}\dot{\phi}^2\ddot{\phi} - 2F_{,\mathcal{X}}\Omega\dot{\phi}^2 + F\Omega, \tag{5.34h}$$

$$B_{\mathcal{W}} = F_{,\mathcal{X}}\dot{\phi}^2\left(\ddot{\phi} + \frac{1}{3}K\dot{\phi}\right) + F\left(\ddot{\phi} - \frac{1}{3}K\dot{\phi}\right) \tag{5.34i}$$

$$C_{\mathcal{W}} = F_{,\mathcal{X}}\Omega\dot{\phi}^2 - F\left(\ddot{\phi} - \frac{1}{3}K\dot{\phi}\right), \tag{5.34j}$$

$$D_{\mathcal{W}} = F\left(\ddot{\phi} - \frac{1}{3}K\dot{\phi}\right), \qquad E_{\mathcal{W}} = -D_{\mathcal{W}}. \tag{5.34k}$$

There are now 13 independent parameters.

An explicit example of a theory of this class is the "reduced" Galileon model,

$$S = \int d^4x\sqrt{-g}\left[\frac{1}{2}M_{\mathrm{pl}}^2 R + c_2\mathcal{X} + c_3\mathcal{X}\Box\phi - \mathcal{L}_{\mathrm{m}}\right], \tag{5.35}$$

where c_2, c_3 are constants. In this reduced Galileon theory, the effective metric (5.27) becomes

$$G^{\mu\nu}_{\mathrm{eff}} = -c_2\left(1 + \frac{2c_3}{c_2}\Box\phi\right)g^{\mu\nu}, \tag{5.36}$$

and the energy-momentum tensor (5.24) becomes

$$U^{\mu\nu} = (c_2 + c_3\Box\phi)\nabla^\mu\phi\nabla^\nu\phi - 2c_3\nabla^\alpha\phi\nabla^{(\mu}\phi\nabla^{\nu)}\nabla_\alpha\phi + g^{\mu\nu}(c_2\mathcal{X} + c_3\nabla^\alpha\phi\nabla^\beta\phi\nabla_\alpha\nabla_\beta\phi). \tag{5.37}$$

For this theory, the tensors (5.26) become

$$\mathcal{A} = 0, \qquad \mathcal{B}^{\mu} = 0, \qquad \mathcal{C}^{\mu\nu} = (c_2 + c_3\Box\phi)g^{\mu\nu}, \tag{5.38a}$$

$$\mathcal{D}^{\mu\nu} = 0, \qquad \mathcal{E}^{\mu\alpha\beta} = c_3\nabla^{\mu}\phi g^{\alpha\beta}, \qquad \mathcal{F}^{\mu\nu\alpha\beta} = 0, \qquad \mathcal{V}^{\mu\nu} = 0 \tag{5.38b}$$

$$\mathcal{Y}^{\alpha\mu\nu} = -2\left[2c_3\nabla^{\alpha}\phi\nabla^{\mu}\nabla^{\nu}\phi - (c_2 + c_3\Box\phi)(g^{\mu\nu}\nabla^{\alpha}\phi - 2g^{\alpha(\mu}\nabla^{\nu)}\phi)\right], \tag{5.38c}$$

$$\mathcal{Z}^{\mu\nu\alpha\beta} = -2c_3\left[g^{\mu\nu}\nabla^{\alpha}\phi\nabla^{\beta}\phi + \mathcal{X}(g^{\mu\nu}g^{\alpha\beta} - 2g^{\mu(\alpha}g^{\beta)\nu})\right], \tag{5.38d}$$

$$\begin{aligned}
\mathcal{W}^{\mu\nu\alpha\beta} = &-(c_2 + c_3\Box\phi)\left(g^{\alpha\beta}\nabla^{\mu}\phi\nabla^{\nu}\phi + g^{\mu\nu}\nabla^{\alpha}\phi\nabla^{\beta}\phi\right)\\
&+2c_3\mathcal{X}\left(g^{\mu\nu}\nabla^{\alpha}\nabla^{\beta}\phi + g^{\alpha\beta}\nabla^{\mu}\nabla^{\nu}\phi\right)\\
&+2(c_2 + c_3\Box\phi)\left(g^{\mu(\alpha}\nabla^{\beta)}\phi\nabla^{\nu}\phi + g^{\nu(\alpha}\nabla^{\beta)}\phi\nabla^{\mu}\phi\right)\\
&-4c_3\mathcal{X}\left(g^{\mu(\alpha}\nabla^{\beta)}\nabla^{\nu}\phi + g^{\nu(\alpha}\nabla^{\beta)}\nabla^{\mu}\phi\right)\\
&+2c_3\left(\nabla^{\mu}\phi\nabla^{\nu}\phi\nabla^{\alpha}\nabla^{\beta}\phi + \nabla^{\alpha}\phi\nabla^{\beta}\phi\nabla^{\mu}\nabla^{\nu}\phi\right)\\
&-\mathcal{L}\left(g^{\mu\nu}g^{\alpha\beta} - 2g^{\mu(\alpha}g^{\beta)\nu}\right),
\end{aligned} \tag{5.38e}$$

$$\mathcal{I}^{\epsilon\rho\sigma} = 0, \qquad \mathcal{J}^{\epsilon\rho\sigma\alpha} = \nabla_{\lambda}\phi \mathsf{S}^{\epsilon\rho\sigma\lambda}{}_{\mu\nu}\mathcal{E}^{\alpha\mu\nu}, \tag{5.38f}$$

$$\mathcal{N}^{\epsilon\rho\sigma\alpha\beta} = 0, \qquad \mathcal{U}^{\rho\mu\nu\alpha\beta} = \nabla_{\lambda}\phi \mathsf{S}^{\rho\mu\nu\lambda}{}_{\epsilon\gamma}\mathcal{Z}^{\epsilon\gamma\alpha\beta}, \tag{5.38g}$$

$$\mathcal{M}^{\rho\mu\nu\sigma\alpha\beta} = 0. \tag{5.38h}$$

5.4 $F(R)$ and Gauss–Bonnet Gravities

We will compute the field equations for a generalized modified gravity theory, and in particular, identify the dark energy-momentum tensor $U^{\mu\nu}$. The class of theories we will consider is defined by the action

$$S = \int \mathrm{d}^4x\,\sqrt{-g}\left[R + 2\mathcal{L}_{\mathrm{mg}}(g_{\mu\nu}, R^{\alpha}{}_{\mu\beta\nu}) - 16\pi G\mathcal{L}_{\mathrm{m}}\right]. \tag{5.39}$$

We will now calculate $U^{\mu\nu}$ for this theory. To do so, it is useful to define the derivatives of the Lagrangian to be

$$A^{\mu\nu} \equiv \frac{\delta\mathcal{L}_{\mathrm{mg}}}{\delta g_{\mu\nu}}, \qquad B_{\mu}{}^{\alpha\nu\beta} \equiv \frac{\delta\mathcal{L}_{\mathrm{mg}}}{\delta R^{\mu}{}_{\alpha\nu\beta}}, \tag{5.40}$$

and note that the variations of the Riemann and Ricci tensor and scalar can be written as

$$\delta R_{\mu\nu} = g^{\beta\sigma} g_{\sigma\alpha} \delta R^{\alpha}{}_{\mu\beta\nu}, \qquad \delta R = g^{\alpha\beta} g^{\rho\sigma} g_{\sigma\pi} \delta R^{\pi}{}_{\alpha\rho\beta} - R^{\alpha\beta} \delta g_{\alpha\beta}, \tag{5.41a}$$

$$\delta R^{\alpha}{}_{\mu\beta\nu} = \Theta^{\alpha\xi\rho\sigma\pi}{}_{\mu\beta\nu} \nabla_{\xi} \nabla_{\rho} \delta g_{\sigma\pi}, \tag{5.41b}$$

where, for convenience we have defined

$$\Theta^{\alpha\xi\rho\sigma\pi}{}_{\mu\beta\nu} \equiv g^{\alpha\rho} \delta^{\sigma}_{\mu} \delta^{[\pi}_{\beta} \delta^{\xi]}_{\nu} + g^{\alpha\pi} \delta^{\sigma}_{\mu} \delta^{[\rho}_{\nu} \delta^{\xi]}_{\beta} + g^{\alpha\pi} \delta^{\rho}_{\mu} \delta^{[\sigma}_{\nu} \delta^{\xi]}_{\beta}. \tag{5.42}$$

Thus, varying the Lagrangian yields

$$\delta \mathcal{L}_{\rm mg} = \left(A^{\sigma\pi} + \Theta^{\alpha\xi\rho\sigma\pi}{}_{\mu\beta\nu} \nabla_{\rho} \nabla_{\xi} B_{\alpha}{}^{\mu\beta\nu} \right) \delta g_{\sigma\pi} + \nabla_{\mu} \delta S^{\mu}, \tag{5.43}$$

where the term which will only contribute to a surface integral is given by

$$\delta S^{\rho} \equiv B_{\alpha}{}^{\mu\beta\nu} \Theta^{\alpha\rho\xi\sigma\pi}{}_{\mu\beta\nu} \nabla_{\xi} \delta g_{\sigma\pi} - \Theta^{\alpha\xi\rho\sigma\pi}{}_{\mu\beta\nu} \nabla_{\xi} (\delta g_{\sigma\pi} B_{\alpha}{}^{\mu\beta\nu}). \tag{5.44}$$

The field equations are given by $G^{\mu\nu} = 8\pi G T^{\mu\nu} + U^{\mu\nu}$, where, under the usual definition, the dark energy momentum tensor is given by

$$U^{\sigma\pi} = -\left[g^{\sigma\pi} \mathcal{L}_{\rm mg} + 2A^{\sigma\pi} + 2\Theta^{\alpha\xi\rho\sigma\pi}{}_{\mu\beta\nu} \nabla_{\rho} \nabla_{\xi} B_{\alpha}{}^{\mu\beta\nu} \right]. \tag{5.45}$$

For an $F(R)$ theory we can use (5.41a) to obtain

$$A^{\mu\nu} = -\mathcal{L}_{,R} R^{\mu\nu}, \qquad B_{\pi}{}^{\alpha\rho\beta} = \mathcal{L}_{,R} g^{\alpha\beta} \delta^{\rho}_{\pi}. \tag{5.46}$$

For a consistency check, an $F(R)$ theory has $\mathcal{L}_{\rm mg} = \frac{1}{2} f(R)$. Using (5.46) to calculate the final term in (5.45) yields

$$\Theta^{\alpha\xi\rho\sigma\pi}{}_{\mu\beta\nu} \delta^{\beta}_{\alpha} g^{\mu\nu} = g^{\rho\sigma} g^{\xi\pi} - g^{\rho\xi} g^{\pi\sigma}, \tag{5.47}$$

so that we obtain

$$U^{\sigma\pi} = (R^{\sigma\pi} + g^{\sigma\pi} \Box - \nabla^{\sigma} \nabla^{\pi}) f_{,R} - \frac{1}{2} g^{\sigma\pi} f, \tag{5.48}$$

which is identical to the expression one finds through direct calculation.

For a (generalized) Gauss-Bonnet theory, the action is given in terms of the Gauss-Bonnet term, $\mathcal{G}$,

$$S = \int \mathrm{d}^4 x \sqrt{-g} \mathcal{L}(\mathcal{G}), \qquad \mathcal{G} \equiv R^2 - 4R^{\mu\nu} R_{\mu\nu} + 4R^{\mu\nu\alpha\beta} R_{\mu\nu\alpha\beta}. \tag{5.49}$$

It is useful to realize that the variation of the Gauss-Bonnet term $\delta\mathcal{G}$ can be written as

$$\delta\mathcal{G} = \mathfrak{A}^{\epsilon\kappa}\delta g_{\epsilon\kappa} + \mathfrak{B}_{\lambda}{}^{\gamma\epsilon\kappa}\delta R^{\lambda}{}_{\gamma\epsilon\kappa}, \tag{5.50}$$

where

$$\begin{aligned} \mathfrak{A}^{\epsilon\kappa} &= -2RR^{\epsilon\kappa} + 8R^{\alpha\epsilon}R^{\kappa}{}_{\alpha} + R^{\epsilon\nu\alpha\beta}R^{\kappa}{}_{\nu\alpha\beta} - R_{\xi}{}^{\nu\alpha\kappa}R^{\xi}{}_{\nu\alpha}{}^{\epsilon} \\ &\quad - R_{\xi}{}^{\nu\kappa\alpha}R^{\xi}{}_{\nu}{}^{\epsilon}{}_{\alpha} - R_{\xi}{}^{\kappa\alpha\beta}R^{\xi\epsilon}{}_{\alpha\beta}, \end{aligned} \tag{5.51a}$$

$$\mathfrak{B}_{\lambda}{}^{\gamma\epsilon\kappa} = 2\Big(Rg^{\gamma\kappa} - 4R^{\gamma\kappa}\Big)\delta^{\epsilon}{}_{\lambda} + 2R_{\lambda}{}^{\gamma\epsilon\kappa}. \tag{5.51b}$$

Hence, for a Gauss-Bonnet theory, we find that

$$A^{\mu\nu} = \mathcal{L}_{,\mathcal{G}}\mathfrak{A}^{\mu\nu}, \qquad B_{\mu}{}^{\alpha\nu\beta} = \mathcal{L}_{,\mathcal{G}}\mathfrak{B}_{\mu}{}^{\alpha\nu\beta}. \tag{5.52}$$

To calculate $\delta U^{\mu\nu}$ in generality is complicated and cumbersome to write down.

In general it is possible to write $\delta U^{\mu\nu}$ as a single rank-4 pseudo-tensor operator acting on $\delta g^{\alpha\beta}$, so that

$$\delta U^{\mu\nu} = \hat{\mathbb{W}}^{\mu\nu\alpha\beta}\delta g_{\alpha\beta}, \tag{5.53a}$$

where $\hat{\mathbb{W}}^{\mu\nu\alpha\beta}$ is an operator given by

$$\begin{aligned} \hat{\mathbb{W}}^{\mu\nu\rho\sigma} &= \mathbb{A}^{\mu\nu\rho\sigma} + \mathbb{B}^{\mu\nu\rho\sigma\alpha}\nabla_{\alpha} + \mathbb{C}^{\mu\nu\rho\sigma\alpha\beta}\nabla_{\alpha}\nabla_{\beta} \\ &\quad + \mathbb{D}^{\mu\nu\rho\sigma\lambda\alpha\beta}\nabla_{\lambda}\nabla_{\alpha}\nabla_{\beta} + \mathbb{E}^{\mu\nu\rho\sigma\zeta\lambda\alpha\beta}\nabla_{\zeta}\nabla_{\lambda}\nabla_{\alpha}\nabla_{\beta}. \end{aligned} \tag{5.53b}$$

The coefficients $\mathbb{A}\cdots\mathbb{E}$ can be written using the (3+1) split.

We will explicitly calculate $\delta U^{\mu\nu}$ for $F(R)$ gravities, and show that $\delta U^{\mu\nu}$ is indeed of this form. The action we study is given by

$$S = \int \mathrm{d}^4x\,\sqrt{-g}\Big[R + f(R) - 16\pi G\mathcal{L}_{\mathrm{m}}\Big]. \tag{5.54}$$

The dark energy-momentum tensor is given by

$$U^{\mu\nu} = \Big(R^{\mu\nu} + g^{\mu\nu}\Box - \nabla^{\mu}\nabla^{\nu}\Big)f' - \frac{1}{2}fg^{\mu\nu}. \tag{5.55}$$

Perturbing this yields

$$\delta U^{\mu\nu} = \mathfrak{A}^{\mu\nu\rho\sigma}\delta g_{\rho\sigma} + \mathfrak{B}^{\mu\nu}\delta R + \mathfrak{C}^{\alpha\mu\nu}\nabla_{\alpha}\delta R + \mathfrak{D}^{\alpha\beta\rho\sigma}\nabla_{\alpha}\nabla_{\beta}\delta R + \mathfrak{E}_{\alpha}{}^{\rho\sigma\mu\nu}\delta\Gamma^{\alpha}{}_{\rho\sigma}, \tag{5.56}$$

where we have defined

$$\mathfrak{A}^{\mu\nu\rho\sigma} \equiv -2f'g^{\rho(\mu}R^{\nu)\sigma} - g^{\mu\rho}g^{\nu\sigma}\Box f' - f'''g^{\mu\nu}\nabla^{\rho}R\nabla^{\sigma}R + 2f'''g^{\rho(\mu}\nabla^{\nu)}R\nabla^{\sigma}R$$
$$+\frac{1}{2}fg^{\mu\rho}g^{\nu\sigma} - f''g^{\mu\nu}\nabla^{\rho}\nabla^{\sigma}R + 2f''g^{\rho(\mu}\nabla^{\nu)}\nabla^{\sigma}R, \quad (5.57a)$$

$$\mathfrak{B}^{\mu\nu} \equiv f''g^{\mu\nu} + f'''g^{\mu\nu}\Box R + f''''g^{\mu\nu}\nabla^{\alpha}R\nabla_{\alpha}R - f''''\nabla^{\mu}R\nabla^{\nu}R - \frac{1}{2}f'g^{\mu\nu}, \quad (5.57b)$$

$$\mathfrak{C}^{\alpha\mu\nu} \equiv 2f'''\Big(g^{\mu\nu}\nabla^{\alpha}R - g^{\alpha(\mu}\nabla^{\nu)}R\Big), \quad (5.57c)$$

$$\mathfrak{D}^{\alpha\beta\mu\nu} \equiv f''\Big(g^{\mu\nu}g^{\alpha\beta} - g^{\mu\alpha}g^{\nu\beta}\Big), \quad (5.57d)$$

$$\mathfrak{E}_{\alpha}{}^{\rho\sigma\mu\nu} \equiv f''\Big(g^{\mu\rho}g^{\nu\sigma} - g^{\mu\nu}g^{\rho\sigma}\Big)\nabla_{\alpha}R \quad (5.57e)$$

To actually compute the explicit form of $\hat{\mathbb{W}}^{\mu\nu\alpha\beta}$ is a rather convoluted task and to do so we use the fact that perturbations such as δR, $\delta R_{\mu\nu}$, $\delta\Gamma^{\lambda}{}_{\alpha\beta}$ can be written entirely in terms of $\delta g^{\alpha\beta}$. For example, we have the identities

$$\delta\Gamma^{\rho}{}_{\mu\nu} = \frac{1}{2}g^{\rho\sigma}\Big(\nabla_{\mu}\delta g_{\nu\sigma} + \nabla_{\nu}\delta g_{\mu\sigma} - \nabla_{\sigma}\delta g_{\mu\nu}\Big), \quad (5.58)$$
$$\delta R_{\alpha\beta} = \nabla_{\rho}\delta\Gamma^{\rho}{}_{\alpha\beta} - \nabla_{\beta}\delta\Gamma^{\rho}{}_{\alpha\rho}, \quad (5.59)$$

which can be rewritten into the more useful form

$$\delta\Gamma^{\rho}{}_{\mu\nu} = \Big(g^{\rho\pi}\delta^{\xi}_{(\mu}\delta^{\lambda}_{\nu)} - \frac{1}{2}g^{\rho\xi}\delta^{\lambda}_{\mu}\delta^{\pi}_{\nu}\Big)\nabla_{\xi}\delta g_{\lambda\pi}, \quad (5.60a)$$
$$\delta R_{\alpha\beta} = \Theta^{\gamma\lambda\xi\epsilon\pi}{}_{\alpha\gamma\beta}\nabla_{\lambda}\nabla_{\xi}\delta g_{\epsilon\pi}, \quad (5.60b)$$

where $\Theta^{\gamma\lambda\xi\epsilon\pi}{}_{\alpha\gamma\beta}$ is defined in (5.42). By writing $\delta R = g^{\mu\nu}\delta R_{\mu\nu} - R^{\mu\nu}\delta g_{\mu\nu}$ we find that from (5.56) we can obtain

$$\delta U^{\mu\nu} = \Big[\mathfrak{A}^{\mu\nu\rho\sigma} - \mathfrak{B}^{\mu\nu}R^{\rho\sigma} - \mathfrak{C}^{\alpha\mu\nu}\nabla_{\alpha}R^{\rho\sigma} - \mathfrak{D}^{\alpha\beta\mu\nu}\nabla_{\alpha}\nabla_{\beta}R^{\rho\sigma}\Big]\delta g_{\rho\sigma} + \Big[\mathfrak{E}_{\alpha}{}^{\rho\sigma\mu\nu}\Big]\delta\Gamma^{\alpha}{}_{\rho\sigma}$$
$$+\Big[-\mathfrak{C}^{\alpha\mu\nu}R^{\rho\sigma} - 2\mathfrak{D}^{(\alpha\beta)\mu\nu}\nabla_{\beta}R^{\rho\sigma}\Big]\nabla_{\alpha}\delta g_{\rho\sigma} + \Big[-\mathfrak{D}^{\alpha\beta\mu\nu}R^{\rho\sigma}\Big]\nabla_{\alpha}\nabla_{\beta}\delta g_{\rho\sigma}$$
$$+\Big[\mathfrak{B}^{\mu\nu}g^{\alpha\beta}\Big]\delta R_{\alpha\beta} + \Big[\mathfrak{C}^{\zeta\mu\nu}g^{\alpha\beta}\Big]\nabla_{\zeta}\delta R_{\alpha\beta} + \Big[\mathfrak{D}^{\zeta\kappa\mu\nu}g^{\alpha\beta}\Big]\nabla_{\zeta}\nabla_{\kappa}\delta R_{\alpha\beta}, \quad (5.61)$$

and if we insert (5.60) into this we obtain

$$\delta U^{\mu\nu} = \Big\{\mathfrak{A}^{\mu\nu\rho\sigma} - \mathfrak{B}^{\mu\nu}R^{\rho\sigma} - \mathfrak{C}^{\alpha\mu\nu}\nabla_{\alpha}R^{\rho\sigma} - \mathfrak{D}^{\alpha\beta\mu\nu}\nabla_{\alpha}\nabla_{\beta}R^{\rho\sigma}\Big\}\delta g_{\rho\sigma}$$

$$+\left\{\mathfrak{E}_{\xi}{}^{\lambda\beta\mu\nu}\left[g^{\xi\sigma}\delta^{\alpha}_{(\lambda}\delta^{\rho}_{\pi)}-\frac{1}{2}g^{\xi\alpha}\delta^{\rho}_{\lambda}\delta^{\sigma}_{\pi}\right]-\mathfrak{C}^{\alpha\mu\nu}R^{\rho\sigma}-2\mathfrak{D}^{(\alpha\beta)\mu\nu}\nabla_{\beta}R^{\rho\sigma}\right\}\nabla_{\alpha}\delta g_{\rho\sigma}$$

$$+\left\{\mathfrak{B}^{\mu\nu}g^{\alpha\beta}\Theta^{\gamma\lambda\xi\epsilon\pi}{}_{\alpha\gamma\beta}-\mathfrak{D}^{\alpha\beta\mu\nu}R^{\rho\sigma}\right\}\nabla_{\lambda}\nabla_{\xi}\delta g_{\epsilon\pi}$$

$$+\left\{\mathfrak{C}^{\zeta\mu\nu}g^{\alpha\beta}\Theta^{\gamma\lambda\xi\epsilon\pi}{}_{\alpha\gamma\beta}\right\}\nabla_{\zeta}\nabla_{\lambda}\nabla_{\xi}\delta g_{\epsilon\pi}$$

$$+\left\{\mathfrak{D}^{\zeta\kappa\mu\nu}g^{\alpha\beta}\Theta^{\gamma\lambda\xi\epsilon\pi}{}_{\alpha\gamma\beta}\right\}\nabla_{\zeta}\nabla_{\kappa}\nabla_{\lambda}\nabla_{\xi}\delta g_{\epsilon\pi}, \tag{5.62}$$

which is clearly of the form (5.53).

5.5 Summary

In this chapter we have shown that a mapping exists between the generalized theories we introduced and studied in Chaps. 3 and 4, and explicit theories constructed from a background Lagrangian. We identified the tensors and coefficients which appear in a (3+1) decomposition of the Lagrangian for perturbations for first and second order scalar field theories (where we had to pick a particular scalar for the second order scalar field theory) by computing the Lagrangian for perturbations for explicit theories and comparing with the generalized expression. We also sketched out how generalized "tensor" theories fit into our formalism, where we computed the perturbed dark energy momentum tensor for an $F(R)$ theory.

We have shown that it is easier to work with the generalized theory than explicit theories. This is due to the more compact appearance of expressions, particularly the way in which the tensors in the Lagrangian for perturbations combine to construct the perturbed energy momentum tensor in the generalized theories.

The mere existence of the mapping significantly enhances the case for further study and development of generalized theories for the dark sector. Of course, it is key that the mapping between the explicit and generalized theories should be further developed.

Chapter 6
Connections to Massive Gravity

6.1 Introduction

The realization that the Universe appears to be accelerating has fuelled the search to alternative theories of gravity [1] as a possible explanation for what has become called *the dark sector*. In this paper we will focus on what is the simplest subset of such theories, in which the dark sector Lagrangian is only a function of the metric (and no extra derivatives thereof) following the approach discussed in [2]. The action for this type of theory is given by

$$S = \int d^4x\sqrt{-g}\left[R + 16\pi G\mathcal{L}_{\rm m} - 2\mathcal{L}_{\rm d}(g_{\mu\nu})\right]. \tag{6.1}$$

If $T_{\mu\nu}$ is the energy-momentum tensor for the matter sector Lagrangian $\mathcal{L}_{\rm m}$ and $U_{\mu\nu}$ is that associated with the dark sector Lagrangian $\mathcal{L}_{\rm d}$, then the Einstein equation is $G_{\mu\nu} = 8\pi GT_{\mu\nu} + U_{\mu\nu}$. Typically, we will be interested in spacetimes which are isotropic where $U_{\mu\nu} = \rho u_\mu u_\nu + P\gamma_{\mu\nu}$ can specified in terms of a density, ρ, and pressure, $P = w\rho$. There are two classes of theories which can be described by (6.1): elastic dark energy and theories of massive gravitons. Typically, theories of massive gravitons have been considered as a fundamental theory around Minkowski space-time, but it is also possible to think about masses for the gravitons as being induced by a some unknown effective physics encoded by $\mathcal{L}_{\rm d}$.

The study of massive gravity theories (see, for example, [3, 4]) has a long history. This started with the linearized theories of Pauli and Feirz [5], progressing to the studies of Boulware and Deser [6]. It has received a new lease of life in recent times with the proposal of non-linear dRGT massive gravity theory [7–10] and its connections to the Vainshtein screening mechanism [11–14]. Massive gravity theories are built upon the pretext that the resulting theory should be "ghost-free" [9, 15–21], and have begun to be studied in cosmological backgrounds [22–36]. Such theories are usually presumed to be Lorentz invariant which leads to the Pauli-Fierz tuning.

J. Pearson, *Generalized Perturbations in Modified Gravity and Dark Energy*, Springer Theses, DOI: 10.1007/978-3-319-01210-0_6,

Giving up Lorentz-invariance is another way to remove ghosts from the theory, as pointed out by [37, 38] and further studied in [3, 39–42].

Elastic dark energy (EDE) is an idea which has been developed from relativistic elasticity theory [43–49]. The basic concept is that the stress-energy component responsible for the dark energy has rigidity which stabililizes perturbations that would, if modelled as those of a perfect fluid, give rise to exponential growth in the density contrast. The framework was adapted for cosmological purposes in [50, 51] in order to provide a phenomenological model for domain wall as an explanation for accelerated expansion which have $P/\rho = w_{\rm dw} = -2/3$. However, in principle the equation of state parameter, w, is allowed to take "any" value so long as the rigidity modulus, μ, is sufficiently large. Indeed, the theory is well-defined in the limit $w \to -1$ where the elastic medium becomes a "cosmological constant" and $w \to 0$ and $\mu \to 0$ which corresponds to cold dark matter. The standard assumption, which we will use in this paper, is that the elasticity tensor is isotropic, but one can also construct anisotropic models [52, 53].

The aim of this chapter is to point out the connections between linearized massive gravity theories, EDE and the framework for linearized perturbations in the generalized models for the dark sector discussed in [2, 54]. The reason for this connection is that, at linearized order, one can represent all possible Lagrangians by a generalized function which is quadratic in the fields. In the specific case we are concerned with here, this is just a quadratic function of the metric which is parameterized by a rank-4 tensor. This has the same symmetries as an elasticity tensor, suggesting an interpretation of massive gravitons as creating an effective rigidity of spacetime. This tensor can split in a way which is compatible with the symmetries of the FRW spacetime, that is more general than the usual Pauli-Fierz case. We will make a survey of the possible mechanisms by which ghost modes can be removed, both Lorentz invariant and violating. The ghosts are associated with a breaking of reparameterization invariance and we show that its re-imposition leads to 3 interesting sub-cases, one of which is compatible with Lorentz invariance which is a cosmological constant and the other two which violate either time or spatial translation invariance. The one which violates spatial translation invariance happens to be the EDE model which we see is a Lorentz-violating ghost-free massive gravity theory.

6.1.1 Second Variation of the Einstein–Hilbert Action

We start by providing a formula for the kinetic term of gravity in General Relativity, and is obtained by computing $\lozenge^2 R$. This is a rather involved calculation, which we do not provide any details of here. When theories for gravitational waves are studied in the literature, the Einstein-Hilbert action is usually written down in flat backgrounds or spacetimes with constant curvature, so that the gravitational field equations at background order can be used replace all Ricci tensors with the Ricci scalar and the metric. We will not introduce either of these simplifications: the variation of the metric is performed about an arbitrary background.

The Einstein-Hilbert action is given by

$$S = \int \mathrm{d}^4x \sqrt{-g}\, R. \tag{6.2}$$

The second variation of the action can be written using the measure weighted variation as

$$\delta^2 S = \int \mathrm{d}^4x \sqrt{-g}\, \lozenge^2 R, \tag{6.3}$$

where the integrand, $\lozenge^2 R$, is given by

$$\begin{aligned}\lozenge^2 R &\equiv \frac{1}{\sqrt{-g}}\delta^2(\sqrt{-g}R)\\ &= \delta^2 R + g^{\mu\nu}\delta g_{\mu\nu}\delta R + \frac{1}{4}R\Big(g^{\mu\nu}g^{\alpha\beta} - 2g^{\mu(\alpha}g^{\beta)\nu}\Big)\delta g_{\mu\nu}\delta g_{\alpha\beta}.\end{aligned} \tag{6.4}$$

The quantity $\lozenge^2 R$ has the interpretation of being the Lagrangian for gravitational waves; because we used the Einstein-Hilbert action, these are waves in Einstein gravity only (later on we will give an expression for the Lagrangian for gravitational waves in $F(R)$ theories). Our metric is

$$g_{\mu\nu} + \delta g_{\mu\nu}, \qquad g^{\mu\nu} + \delta g^{\mu\nu}, \qquad \delta g^{\mu\nu} = -g^{\mu(\alpha}g^{\beta)\nu}\delta g_{\alpha\beta}. \tag{6.5}$$

We will also write $\delta g^{\mu}{}_{\nu} = g^{\mu\alpha}\delta g_{\alpha\nu}$. We must be careful when comparing results to the literature: sometimes $h_{\mu\nu} = \delta g_{\mu\nu}$ is written but they also set $h^{\mu\nu} = -\delta g^{\mu\nu}$, which will induce some minus-sign discrepancies. By careful calculation, one finds that

$$\begin{aligned}\lozenge^2 R = {}& \nabla^{\lambda}\delta d g_{\mu\nu}\left[\frac{1}{2}g^{\mu\nu}\nabla_{\lambda}\delta g^{\alpha}{}_{\alpha} + \frac{1}{2}\nabla_{\lambda}\delta g^{\mu\nu} + \nabla^{\mu}\delta g^{\nu}{}_{\lambda} - g^{\mu\nu}\nabla^{\alpha}\delta g_{\alpha\lambda}\right]\\ &+\frac{1}{4}\Big[4R^{\alpha(\mu}g^{\nu)\beta} + 4R^{\beta(\mu}g^{\nu)\alpha} - 2(g^{\mu\nu}R^{\alpha\beta} + g^{\alpha\beta}R^{\mu\nu})\\ &\qquad + R(g^{\mu\nu}g^{\alpha\beta} - 2g^{\mu(\alpha}g^{\beta)\nu})\Big]\delta g_{\mu\nu}\delta g_{\alpha\beta}.\end{aligned} \tag{6.6}$$

It is to be noted that the final line is a mass-term for the gravitational fluctuation; it is rather trivial to realise that the mass-term may be space and/or time dependent (obviously, depending on the symmetry of the background). The expression (6.6) is clearly encompassed by the second order Lagrangian we wrote down for the metric derivative theory (4.22) where, by comparison with (6.6), we see that the mixing term vanishes, $\mathcal{P} = 0$.

We will now write $\lozenge^2 R$ for two special cases of the background spacetime, to check that our general formula reproduces simpler formulae usually presented in the literature.

6.1.1.1 Minkowski Background

In a Minkowski background t he Ricci scalar and tensor vanish so that (6.6) simplifies to become

$$\lozenge^2 R = \partial^\lambda \delta g_{\mu\nu}\left[\frac{1}{2}g^{\mu\nu}\partial_\lambda \delta g^\alpha{}_\alpha + \frac{1}{2}\partial_\lambda \delta g^{\mu\nu} + \partial^\mu \delta g^\nu{}_\lambda - g^{\mu\nu}\partial^\alpha \delta g_{\alpha\lambda}\right]. \quad (6.7)$$

Notice that the gravitational fluctuations are now mass-less; they can become massive when, for example, the Pauli-Feirz mass-term is inserted, although this is a deviation from Einstein gravity. We now set $\delta g_{\mu\nu} = h_{\mu\nu}, \delta g^{\mu\nu} = -h^{\mu\nu}$ to write this in the more familiar form,

$$\lozenge^2 R = \partial^\lambda h_{\mu\nu}\left[\frac{1}{2}g^{\mu\nu}\partial_\lambda h - \frac{1}{2}\partial_\lambda h^{\mu\nu} + \partial^\mu h^\nu{}_\lambda - g^{\mu\nu}\partial^\alpha h_{\alpha\lambda}\right]. \quad (6.8)$$

This agrees with the formula given by [4, 55].

6.1.1.2 Universe with Constant Curvature

In the presence of a cosmological constant and absence of matter sources, the gravitational Lagrangian is $\mathcal{L} = R - 2\Lambda$, and the field equations are $R_{\mu\nu} - \frac{1}{2}Rg_{\mu\nu} + \Lambda g_{\mu\nu} = 0$. The trace of the field equation yields

$$\Lambda = \frac{1}{4}R, \quad (6.9)$$

so that the field equations become

$$R_{\mu\nu} = \frac{1}{4}g_{\mu\nu}R. \quad (6.10)$$

The second variation of the gravitational action (with a cosmological constant) is given by

$$\delta^2 S = \int \mathrm{d}^4x\sqrt{-g}\left[\lozenge^2 R - 2\lozenge^2\Lambda\right]. \quad (6.11)$$

Inserting the field equation (6.10) into the general formula (6.6) yields the second measure weighted variation of the Ricci scalar for a Universe with constant curvature:

$$\lozenge^2 R = \nabla^\lambda \delta g_{\mu\nu}\left[\frac{1}{2}g^{\mu\nu}\nabla_\lambda \delta g + \frac{1}{2}\nabla_\lambda \delta g^{\mu\nu} + \nabla^\mu \delta g^\nu{}_\lambda - g^{\mu\nu}\nabla^\alpha \delta g_{\alpha\lambda}\right]. \quad (6.12)$$

Notice that the specification (6.10) caused the entire mass term from (6.6) to vanish: the Ricci scalar itself does not contribute to the mass of the graviton in a universe with constant curvature. It is a simple matter to compute the second measure weighted variation of a constant:

$$\begin{aligned}\lozenge^2\Lambda &= \frac{1}{4}\Lambda\left(g^{\mu\nu}g^{\alpha\beta} - 2g^{\mu(\alpha}g^{\beta)\nu}\right)\delta g_{\mu\nu}g_{\alpha\beta} \\ &= \frac{1}{16}R\left(g^{\mu\nu}g^{\alpha\beta} - 2g^{\mu(\alpha}g^{\beta)\nu}\right)\delta g_{\mu\nu}\delta g_{\alpha\beta},\end{aligned} \tag{6.13}$$

where in the last equality we used (6.9). We now combine (6.12) and (6.13) to yield

$$\begin{aligned}\delta^2 S = \int d^4x\sqrt{-g}\,\Bigg\{&\nabla^\lambda \delta g_{\mu\nu}\left(\frac{1}{2}g^{\mu\nu}\nabla_\lambda \delta g + \frac{1}{2}\nabla_\lambda \delta g^{\mu\nu} + \nabla^\mu \delta g^\nu{}_\lambda - g^{\mu\nu}\nabla^\alpha \delta g_{\alpha\lambda}\right) \\ &+\frac{1}{4}R\left(g^{\mu(\alpha}g^{\beta)\nu} - \frac{1}{2}g^{\mu\nu}g^{\alpha\beta}\right)\delta g_{\mu\nu}\delta g_{\alpha\beta}\Bigg\}.\end{aligned} \tag{6.14}$$

Notice that the gravitational waves are now massive, with their mass being given by the value of the (constant) curvature.

6.1.2 Generalization to $F(R)$ Gravity

We now generalize to an $F(R)$ theory; this severely complicates the calculation. The Lagrangian density for gravitational waves in an $F(R)$ theory is given by

$$\begin{aligned}\lozenge^2 F(R) = {}& F'\delta^2 R + F''\delta R\delta R + F' g^{\mu\nu}\delta R\delta g_{\mu\nu} \\ &+\frac{1}{4}F\left(g^{\mu\nu}g^{\alpha\beta} - 2g^{\mu(\alpha}g^{\beta)\nu}\right)\delta g_{\mu\nu}\delta g_{\alpha\beta}.\end{aligned} \tag{6.15}$$

The presence of the functions F', F'' infront of the variations will not allow an "easy" removal of total derivatives. Also notice that the F'' represents a "new" term in the action density. Inserting expressions to the point of writing down variations in the Ricci tensor, we obtain

$$\begin{aligned}\lozenge^2 F(R) = {}& F'\left[g^{\mu\nu}\delta^2 R_{\mu\nu} - 2g^{\mu(\alpha}g^{\beta)\nu}\delta g_{\alpha\beta}\delta R_{\mu\nu} + g^{\mu\nu}g^{\alpha\beta}\delta R_{\alpha\beta}\delta g_{\mu\nu}\right] \\ &+F''\left[g^{\mu\nu}g^{\alpha\beta}\delta R_{\alpha\beta}\delta R_{\mu\nu} - R^{\mu\nu}g^{\alpha\beta}\delta R_{\alpha\beta}\delta g_{\mu\nu} - g^{\mu\nu}R^{\alpha\beta}\delta R_{\mu\nu}\delta g_{\alpha\beta}\right] \\ &+\Bigg[\tfrac{1}{2}F'\left(2R^{\alpha(\mu}g^{\nu)\beta} + 2R^{\beta(\mu}g^{\nu)\alpha} - g^{\mu\nu}R^{\alpha\beta} - g^{\alpha\beta}R^{\mu\nu}\right) \\ &\qquad+\tfrac{1}{4}F\left(g^{\mu\nu}g^{\alpha\beta} - 2g^{\mu(\alpha}g^{\beta)\nu}\right) + F''R^{\mu\nu}R^{\alpha\beta}\Bigg]\delta g_{\alpha\beta}\delta g_{\mu\nu}.\end{aligned} \tag{6.16}$$

Whilst we have not explicitly written out the variations of the Ricci tensor, it is clear that they will not contribute to the mass term. We have obtained an expression for the mass of gravitational waves in $F(R)$ gravity.

6.2 The General "Metric Only" Theory

The action which will give linearized field equations for the perturbed field variables is given by

$$S_{\{2\}} = \int \mathrm{d}^4x\sqrt{-g}\left[\Diamond^2 R + 16\pi G\Diamond^2\mathcal{L}_{\mathrm{m}} - 2\mathcal{L}_{\{2\}}\right]. \tag{6.17}$$

We use $\Diamond^2 Q$ to denote the second measure-weighted variation of the quantity Q, defined as $\Diamond^2 Q \equiv \frac{1}{\sqrt{-g}}\delta^2(\sqrt{-g}Q)$. $\mathcal{L}_{\{2\}}$ is the Lagrangian for dark sector perturbations, given by

$$\mathcal{L}_{\{2\}} = \frac{1}{8}\mathcal{W}^{\mu\nu\alpha\beta}\delta g_{\mu\nu}\delta g_{\alpha\beta}, \tag{6.18}$$

where $\delta g_{\mu\nu}$ is the metric fluctuation. This clearly looks like a mass term for the metric perturbations. The tensor

$$\mathcal{W}^{\mu\nu\alpha\beta} = \mathcal{W}^{(\mu\nu)(\alpha\beta)} = \mathcal{W}^{\alpha\beta\mu\nu} \tag{6.19}$$

is the *mass matrix* determining how the components of $\delta g_{\mu\nu}$ mix to provide the mass; all linearized massive gravities are encoded by choices of $\mathcal{W}$. Hence, the complete linearized theory we study is

$$S_{\{2\}} = \int \mathrm{d}^4x\sqrt{-g}\left[\Diamond^2 R + 16\pi G\Diamond^2\mathcal{L}_{\mathrm{m}} - \tfrac{1}{4}\mathcal{W}^{\mu\nu\alpha\beta}\delta g_{\mu\nu}\delta g_{\alpha\beta}\right]. \tag{6.20}$$

This theory contains metric fluctuations which have a kinetic term, and a mass term. We will see that the spatial components of $\mathcal{W}$ can be interpreted as an elasticity tensor.

To isolate the degrees of freedom in the theory, $\delta g_{\mu\nu}$ can be decomposed as

$$\delta g_{\mu\nu} = h_{\mu\nu} + 2\nabla_{(\mu}\xi_{\nu)}. \tag{6.21}$$

In the parlance of [46, 47, 56, 57] $h_{\mu\nu}$ is the Eulerian metric perturbation and ξ_μ is a vector field representing possible coordinate transformations. In standard General Relativity, the action is independent of ξ_μ, but in more general theories this can become a physical field. This formulation is equivalent to what is sometimes called the Stuckelberg trick [4, 21, 58]. Inserting (6.21) into the action (6.20) and integrating by parts reveals that

$$S_{\{2\}} = \int \mathrm{d}^4x\sqrt{-g}\left[\left(\delta_{\mathrm{E}}G^{\mu\nu} - 8\pi G\delta_{\mathrm{E}}T^{\mu\nu} - \delta_{\mathrm{E}}U^{\mu\nu}\right)h_{\mu\nu} + 2\xi_{(\mu}\delta_{\mathrm{E}}(\nabla_{\nu)}U^{\mu\nu})\right], \tag{6.22}$$

where the variational operator "δ_{E}" denotes that the quantity is evaluated with the metric perturbation variable $h_{\mu\nu}$ (rather than $\delta g_{\mu\nu}$), and where we have defined the perturbed dark energy-momentum tensor

$$\delta_{\mathrm{E}}U^{\mu\nu} = -\tfrac{1}{2}(\mathcal{W}^{\mu\nu\alpha\beta} + U^{\mu\nu}g^{\alpha\beta})\delta g_{\alpha\beta} - \xi^{\alpha}\nabla_{\alpha}U^{\mu\nu} + 2U^{\alpha(\mu}\nabla_{\alpha}\xi^{\nu)}. \tag{6.23}$$

It is now a simple matter to obtain the functional derivatives of the action with respect to the perturbed metric $h_{\mu\nu}$ and ξ^{μ}-fields,

$$\frac{\hat{\delta}}{\hat{\delta}h_{\mu\nu}}S_{\{2\}} = \delta_{\mathrm{E}}G^{\mu\nu} - 8\pi G\delta_{\mathrm{E}}T^{\mu\nu} - \delta_{\mathrm{E}}U^{\mu\nu} = 0, \tag{6.24a}$$

$$\frac{\hat{\delta}}{\hat{\delta}\xi_{\mu}}S_{\{2\}} = \delta_{\mathrm{E}}(\nabla_{\nu}U^{\mu\nu}) = 0. \tag{6.24b}$$

The variational principle was used to demand that these expressions vanish, yielding the perturbed gravitational field equations and perturbed conservation equation respectively. Using (6.23) to evaluate (6.24b) yields

$$\begin{aligned} L^{\mu\nu\alpha\beta}\nabla_{\mu}\nabla_{\alpha}\xi_{\beta} + (\nabla_{\mu}\mathcal{W}^{\mu\nu\alpha\beta})\nabla_{\alpha}\xi_{\beta} + (\nabla_{\mu}\nabla_{\alpha}U^{\mu\nu})\xi^{\alpha} \\ = -\tfrac{1}{2}\left((\nabla_{\mu}\mathcal{W}^{\mu\nu\alpha\beta})h_{\alpha\beta} + P^{\mu\nu\alpha\beta}\nabla_{\mu}h_{\alpha\beta}\right), \end{aligned} \tag{6.25}$$

where we defined the *effective metric* $L^{\mu\nu\alpha\beta}$ and *derivative-coupling* $P^{\mu\nu\alpha\beta}$ terms,

$$L^{\mu\nu\alpha\beta} \equiv \mathcal{W}^{\mu\nu\alpha\beta} + U^{\mu\nu}g^{\alpha\beta} - 2U^{\alpha(\mu}g^{\nu)\beta}, \tag{6.26a}$$

$$P^{\mu\nu\alpha\beta} \equiv \mathcal{W}^{\mu\nu\alpha\beta} + U^{\alpha\beta}g^{\mu\nu} - 2g^{\nu\beta}U^{\alpha\mu}. \tag{6.26b}$$

We impose spatial isotropy upon the background with the $(3+1)$ decomposition and in doing so we will obtain the most general linearized massive gravity Lagrangian compatible with spatial isotropy of the background. We foliate the 4D spacetime by 3D sheets with a time-like unit vector, u_{μ}, being everywhere orthogonal to the sheets. The 4D spacetime has metric $g_{\mu\nu}$, and the 3D sheets have metric $\gamma_{\mu\nu}$. The $(3+1)$ decomposition of the 4D metric is $g_{\mu\nu} = \gamma_{\mu\nu} - u_{\mu}u_{\nu}$, where $u^{\mu}u_{\mu} = -1$, $\gamma^{\mu\nu}u_{\mu} = 0$. This structure provides an extrinsic curvature tensor $K_{\mu\nu} = K_{(\mu\nu)}$ on the 3D sheets, given by $K_{\mu\nu} = \nabla_{\mu}u_{\nu}$ and satisfying $u^{\mu}K_{\mu\nu} = 0$ (the extrinsic curvature tensor is given by $K_{\mu\nu} = \frac{1}{3}K\gamma_{\mu\nu}$). We define "time" and "space" differentiation as the derivative operator projected along the time and space directions,

$$\dot{\psi} \equiv u^{\mu}\nabla_{\mu}\psi, \qquad \overline{\nabla}_{\mu}\psi \equiv \gamma^{\nu}{}_{\mu}\nabla_{\nu}\psi. \tag{6.27}$$

Using this technology we decompose the gradient of a scalar into two orthogonal terms,

$$\nabla_{\mu}\psi = -\dot{\psi}u_{\mu} + \overline{\nabla}_{\mu}\psi. \tag{6.28}$$

This enables us to find the values of two useful "kinetic scalars",

$$\nabla^{\mu}\psi\nabla_{\mu}\psi = -\dot{\psi}^2 + \overline{\nabla}^{\mu}\psi\overline{\nabla}_{\mu}\psi, \tag{6.29a}$$

$$\Box\psi \equiv \nabla^{\mu}\nabla_{\mu}\psi = -\ddot{\psi} + \overline{\nabla}^{\mu}\overline{\nabla}_{\mu}\psi. \tag{6.29b}$$

The last term of each expression simply selects out the spatial derivatives of the scalar field. Another useful application of the (3 + 1) decomposition is to find all the freedom in a tensor which is compatible with spatial isotropy of the background spacetime.

We use the (3 + 1) decomposition to isolate the components of the perturbed metrics by writing

$$\delta g_{\mu\nu} = 2\Phi u_{\mu}u_{\nu} + 2N_{(\mu}u_{\nu)} + \bar{H}_{\alpha\beta}\gamma^{\alpha}{}_{\mu}\gamma^{\beta}{}_{\nu}, \tag{6.30a}$$

$$h_{\mu\nu} = 2\phi u_{\mu}u_{\nu} + 2n_{(\mu}u_{\nu)} + \bar{h}_{\alpha\beta}\gamma^{\alpha}{}_{\mu}\gamma^{\beta}{}_{\nu}, \tag{6.30b}$$

and we isolate the time-like and space-like components of the vector field via

$$\xi_{\mu} = -\chi u_{\mu} + \omega_{\mu}, \tag{6.30c}$$

where $N^{\mu}u_{\mu} = n^{\mu}u_{\mu} = 0$, $u^{\mu}\bar{H}_{\mu\nu} = u^{\mu}\bar{h}_{\mu\nu} = 0$ and $u^{\mu}\omega_{\mu} = 0$.

In [2] we showed that the general decomposition of the mass-matrix $\mathcal{W}$ compatible with spatial isotropy is

$$\begin{aligned} \mathcal{W}^{\mu\nu\alpha\beta} &= A_{\mathcal{W}}u^{\mu}u^{\nu}u^{\alpha}u^{\beta} + B_{\mathcal{W}}\left(u^{\mu}u^{\nu}\gamma^{\alpha\beta} + u^{\alpha}u^{\beta}\gamma^{\mu\nu}\right) \\ &\quad +2C_{\mathcal{W}}\left(\gamma^{\alpha(\mu}u^{\nu)}u^{\beta} + \gamma^{\beta(\mu}u^{\nu)}u^{\alpha}\right) \\ &\quad +D_{\mathcal{W}}\gamma^{\mu\nu}\gamma^{\alpha\beta} + 2E_{\mathcal{W}}\gamma^{\mu(\alpha}\gamma^{\beta)\nu}, \end{aligned} \tag{6.31}$$

where there are only 5 free functions which only depend on time. Using the mass matrix (6.31) and (6.30a) in the Lagrangian (6.18) yields

$$8\mathcal{L}_{\{2\}} = 4A_{\mathcal{W}}\Phi^2 + 4B_{\mathcal{W}}\Phi\bar{H} + 2C_{\mathcal{W}}N_\alpha N^\alpha + D_{\mathcal{W}}\bar{H}^2 + 2E_{\mathcal{W}}\bar{H}^{\alpha\beta}\bar{H}_{\alpha\beta}. \quad (6.32)$$

This can be written in terms of "graviton masses" (see e.g., [3]) where the free functions $\{A_{\mathcal{W}}, \ldots, E_{\mathcal{W}}\}$ are given by

$$2\mathcal{L}_{\{2\}} = m_0^2(\delta g_{00})^2 + 2m_1^2(\delta g_{0i})^2 - m_2^2(\delta g_{ij})^2 + m_3^2(\delta g_{ii})^2 - 2m_4^2\delta g_{00}\delta g_{ii} \quad (6.33)$$

with $A_{\mathcal{W}} = m_0^2$, $B_{\mathcal{W}} = -2m_4^2$, $C_{\mathcal{W}} = 4m_1^2$, $D_{\mathcal{W}} = 4m_3^2$ and $E_{\mathcal{W}} = -2m_2^2$. One should keep in mind, therefore, that when we talk about the $\{A_{\mathcal{W}}, \ldots, E_{\mathcal{W}}\}$ we are actually talking about the graviton masses m_i^2, albeit ones that depend on time. The values of these masses for a "Goldstone" theory are given in [59] and those which are induced by perturbations in scalar fields in [2].

Using (6.30) to evaluate (6.21) yields

$$\Phi = \phi + \dot{\chi}, \qquad N_\alpha = n_\alpha - \dot{\varpi}_\alpha - \overline{\nabla}_\alpha\chi, \quad (6.34a)$$

$$\bar{H}_{\alpha\beta} = \bar{h}_{\alpha\beta} + 2\overline{\nabla}_{(\alpha}\omega_{\beta)} - \tfrac{2}{3}K\gamma_{\alpha\beta}\chi - \tfrac{2}{3}K\omega_{(\alpha}u_{\beta)}, \quad (6.34b)$$

where $\dot{\varpi}_\alpha \equiv \left(\dot{\omega}_\alpha - \frac{1}{3}K\omega_\alpha\right)$. Substituting (6.30a) into the action (6.20) one finds the absence of $\dot{\phi}^2$ and $\dot{n}_\alpha^2$ terms in the kinetic part of the Einstein-Hilbert Lagrangian; this can also be seen in results given by [3, 37, 41] and in the ADM formulation [60]. ϕ and n_α are now Lagrange multipliers whose equations of motion are constraint equations, allowing them to be eliminated. Using re-definitions of the coefficients, we can effectively set $\phi = 0$ and $n_\alpha = 0$ which is equivalent to choice of the synchronous gauge. We will make this choice in what follows.

The two independent components of the equation of motion (6.25), after inserting the $(3+1)$-decomposition, are given by

$$\begin{aligned}
&\left[A_{\mathcal{W}} + \rho\right]\ddot{\chi} + \left[\dot{A}_{\mathcal{W}} + \mathcal{H}(4A_{\mathcal{W}} + \rho - 3P)\right]\dot{\chi} + \left[P + C_{\mathcal{W}}\right]\nabla^2\chi \\
&\quad -\big[\mathcal{H}(3\dot{P} + 2\dot{\rho} - \dot{A}_{\mathcal{W}} + 3\dot{B}_{\mathcal{W}}) - (2A_{\mathcal{W}} + 5\rho + 3P - 3B_{\mathcal{W}} - 9D_{\mathcal{W}} - 6E_{\mathcal{W}})\mathcal{H}^2 \\
&\qquad + (2\rho + 3B_{\mathcal{W}} - A_{\mathcal{W}})\tfrac{\ddot{a}}{a} + \ddot{\rho}\big]\chi \\
&\quad -\left[\dot{B}_{\mathcal{W}} + \mathcal{H}(3B_{\mathcal{W}} + 3D_{\mathcal{W}} + 2E_{\mathcal{W}} - 2P)\right]\partial_i\omega^i + \left[B_{\mathcal{W}} + C_{\mathcal{W}}\right]\partial_i\dot{\omega}^i \\
&\qquad = \frac{1}{2}\left[\dot{B}_{\mathcal{W}} + \mathcal{H}(3B_{\mathcal{W}} + 3D_{\mathcal{W}} + 2E_{\mathcal{W}} - 2P)\right]h + \frac{1}{2}\left[B_{\mathcal{W}} - P\right]\dot{h},
\end{aligned} \quad (6.35a)$$

$$\begin{aligned}
&\left[\rho - C_{\mathcal{W}}\right]\ddot{\omega}^i - \left[\dot{C}_{\mathcal{W}} + \mathcal{H}(4C_{\mathcal{W}} - \rho + 3P)\right]\dot{\omega}^i - \left[E_{\mathcal{W}} - P\right]\nabla^2\omega^i - \left[E_{\mathcal{W}} + D_{\mathcal{W}}\right]\partial^i\partial_k\omega^k \\
&\quad + \left[\dot{C}_{\mathcal{W}} - \mathcal{H}(3D_{\mathcal{W}} - B_{\mathcal{W}} + 2E_{\mathcal{W}} - 4C_{\mathcal{W}} - 2P)\right]\partial^i\chi + \left[B_{\mathcal{W}} + C_{\mathcal{W}}\right]\partial^i\dot{\chi} \\
&\qquad = -\left[P - E_{\mathcal{W}}\right]\partial_j h^{ij} + \frac{1}{2}\left[P + D_{\mathcal{W}}\right]\partial^i h.
\end{aligned} \quad (6.35b)$$

ρ and P are the density and pressure coming from the dark fluid (i.e. the components of the background dark energy momentum tensor $U_{\mu\nu}$). The benefit of using the (3+1) decomposition has become apparent: we are able to identify the degrees of freedom. There are the tensor degrees of freedom h_+ and $h_\times$ which are present in standard General Relativity, a vector degree of freedom ω^i, that can be split into a longitudinal (scalar) and two transverse (vector) degrees of freedom, and a scalar degree of freedom χ. Therefore, prima facie there are 6 extra degrees of freedom. As we will discuss below either χ or ω^i can be a ghost and therefore the coefficients must be chosen to suppress one or both of them. In the case where χ is the ghost then there are 5 degrees of freedom with those in ω^i being split into a longitudinal (scalar) and two transverse (vector) degrees of freedom. If ω^i is the ghost then there are only 3 degrees of freedom.

6.3 Mechanisms for the Elimination of Ghosts

From (6.22) we see that the kinetic terms of the ω_μ and χ fields enter the theory via

$$\frac{1}{2}S_{\{2\}} \supset \int \mathrm{d}^4x\sqrt{-g}\Big[(A_{\mathcal{W}}+\rho)\dot{\chi}^2 + (C_{\mathcal{W}}+P)\overline{\nabla}_\mu\chi\overline{\nabla}^\mu\chi + (C_{\mathcal{W}}-\rho)\dot{\omega}_\mu\dot{\omega}^\mu + (D_{\mathcal{W}}+P)(\overline{\nabla}_\mu\omega^\mu)^2 + 2(E_{\mathcal{W}}-P)\overline{\nabla}_\mu\omega_\nu\overline{\nabla}^\mu\omega^\nu\Big]. \tag{6.36}$$

Let us now focus on the standard scenario of perturbations around Minkowski space-time when both $\rho = P = 0$. If $A_{\mathcal{W}} > 0$, then $C_{\mathcal{W}} < 0$ is required for χ to have a kinetic term with the "proper sign". But if this is the case then ω_μ has a kinetic term with the "wrong sign" (the same is true if $A_{\mathcal{W}} < 0$). Hence, one of χ, ω_μ must be a ghost. There are a few ways to get out of this.

First, one can make the coefficient of $\dot{\chi}^2$ vanish by setting $A_{\mathcal{W}} = 0$, which removes χ as a propagating mode and the equation of motion is a constraint that can be enforced in a Lorentz invariant theory. Secondly, one could set $C_{\mathcal{W}} = 0$, since that would remove ω_μ as a propagating mode. Finally, one could set $\chi \equiv 0$ directly which requires a breaking of Lorentz invariance (since we will be manually forcing one of the four components of a 4-vector to zero).

When $A_{\mathcal{W}} = 0$, there is no $\dot{\chi}^2$ term in the Lagrangian and the equation of motion simply becomes a constraint equation specifying the value of χ from the other field variables. This can be back-substituted into the action so that the theory explicitly does not contain the χ-field. From our presentation it is clear that in this case the ghost can be identified with the time-like degree of freedom χ. We have been able to deduce this since we used a $(3+1)$ decomposition. Performing a transverse-longitudinal decomposition does not aid the identification of the ghost.

If we choose the 5 parameters in (6.31) to be given by $A_{\mathcal{W}} = X + Y$, $B_{\mathcal{W}} = -X$, $C_{\mathcal{W}} = -\frac{1}{2}Y$, $D_{\mathcal{W}} = X$, $E_{\mathcal{W}} = \frac{1}{2}Y$, then the theory is Lorentz invariant and the mass-matrix is be given by

$$\mathcal{W}^{\mu\nu\alpha\beta} = Xg^{\mu\nu}g^{\alpha\beta} + Yg^{\mu(\alpha}g^{\beta)\nu}, \tag{6.37}$$

where X, Y are two parameters that are dependent on background field variables only. The standard route for isolating the ghost in the Lorentz invariant theory [58] decomposes ξ^{μ} into its transverse and longitudinal modes as

$$\xi_{\mu} = \zeta_{\mu} + \nabla_{\mu}\kappa, \tag{6.38}$$

where $\nabla_{\mu}\zeta^{\mu} = 0$ and κ is a scalar field. For Minkowski background spacetime, inserting (6.21), (6.38) and the mass-matrix (6.37) into the Lagrangian (6.18), whilst assuming that X, Y are constants, yields

$$\begin{aligned}\mathcal{L}_{\{2\}} = \frac{1}{8}(Xh^2 + Yh^{\mu\nu}h_{\mu\nu}) + \frac{1}{2}Yh^{\mu\nu}\partial_{(\mu}\zeta_{\nu)} + \frac{1}{2}(Xh\Box\kappa + Yh^{\mu\nu}\partial_{\mu}\partial_{\nu}\kappa)\\ +\frac{1}{2}Y\partial^{\mu}\zeta^{\nu}\partial_{(\mu}\zeta_{\nu)} + Y\partial^{(\mu}\zeta^{\nu)}\partial_{\mu}\partial_{\nu}\kappa + \frac{1}{2}(X+Y)(\Box\kappa)^2.\end{aligned} \tag{6.39}$$

This expression has made the ghost problem manifest in the Lorentz invariant language. The existence of the last term, $(X+Y)(\Box\kappa)^2$, means that ghosts are inevitable (see e.g. [18, 61, 62]). The cure is to set $X = -Y$, which removes the problematic kinetic term, and leaves the Pauli-Feriz mass-term, $\mathcal{L}_{\{2\}} \supset \mathcal{L}_{\text{PF}} = h^2 - h^{\mu\nu}h_{\mu\nu}$. The parameter choice $X = -Y$ is called the *Pauli-Feirz tuning*, and will render massive gravitons ghost-free at linearized order on Minkowski backgrounds. In this case $A_{\mathcal{W}} = 0$, $B_{\mathcal{W}} = -X$, $C_{\mathcal{W}} = \frac{1}{2}X$, $D_{\mathcal{W}} = X$, $E_{\mathcal{W}} = -\frac{1}{2}X$ which is a special case of the more general situation discussed earlier.

Rather than retain Lorentz invariance and be forced to use the Pauli-Feirz tuning to remove the ghost, it has been suggest that one can just fix the field $u^{\mu}\xi_{\mu} = 0$ which implies that $\chi = 0$, removing it as a physical degree of freedom. Of course, this is not really a solution to the problem of the ghost, since we have just set the field to zero. However, as we will see in the next section that it is possible to impose a symmetry which is equivalent to this. The condition $u^{\mu}\xi_{\mu} = 0$ imposes an interesting structure upon the fields in theory when we use the transverse-longitudinal split language. Using this and (6.38) implies that $\dot{\kappa} = -u^{\mu}\zeta_{\mu}$ which can be differentiated to yield $\ddot{\kappa} = -u^{\mu}\dot{\zeta}_{\mu}$. This shows us that the κ-field (i.e. the longitudinal component of the ξ^{μ}-field) does not propagate. Instead, the nth time derivative of κ is replaced by the $(n-1)^{\text{th}}$ time derivative of ζ_{μ}. Evaluating the "kinetic scalars" (6.29) for the scalar κ, yields

$$\nabla^{\mu}\kappa\nabla_{\mu}\kappa = -u^{\mu}u^{\nu}\zeta_{\mu}\zeta_{\nu} + \overline{\nabla}^{\mu}\kappa\overline{\nabla}_{\mu}\kappa, \tag{6.40a}$$

$$\Box\kappa = u^{\mu}\dot{\zeta}_{\mu} + \overline{\nabla}^{\mu}\overline{\nabla}_{\mu}\kappa. \tag{6.40b}$$

Using (6.40b), the previously offensive term in (6.39) becomes

$$(X + Y)(\Box\kappa)^2 = (X + Y)(u^{\mu}u^{\nu}\dot{\zeta}_{\mu}\dot{\zeta}_{\nu} + 2u^{\alpha}\dot{\zeta}_{\alpha}\overline{\nabla}^{\mu}\overline{\nabla}_{\mu}\kappa + \overline{\nabla}^{\mu}\overline{\nabla}_{\mu}\kappa\overline{\nabla}^{\alpha}\overline{\nabla}_{\alpha}\kappa), \tag{6.41}$$

and we observe that the multiple derivatives of κ that are present are entirely spatial. The upshot is that there are no time-derivatives of the scalar κ left, and, *crucially* no second time-derivatives of κ in $\Box\kappa$. The term $(X + Y)(\Box\kappa)^2$ in (6.39) is now longer problematic, and does not require removal.

6.4 Imposing Reparameterization Invariance

A key aspect of the theories under consideration here is the spontaneous violation of reparameterization invariance. It is interesting to see under what conditions it can be be reimposed on the theory. Therefore, we consider how the vector field ξ^{μ} sources the perturbed gravitational field equations, and under what circumstances its components decouples from the field equations. From (6.17) the field equations for the metric are $\delta_{\mathrm{E}}G^{\mu\nu} = 8\pi G\delta_{\mathrm{E}}T^{\mu\nu} + \delta_{\mathrm{E}}U^{\mu\nu}$, where $\delta_{\mathrm{E}}U^{\mu\nu}$ is the dark energy momentum tensor and contains contributions to the field equations from the dark sector. In [2] we showed that

$$\delta_{\mathrm{E}}U^{\mu\nu} = -\frac{1}{2}(\mathcal{W}^{\mu\nu\alpha\beta} + U^{\mu\nu}g^{\alpha\beta})\delta g_{\alpha\beta} - \xi^{\alpha}\nabla_{\alpha}U^{\mu\nu} + 2U^{\alpha(\mu}\nabla_{\alpha}\xi^{\nu)}. \tag{6.42}$$

It is useful to note the terms in $\delta_{\mathrm{E}}U^{\mu\nu}$ which are present due to the background dark energy-momentum tensor $U^{\mu\nu}$. The components of $\delta_{\mathrm{E}}U^{\mu\nu}$ are written as perturbed fluid variables,

$$\delta_{\mathrm{E}}U^{\mu}{}_{\nu} = \delta\rho u^{\mu}u_{\nu} + 2(\rho + P)v^{(\mu}u_{\nu)} + \delta P\gamma^{\mu}{}_{\nu} + P\Pi^{\mu}{}_{\nu}, \tag{6.43}$$

where one can obtain

$$\delta\rho = \left[\dot{\rho} + \mathcal{H}(2\rho - A_{\mathcal{W}} + 3B_{\mathcal{W}})\right]\chi - (A_{\mathcal{W}} + \rho)\dot{\chi} + (\rho + B_{\mathcal{W}})\left(\frac{1}{2}h + \partial_i\omega^i\right), \tag{6.44a}$$

$$\delta P = -\left[\dot{P} + \mathcal{H}(2P - B_{\mathcal{W}} + 3D_{\mathcal{W}} + 2E_{\mathcal{W}})\right]\chi + (B_{\mathcal{W}} - P)\dot{\chi}$$
$$-\tfrac{1}{3}(P + 3D_{\mathcal{W}} + 2E_{\mathcal{W}})\left(\tfrac{1}{2}h + \partial_i\omega^i\right), \tag{6.44b}$$

$$(\rho + P)v^i = (P + C_{\mathcal{W}})\partial^i\chi + (\rho - C_{\mathcal{W}})\dot{\omega}^i, \tag{6.44c}$$

$$P\Pi^i{}_j = 2(P - E_{\mathcal{W}})\left(\tfrac{1}{2}h^i{}_j + \partial^{(i}\omega_{j)} - \tfrac{1}{3}\delta^i{}_j(\tfrac{1}{2}h + \partial_k\omega^k)\right). \tag{6.44d}$$

These effective fluid variables define how the components of the vector field ξ^μ sources the gravitational field equations. If one, or both, of the fields χ and ω^i does not appear in (6.44) then that field is not dynamical and hence can be completely ignored. It is clear that particular choices of the free functions in the mass-matrix it will be possible to achieve this. When one or both does not appear, it means that the theory is invariant under the symmetry associated with that field. Therefore, we can impose reparameterization invariance in three natural ways:

- ξ^μ-field decouples from the system when $\dot{\rho} + 3\mathcal{H}(\rho + P) = 0$, $\dot{P} + \mathcal{H}(P + 3D_{\mathcal{W}} + 2E_{\mathcal{W}}) = 0$, $A_{\mathcal{W}} = -\rho$, $B_{\mathcal{W}} = P$, $C_{\mathcal{W}} = -P$, $\rho = -B_{\mathcal{W}}$, $\rho = C_{\mathcal{W}}$, $P = E_{\mathcal{W}}$, $D_{\mathcal{W}} = -P$. Hence, in the "fully" reparameterization invariant case, where the theory is forced to be invariant under $x^\mu \to x^\mu + \xi^\mu$, the only values of ρ, P that are allowed are those which are provided by a cosmological constant, $\rho = -P$, and all perturbed fluid variables vanish. Neither the χ- nor the ω^i-field propagate.
- $\chi = u_\mu\xi^\mu$ field decouples from the system when the parameters satisfy $A_{\mathcal{W}} = -\rho$, $B_{\mathcal{W}} = P$, $C_{\mathcal{W}} = -P$, $\dot{\rho} + \mathcal{H}(2\rho - A_{\mathcal{W}} + 3B_{\mathcal{W}}) = 0$, $\dot{P} + \mathcal{H}(2P - B_{\mathcal{W}} + 3D_{\mathcal{W}} + 2E_{\mathcal{W}}) = 0$ from which we can deduce that $\dot{\rho} + 3\mathcal{H}(\rho + P) = 0$ and $\dot{P} + \mathcal{H}(P + 3D_{\mathcal{W}} + 2E_{\mathcal{W}}) = 0$. These equations appear to leave two coefficients, $D_{\mathcal{W}}$ and $E_{\mathcal{W}}$, unspecified. If we now define two parameters β and μ via $D_{\mathcal{W}} = \beta - P - \frac{2}{3}\mu$ and $E_{\mathcal{W}} = \mu + P$ then we find that $\beta = (\rho + P)\frac{\mathrm{d}P}{\mathrm{d}\rho}$ which is the definition of the relativistic bulk modulus and μ can then be interpreted as a rigidity modulus of an elastic medium. Hence, in the case where we impose time translation invariance, $t \to t + \chi$, but not spatial translation invariance, then we find that the theory must be EDE. The equations of motion (6.35) become

$$-3\mathcal{H}\big[\dot{P} + 3\beta\mathcal{H}\big]\chi = 0, \tag{6.45a}$$

$$\big[\rho + P\big]\ddot{\omega}^i + \big[\dot{P} + \mathcal{H}(P + \rho)\big]\dot{\omega}^i - \big[E_{\mathcal{W}} - P\big]\nabla^2\omega^i - \big[E_{\mathcal{W}} + D_{\mathcal{W}}\big]\partial^i\partial_k\omega^k$$
$$= -\big[P - E_{\mathcal{W}}\big]\partial_j h^{ij} + \tfrac{1}{2}\big[P + D_{\mathcal{W}}\big]\partial^i h. \tag{6.45b}$$

Note that (6.45a) vanishes for arbitrary values of χ since $\beta = (\rho + P)\dot{P}/\dot{\rho}$, and that there is a propagating vector degree of freedom, ω^i. (6.45b) is the equation of motion presented in [63]. In this case the mass-term for the gravitons is

$$\mathcal{L}_{\{2\}} = \tfrac{1}{8}\rho\left[(w^2 - \tfrac{2}{3}\hat{\mu})\bar{h}^2 + 2(w + \hat{\mu})\bar{h}_{\mu\nu}\bar{h}^{\mu\nu}\right], \tag{6.46}$$

where $w = P/\rho$, $\hat{\mu} \equiv \mu/\rho$. This case is equivalent to setting $A_{\mathcal{W}} = 0$ in the Minkowski space case. Since χ does not appear in (6.44) it is no longer a physical degree and there is no ghost.

- ω^i decouples when $\rho + B_{\mathcal{W}} = 0$, $P + 3D_{\mathcal{W}} + 2E_{\mathcal{W}} = 0$, $\rho = C_{\mathcal{W}}$, $P = E_{\mathcal{W}}$ from which we can deduce that $B_{\mathcal{W}} = -C_{\mathcal{W}} = -\rho$ and $D_{\mathcal{W}} = -E_{\mathcal{W}} = -P$. Therefore, we see that in the case where we impose spatial translation invariance, $x^i \rightarrow x^i + \xi^i$, but not time translation invariance, then the perturbations have some of the characteristics of massive scalar field theory as explained in [2]. The equations of motion (6.35) become

$$\begin{aligned}
&\big[A_{\mathcal{W}} + \rho\big]\ddot{\chi} + \big[\dot{A}_{\mathcal{W}} + \mathcal{H}(4A_{\mathcal{W}} + \rho - 3P)\big]\dot{\chi} + \big[\rho + P\big]\nabla^2\chi \\
&\quad + \big[(\dot{A}_{\mathcal{W}} + 3(A_{\mathcal{W}} - \rho - 2P)\mathcal{H})\mathcal{H} + (A_{\mathcal{W}} + 4\rho + 3P)\dot{\mathcal{H}}\big]\chi \\
&\quad = -\tfrac{1}{2}\big[\rho + P\big]\dot{h},
\end{aligned} \tag{6.47a}$$

$$\big[\dot{\rho} + 3\mathcal{H}(\rho + P)\big]\partial^i\chi = 0. \tag{6.47b}$$

Note that (6.47b) vanishes for arbitrary values of χ due to the background conservation equation. If we define the entropy

$$w\Gamma = \left(\frac{\delta P}{\delta\rho} - w\right)\delta\,, \tag{6.48}$$

and set $A_{\mathcal{W}} = [w + \epsilon(1 + w)]\rho$ we find that

$$w\Gamma = \left(\frac{1}{1+\epsilon} - w\right)\left[\delta - 3\mathcal{H}\left(\frac{w(1+\epsilon)+1}{w(1+\epsilon)-1}\right)(1+w)\theta\right]. \tag{6.49}$$

This implies that this theory is a scalar field theory with a non-standard kinetic term. The mass-term is given by

$$\mathcal{L}_{\{2\}} = -\tfrac{1}{8}w\rho\left[\bar{h}^2 - 2\bar{h}_{\mu\nu}\bar{h}^{\mu\nu}\right], \tag{6.50}$$

which we note does not satisfy the Pauli-Feirz tuning. This case is equivalent to setting $C_{\mathcal{W}} = 0$ in the Minkowski space case. Since ω^i does not appear in (6.44) it is no longer a physical degree and there is no ghost.

6.5 Discussion

It is well established in the literature, that can be ghosts in *general* massive gravity theories, and various methods have been devised to remove them. If one imposes Lorentz invariance, then one is forced to use the Pauli-Feirz tuning to excise the ghost. If one is willing to give up Lorentz invariance, then certain parameter choices allow for ghost-free massive gravity theories. We have shown that the theory (6.18) with the $(3+1)$-decomposition of the mass-matrix (6.31) imposed with just time-translation invariance constitutes a linearized theory with healthy Lorentz-violating massive gravitons with 5 physical degrees of freedom. That theory *is* exactly the EDE model previously discussed in the literature [63]. In addition, if one imposes spatial translation invariance then there is another ghost-free massive gravity theory with 3 degrees of freedom.

In terms of the EDE parameters and graviton masses, $A_{\mathcal{W}} = m_0^2 = -\rho$, $B_{\mathcal{W}} = -2m_4^2 = P$, $C_{\mathcal{W}} = 4m_1^2 = -P$, $D_{\mathcal{W}} = 4m_3^2 = \beta - P - \frac{2}{3}\mu$, $E_{\mathcal{W}} = -2m_2^2 = \mu + P$. The masses can be conveniently parameterized by some overall mass scale $M^2 \equiv \rho \sim H_0^2$ (since we wrote $U_{\mu\nu} = \rho u_\mu u_\nu + P\gamma_{\mu\nu}$, ρ has units of mass squared),

$$m_0^2 = -M^2, \quad m_1^2 = -\tfrac{1}{4}wM^2, \quad m_2^2 = -\tfrac{1}{2}(\hat{\mu} + w)M^2, \tag{6.51a}$$

$$m_3^2 = \tfrac{1}{4}(w^2 - \tfrac{2}{3}\hat{\mu})M^2, \quad m_4^2 = -\tfrac{1}{2}wM^2. \tag{6.51b}$$

We defined $\hat{\mu} \equiv \mu/\rho$ in analogy with $w = P/\rho$.

To connect to dark energy, we note that the fraction of the total energy density that is dark energy is linked to the mass scale M^2 via $\Omega_{\rm de} = M^2/(3H_0^2)$. Hence, we see that the "natural" scale for the masses in order for the modification of gravity to act as a source of cosmic acceleration is of the order the Hubble parameter, and are all multiplied by order unity "corrections" defined by two parameters which encode the properties of the elastic medium: its equation of state parameter w and shear modulus $\hat{\mu}$. The longitudinal and transverse sound speeds of EDE are [63]

$$c_{\rm s}^2 = w + \frac{\frac{4}{3}\hat{\mu}}{1+w}, \qquad c_{\rm v}^2 = \frac{\hat{\mu}}{1+w}. \tag{6.52}$$

Stability and subluminality require that $0 \le c_{\rm i}^2 \le 1$, so that we have the following constraints on the possible values of $\hat{\mu}$

$$-\tfrac{3}{4}w(1+w) \le \hat{\mu} \le \tfrac{3}{4}(1-w^2), \qquad 0 \le \hat{\mu} \le 1 + w. \tag{6.53}$$

In Fig. 6.1 we plot the allowed values of $(w, \hat{\mu})$ which satisfy (6.53), and some lines of constant m_2^2 and m_3^2. Observationally the values of w, M^2 and $\hat{\mu}$ (only $\hat{\mu}$ is the

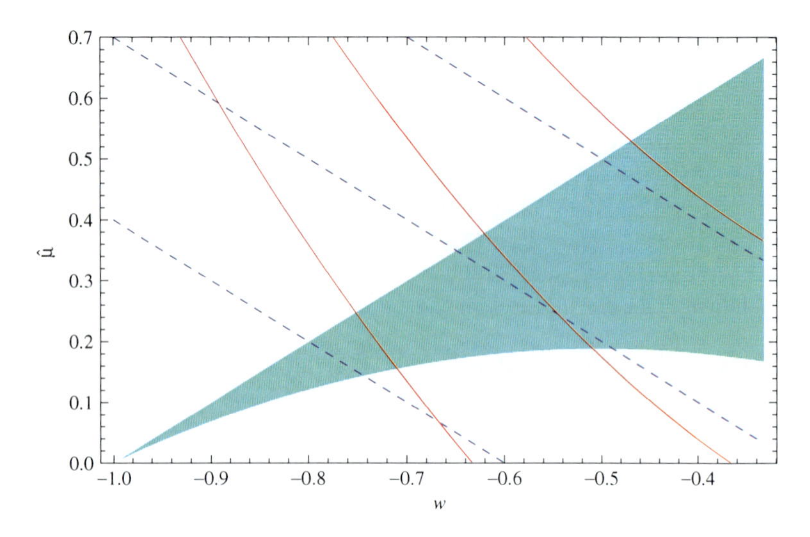

Fig. 6.1 The shaded region denotes the range of values of the equation of state w and shear modulus $\hat{\mu}$ which yield sound speeds less than unity, which is where the inequalities (6.53) are satisfied. The red (solid) lines denote lines of constant $m_2^2 = \frac{1}{2}M^2(0.6, 0.3, 0)$ from left to right and the blue (dashed) lines of constant $m_3^2 = \frac{1}{6}M^2(0.6, 0.2, -0.2)$, again from left to right

"new" parameter) can be constrained and then (6.51a) used to obtain the graviton masses.

Elastic dark energy and massive gravity share two common features. First, they are both constructed from rank-4 tensors, (the elasticity tensor and mass-matrix, respectively) and these tensors have identical symmetries in their indices. Secondly, they both have five propagating degrees of freedom. The extra degrees of freedom in elastic dark energy may have a different fundamental origin to those in massive gravity, but they enable an interesting interpretation to be extracted from massive gravities. Our interpretation is that *massive gravity is the manifestation of rigidity of spacetime*.

References

1. T. Clifton, P.G. Ferreira, A. Padilla, C. Skordis, Modified gravity and cosmology. Phys. Rept. **513**, 1–189 (2012) [arXiv:1106.2476]
2. R.A. Battye, J.A. Pearson, Effective action approach to cosmological perturbations in dark energy and modified gravity. JCAP **1207**, 019 (2012) [arXiv:1203.0398]
3. V.A. Rubakov, P.G. Tinyakov, Infrared-modified gravities and massivegravitons. Phys. Usp. **51**, 759–792 (2008) [arXiv:0802.4379]
4. K. Hinterbichler, Theoretical aspects of massive gravity. Rev. Mod. Phys. **84**, 671–710 (2012) [arXiv:1105.3735]
5. M. Fierz, W. Pauli, On relativistic wave equations for particles of arbitrary spin in an electromagnetic field. Proc. Roy. Soc. Lond. **A173**, 211–232 (1939)

6. D. Boulware, S. Deser, Can gravitation have a finite range?. Phys. Rev. **D6**, 3368–3382 (1972)
7. C. de Rham, G. Gabadadze, A. J. Tolley, Resummation of massive gravity. Phys. Rev. Lett. **106**, 231101 (2011) [arXiv:1011.1232]
8. C. de Rham, G. Gabadadze, Generalization of the fierz-pauli action. Phys. Rev. **D82**, 044020 (2010) [arXiv:1007.0443]
9. S. Hassan, R.A. Rosen, Resolving the ghost problem in non-linear massive gravity. Phys. Rev. Lett. **108**, 041101 (2012) [arXiv:1106.3344]
10. S. Hassan, R.A. Rosen, On non-linear actions for massive gravity. JHEP **1107**, 009 (2011) [arXiv:1103.6055]
11. A. Vainshtein, To the problem of nonvanishing gravitation mass. Phys. Lett. B **39**, 393–394 (1972)
12. E. Babichev, C. Deffayet, R. Ziour, The recovery of general relativity in massive gravity via the vainshtein mechanism. Phys. Rev. **D82**, 104008 (2010) [arXiv:1007.4506]
13. A. De Felice, R. Kase, S. Tsujikawa, Vainshtein mechanism in second-order scalar-tensor theories. Phys. Rev. **D85**, 044059 (2012) [arXiv:1111.5090]
14. F. Sbisa, G. Niz, K. Koyama, G. Tasinato, Characterising vainshtein solutions in massive gravity. Phys. Rev. **D86**, 024033 (2012) [arXiv:1204.1193]
15. S. Hassan, R.A. Rosen, A. Schmidt-May, ghost-free massive gravity with a general reference metric. JHEP **1202**, 026 (2012) [arXiv:1109.3230]
16. S.F. Hassan, R.A. Rosen, Bimetric gravity from ghost-free massive gravity. JHEP **02**, 126 (2012) [arXiv:1109.3515]
17. M.F. Paulos, A.J. Tolley, Massive Gravity theories and limits of ghost-free bigravitymodels. JHEP **1209**, 002 (2012). doi:10.1007/JHEP09(2012)002 [arXiv:1203.4268]
18. P. Creminelli, A. Nicolis, M. Papucci, E. Trincherini, Ghosts in massive gravity. JHEP **09**, 003 (2005) [hep-th/0505147]
19. L. Alberte, A.H. Chamseddine, V. Mukhanov, Massive gravity: Exorcising the ghost. JHEP **1104**, 004 (2011) [arXiv:1011.0183]
20. L. Alberte, A.H. Chamseddine, V. Mukhanov, Massive gravity: Resolving the puzzles. JHEP **1012**, 023 (2010) [arXiv:1008.5132]
21. C. de Rham, G. Gabadadze, A. Tolley, Ghost free massive gravity in the Stúckelberg language. Phys. Lett. **B711**, 190–195 (2012) [arXiv:1107.3820]
22. L. Grisa, L. Sorbo, Pauli-Fierz gravitons on friedmann-robertson-walker background. Phys. Lett. **B686**, 273–278 (2010) [arXiv:0905.3391]
23. F. Berkhahn, D.D. Dietrich, S. Hofmann, Self-protection of massive cosmological gravitons. JCAP **1011**, 018 (2010) [arXiv:1008.0644]
24. F. Berkhahn, D.D. Dietrich, S. Hofmann, Consistency of relevant cosmological deformations on all Scales. JCAP **1109**, 024 (2011) [arXiv:1104.2534]
25. F. Berkhahn, D.D. Dietrich, S. Hofmann, Cosmological classicalization: Maintaining unitarity under relevant deformations of the Einstein-Hilbert action. Phys. Rev. Lett. **106**, 191102 (2011) [arXiv:1102.0313]
26. K. Koyama, G. Niz, G. Tasinato, Analytic solutions in non-linear massive gravity. Phys. Rev. Lett. **107**, 131101 (2011) [arXiv:1103.4708]
27. K. Koyama, G. Niz, G. Tasinato, Strong interactions and exact solutions in non-linear massive gravity. Phys. Rev. **D84**, 064033 (2011) [arXiv:1104.2143]
28. G. D'Amico et al. Massive cosmologies. Phys. Rev. **D84**, 124046 (2011) [arXiv:1108.5231]
29. A.E. Gumrukcuoglu, C. Lin, S. Mukohyama, Open FRW universes and self-acceleration from nonlinear massive gravity. JCAP **1111**, 030 (2011) [arXiv:1109.3845]
30. A.E. Gumrukcuoglu, C. Lin, S. Mukohyama, Cosmological perturbations of self-accelerating universe in nonlinear massive gravity. JCAP **1203**, 006 (2012) [arXiv:1111.4107]
31. M. Crisostomi, D. Comelli, L. Pilo, Perturbations in massive gravity cosmology. JHEP **1206**, 085 (2012) [arXiv:1202.1986]
32. P. Gratia, W. Hu, M. Wyman, Self-accelerating massive gravity: Exact solutions for any isotropic matter distribution. Phys. Rev. **D86**, 061504 (2012) [arXiv:1205.4241]

33. A. De Felice, A.E. Gumrukcuoglu, S. Mukohyama, A. E. Gumrukcuoglu, S. Mukohyama, Massive gravity: Nonlinear instability of the homogeneous and isotropic universe. Phys. Rev. Lett. **109**, 171101 (2012) [arXiv:1206.2080]
34. M.S. Volkov, Cosmological solutions with massive gravitons in the bigravity theory. JHEP **1201**, 035 (2012) [arXiv:1110.6153]
35. M.S. Volkov, Exact self-accelerating cosmologies in the ghost-free massive gravity-the detailed derivation. Phys. Rev. **D86**, 104022 (2012). doi:10.1103/PhysRevD.86.104022 [arXiv:1207.3723]
36. G. D'Amico, Cosmology and perturbations in massive gravity. Phys. Rev. **D86**, 124019 (2012). doi:10.1103/PhysRevD.86.124019 [arXiv:1206.3617]
37. V. Rubakov, Lorentz-violating graviton masses: Getting around ghosts, low strong coupling scale and VDVZ discontinuity. [hep-th/0407104]
38. N. Arkani-Hamed, H.-C. Cheng, M.A. Luty, S. Mukohyama, Ghost condensation and a consistent infrared modification of gravity. JHEP **0405**, 074 (2004) [hep-th/0312099]
39. S. Dubovsky, Phases of massive gravity. JHEP **0410**, 076 (2004) [hep-th/0409124]
40. G. Gabadadze, L. Grisa, Lorentz-violating massive gauge and gravitational fields. Phys. Lett. **B617**, 124–132 (2005) [hep-th/0412332]
41. D. Blas, D. Comelli, F. Nesti, L. Pilo, Lorentz breaking massive gravity in curved space. Phys. Rev. **D80**, 044025 (2009) [arXiv:0905.1699]
42. D. Comelli, M. Crisostomi, F. Nesti, L. Pilo, Degrees of freedom in massive gravity. Phys. Rev. **D86**, 101502 (2012). doi:10.1103/PhysRevD.86.101502 [arXiv:1204.1027]
43. B. Carter, H. Quintana, Foundations of general relativistic high-pressure elasticity theory, Proceedings of the Royal Society of London. A. Math. Phys. Sci. **331**(1584), 57–83 (1972)
44. B. Carter, Elastic perturbation theory in general relativity and a variation principle for a rotating solid star. Commun. Math. Phys. **30**, 261–286 (1973). doi:10.1007/BF01645505
45. B. Carter, Speed of sound in a high-pressure general-relativistic solid. Phys. Rev. **D7**, 1590–1593 (1973)
46. J.L. Friedman, B. F. Schutz, Erratum: On the stability of relativistic systems. Astrophys. J. **200**, 204–220 (1975)
47. B. Carter, H. Quintana, Gravitational and acoustic waves in an elastic medium. Phys. Rev **D16**, 1016 (1977)
48. B. Carter, Rheometric structure theory, convective differentiation and continuum electrodynamics, Proceedings of the Royal Society of London. A. Math. Phys. Sci. **372**(1749), 169–200 (1980)
49. B. Carter, Interaction of gravitational waves with an elastic solid medium. gr-qc/0102113
50. M. Bucher, D. N. Spergel, Is the dark matter a solid?. Phys. Rev. **D60**, 043505 (1999) [astro-ph/9812022]
51. R.A. Battye, M. Bucher, and D. Spergel, Domain wall dominated universes. astro-ph/9908047
52. R.A. Battye, A. Moss, Anisotropic perturbations due to dark energy. Phys. Rev. **D74**, 041301 (2006) [astro-ph/0602377]
53. R. Battye, A. Moss, Anisotropic dark energy and CMB anomalies. Phys. Rev. **D80**, 023531 (2009) [arXiv:0905.3403]
54. J.A. Pearson, Effective field theory for perturbations in dark energy and modified gravity. [arXiv:1205.3611]
55. R.A. Battye, B. Carter, A. Mennim, Linearized self-forces for branes. Phys. Rev. **D71**, 104026 (2005) [hep-th/0412053]
56. R.A. Battye, B. Carter, Second order Lagrangian and symplectic current for gravitationally perturbed Dirac-Goto-Nambu strings and branes. Class. Quant. Grav. **17**, 3325–s3334 (2000) [hep-th/9811075]
57. R. Battye, B. Carter, Gravitational perturbations of relativistic membranes and strings. Phys. Lett. **B357**, 29–35 (1995) [hep-ph/9508300]
58. N. Arkani-Hamed, H. Georgi, M. D. Schwartz, Effective field theory for massive gravitons and gravity in theory space. Ann. Phys. **305**, 96–118 (2003) [hep-th/0210184]

59. S. Dubovsky, P. Tinyakov, I. Tkachev, Cosmological attractors in massive gravity. Phys. Rev. **D72**, 084011 (2005) [hep-th/0504067]
60. R.L. Arnowitt, S. Deser, C. W. Misner, The Dynamics of general relativity. gr-qc/0405109
61. F. de Urries, J. Julve, Degrees of freedom of arbitrarily higher derivative field theories. gr-qc/9506009
62. F. de Urries, J. Julve, Ostrogradski formalism for higher derivative scalar field theories. J. Phys.A **A31**, 6949–6964 (1998) [hep-th/9802115]
63. R.A. Battye, A. Moss, Cosmological perturbations in elastic dark energy models. Phys. Rev. **D76**, 023005 (2007) [astro-ph/0703744]

Chapter 7
Generalized Fluid Description

7.1 Introduction

So far in this thesis we have developed and constructed a formalism which allows the perturbed dark energy momentum tensor to be obtained for particular field contents of the dark sector. All modifications to the equations governing perturbations are contained within $\delta_{\rm E}U^{\mu}{}_{\nu}$ and the perturbed fluid equations. In this chapter we rewrite some of these results to obtain what we call *equations of state for dark sector perturbations*. As the name suggests, these equations of state for perturbations play an identical role to the background equation of state $P = w\rho$ used to close the background fluid equation $\dot{\rho} = -3\mathcal{H}(\rho + P)$. The structure of these equations of state for dark sector perturbations will be our window into detecting and discriminating between modified gravity theories.

To achieve this goal we identify the entropy contribution, Γ, and anisotropic stresses, $\Pi^{\rm S}$, $\Pi^{\rm V}$, $\Pi^{\rm T}$, for some of the generalized dark theories we have so far discussed. We will deduce whether or not the form of Γ suggested by [1, 2] is the most general way in which entropy should be specified. Unless explicitly stated, we will continue to work in the synchronous gauge.

7.2 Perturbed Fluid Variables and Fluid Equations

We identify the components $\delta_{\rm E}U^{\mu}{}_{\nu}$ with the density contrast $\delta \equiv \delta\rho/\rho$, perturbed pressure δP, velocity v^{μ} and anisotropic stress $\Pi^{\mu}{}_{\nu}$ of a *generalized fluid*. This allows some sort of physical picture to be attached to the $\delta_{\rm E}U^{\mu}{}_{\nu}$. The fluid decomposition of $\delta_{\rm E}U^{\mu}{}_{\nu}$ is

$$\delta_{\rm E}U^{\mu}{}_{\nu} = \delta\rho u^{\mu}u_{\nu} + \delta P\gamma^{\mu}{}_{\nu} + 2(\rho + P)v^{(\mu}u_{\nu)} + P\Pi^{\mu}{}_{\nu}, \tag{7.1}$$

J. Pearson, *Generalized Perturbations in Modified Gravity and Dark Energy*,
Springer Theses, DOI: 10.1007/978-3-319-01210-0_7,

where ρ, P are the density and pressure of the "dark sector fluid", u^μ is a time-like unit vector, v^μ is a space-like vector (parameterizing distances between flow lines) and $\Pi^\mu{}_\nu$ is the spatial anisotropic stress tensor. These vectors and tensors satisfy

$$u^\mu u_\mu = -1, \quad u^\mu v_\mu = 0, \quad \Pi^\mu{}_\mu = 0, \quad \Pi^\mu{}_\nu u_\mu = 0, \tag{7.2}$$

so that

$$\delta_{\mathrm{E}} U^0{}_0 = -\delta\rho, \quad \delta_{\mathrm{E}} U^i{}_0 = -\rho(1+w)v^i, \quad \delta_{\mathrm{E}} U^i{}_j = \delta P \delta^i{}_j + P\Pi^i{}_j, \tag{7.3a}$$

$$\delta_{\mathrm{E}} U^i{}_j - \frac{1}{3}\delta^i{}_j \delta_{\mathrm{E}} U^k{}_k = P\Pi^i{}_j. \tag{7.3b}$$

As we have made clear, $\delta_{\mathrm{E}} U^\mu{}_\nu$ satisfies the perturbed conservation equation, $\delta_{\mathrm{E}}(\nabla_\mu U^\mu{}_\nu) = 0$. This equation has a time-like and space-like component, which can be extracted by contracting with u_μ and $\gamma_{\mu\nu}$ respectively, yielding

$$u^\nu \delta_{\mathrm{E}}(\nabla_\mu U^\mu{}_\nu) = 0, \quad \gamma^\nu{}_\alpha \delta_{\mathrm{E}}(\nabla_\mu U^\mu{}_\nu) = 0. \tag{7.4}$$

These can be evaluated for general metric perturbations, which we parameterize as

$$\delta_{\mathrm{E}} g_{\mu\nu} = 2\phi u_\mu u_\nu + 2n_{(\mu}u_{\nu)} + \frac{1}{3}h\gamma_{\mu\nu} + h^{(\Pi)}_{\mu\nu}. \tag{7.5}$$

We will use our usual notation for time and space differentiation: $\dot{X} \equiv u^\mu \nabla_\mu X$, $\bar{\nabla}_\mu X \equiv \gamma^\nu{}_\mu \nabla_\nu X$. Inserting (7.1) and (7.5) into (7.4) respectively yields

$$\dot{\delta} = -(1+w)\left(\nabla_\mu v^\mu + \frac{1}{2}\dot{h}\right) - 3\mathcal{H}\left(\frac{\delta P}{\delta\rho} - w\right)\delta, \tag{7.6a}$$

$$\begin{aligned} \dot{v}_\alpha = &-\mathcal{H}(1-3w)v_\alpha - \frac{\dot{w}}{1+w}v_\alpha + \left(\bar{\nabla}_\alpha\phi - \mathcal{H}n_\alpha\right) \\ &- \frac{1}{\rho(1+w)}\bar{\nabla}_\alpha \delta P - \frac{w}{1+w}\nabla_\mu \Pi^\mu{}_\alpha. \end{aligned} \tag{7.6b}$$

These equations can be adapted for use in the conformal Newtonian gauge (by setting $n_\mu = 0$, $h = 6\psi$), the synchronous gauge (by setting $\phi = n_\mu = 0$) or for studying scalar perturbations.

For scalar perturbations in the synchronous gauge [3, 4] the perturbed conservation equation yields

$$\dot{\delta} = -(1+w)\left(kv + \frac{1}{2}\dot{h}\right) - 3\mathcal{H}w\Gamma, \tag{7.7a}$$

$$\dot{v} = -\mathcal{H}\big(1 - 3w\big)v + \frac{w}{1+w}k\delta + \frac{k}{1+w}w\Gamma - \frac{2}{3}\frac{w}{1+w}k\Pi^{\mathrm{S}}, \tag{7.7b}$$

where $w = P/\rho$ is the equation of state, $\delta = \delta\rho/\rho$ is the density contrast, Π^{S} is the scalar anisotropic source, and the entropy contribution is given by

$$w\Gamma \equiv \left(\frac{\delta P}{\delta\rho} - \frac{\mathrm{d}P}{\mathrm{d}\rho}\right)\delta = \left(\frac{\delta P}{\delta\rho} - w + \frac{\dot{w}}{3\mathcal{H}(1+w)}\right)\delta. \tag{7.8}$$

For the subsequent discussion we take $\dot{w} = 0$ for simplicity. The perturbed conservation equations (7.7) provide evolution equations for δ and v but not δP or Π. Once δP and Π are known by some means, the system of equations (7.7) becomes closed and can be solved. Our formalism enables us to obtain general closed forms of δP and Π (we use the entropy $w\Gamma$ instead of δP).

In [1, 2, 5] these equations are modified by introducing a gauge invariant density contrast. Their modification is equivalent to parameterizing the entropy as

$$w\Gamma = (c_{\mathrm{s}}^2 - w)\left(\delta + 3\mathcal{H}(1+w)\frac{v}{k}\right), \tag{7.9}$$

as noted by [3]. The important thing to notice here is that, for a particular value of w there is a single function which specifies the entropy: the sound speed c_{s}^2. One should note that when $w = -1$ the entropy contribution becomes $w\Gamma = (c_{\mathrm{s}}^2 + 1)\delta$.

The scalar decomposition of the vector field and metric perturbation is

$$\xi_i = \partial_i\xi^{\mathrm{S}}, \quad h_{ij} = D_{ij}\eta + \frac{1}{3}h\delta_{ij}, \qquad D_{ij} \equiv \partial_i\partial_j - \frac{1}{3}\delta_{ij}\nabla^2, \tag{7.10a}$$

where D_{ij} is a transverse-traceless spatial derivative operator, ∇^2 is the spatial Laplacian; the trace of the metric perturbation is h and the transverse-traceless part of the metric perturbation is

$$h_{ij} - \frac{1}{3}\delta_{ij}h = D_{ij}\eta. \tag{7.10b}$$

The scalar decomposition of the perturbed fluid variables is

$$v^i = \partial^i\theta, \quad \Pi_{ij} = D_{ij}\Pi^{\mathrm{S}}. \tag{7.10c}$$

We have that Π^{S} is the scalar anisotropic perturbation, and θ is the scalar velocity divergence perturbation.

We can also decompose the vector and tensor pieces of vectorial and tensorial quantities in the usual way; the equations for the vector and tensor sectors are given in [3]. To summarise our decompositions:

$$\xi^{\mu} \to \{\xi^{\mathrm{S}}, \xi^{\mathrm{V}}\}, \quad \delta_{\mathrm{E}}g_{\mu\nu} \to \{h, \eta, H^{\mathrm{V}}, H^{\mathrm{T}}\}, \tag{7.11a}$$

$$\delta_{\mathrm{E}} U^{\mu}{}_{\nu} \rightarrow \{\delta, \theta, \delta P, \Pi^{\mathrm{S}}, \theta^{\mathrm{V}}, \Pi^{\mathrm{V}}, \Pi^{\mathrm{T}}\}. \tag{7.11b}$$

In order to close the equations we need to specify four of (7.11b), for example δP (via Γ), Π^{S}, Π^{V}, Π^{T}, possibly as functions of the other three (which have evolution equations), plus their derivatives and the metric variables h, η, H^{V}, H^{T}.

As a final piece of notation, we will extract some of the time dependance of a quantity by dividing out the density and write

$$\hat{X} \equiv \frac{X}{\rho}. \tag{7.12}$$

A common example of this is using an equation of state to link the pressure and density, $w = P/\rho$. Although we will not *a priori* assume that these "hatted" quantities are constant, it may well turn out that the problem significantly simplifies if they are constant.

7.3 No Extra Fields

For the theory with field content $\mathcal{L} = \mathcal{L}(g_{\mu\nu})$, one can use (3.65) to obtain the perturbed fluid variables

$$\delta = -\big(1 + \widehat{B_{\mathcal{W}}}\big)\left(\nabla^2 \xi^{\mathrm{S}} + \frac{1}{2}h\right), \tag{7.13a}$$

$$\theta = \frac{1 - \widehat{C_{\mathcal{W}}}}{1 + w}\dot{\xi}^{\mathrm{S}}, \tag{7.13b}$$

$$\delta P = -\rho\left(\widehat{D_{\mathcal{W}}} + \frac{2}{3}\widehat{E_{\mathcal{W}}} + \frac{1}{3}w\right)\left(\nabla^2 \xi^{\mathrm{S}} + \frac{1}{2}h\right), \tag{7.13c}$$

$$\Pi^{\mathrm{S}} = 2\frac{w - \widehat{E_{\mathcal{W}}}}{w}\left(\xi^{\mathrm{S}} + \frac{1}{2}\eta\right), \tag{7.13d}$$

$$\theta^{\mathrm{V}} = \frac{1 - \widehat{C_{\mathcal{W}}}}{1 + w}\dot{\xi}^{\mathrm{V}}, \tag{7.13e}$$

$$\Pi^{\mathrm{V}} = \frac{\widehat{E_{\mathcal{W}}} - w}{w}(k\xi^{\mathrm{V}} - H^{\mathrm{V}}), \tag{7.13f}$$

$$\Pi^{\mathrm{T}} = \frac{w - \widehat{E_{\mathcal{W}}}}{w}H^{\mathrm{T}}. \tag{7.13g}$$

The entropy contribution (7.8) can be identified as

$$w\Gamma = \big(C - w\big)\delta, \tag{7.14}$$

where we defined

$$C \equiv \frac{D_{\mathcal{W}} + \frac{2}{3}E_{\mathcal{W}} + \frac{1}{3}P}{\rho + B_{\mathcal{W}}}. \tag{7.15}$$

Hence, we observe that this theory is capable of supporting non-trivial scalar, vector and tensor anisotropic and velocity sources. When one makes the parameter choice corresponding to elastic dark energy (3.12) one obtains $\Gamma = 0$, which is in agreement with the results in [3].

7.4 Scalar Fields

For a theory with field content $\mathcal{L} = \mathcal{L}(g_{\mu\nu}, \phi, \nabla_\mu\phi)$, one can use (3.59) to obtain the perturbed fluid variables

$$\delta = -\big(1 + \widehat{B_{\mathcal{W}}}\big)\left(\nabla^2\xi^{\mathrm{S}} + \frac{1}{2}h\right) - \frac{1}{2}\left(\widehat{A_{\mathcal{V}}}\delta\phi + \frac{1}{a}\widehat{A_{\mathcal{Y}}}\delta\dot{\phi}\right), \tag{7.16a}$$

$$\theta = -\frac{1}{1+w}\left[(\widehat{C_{\mathcal{W}}} - 1)\dot{\xi}^{\mathrm{S}} + \frac{1}{2a}\widehat{C_{\mathcal{Y}}}\delta\phi\right], \tag{7.16b}$$

$$\delta P = -\rho\left[\widehat{D_{\mathcal{W}}} + \frac{2}{3}\widehat{E_{\mathcal{W}}} + \frac{1}{3}w\right]\left(\nabla^2\xi^{\mathrm{S}} + \frac{1}{2}h\right) - \frac{1}{2}\left(B_{\mathcal{V}}\delta\phi + \frac{1}{a}B_{\mathcal{Y}}\delta\dot{\phi}\right), \tag{7.16c}$$

$$\Pi^{\mathrm{S}} = 2\frac{w - \widehat{E_{\mathcal{W}}}}{w}\left(\xi^{\mathrm{S}} + \frac{1}{2}\eta\right), \tag{7.16d}$$

$$\theta^{\mathrm{V}} = \frac{1 - \widehat{C_{\mathcal{W}}}}{1+w}\dot{\xi}^{\mathrm{V}}, \tag{7.16e}$$

$$\Pi^{\mathrm{V}} = \frac{\widehat{E_{\mathcal{W}} - w}}{w}(k\xi^{\mathrm{V}} - H^{\mathrm{V}}), \tag{7.16f}$$

$$\Pi^{\mathrm{T}} = \frac{w - \widehat{E_{\mathcal{W}}}}{w}H^{\mathrm{T}}. \tag{7.16g}$$

One should notice that this theory is capable of supporting anisotropic stresses, with $\Pi^{\mathrm{S}}, \Pi^{\mathrm{V}}, \Pi^{\mathrm{T}} \neq 0$ in general, but these are only sourced by the perturbed metric and ξ^μ.

The entropy contribution (7.8) is found to be

$$\begin{aligned} w\Gamma = {} & \Big((1 + \widehat{B_{\mathcal{W}}})D - \mathcal{Z}\Big)\left(\nabla^2\xi^{\mathrm{S}} + \frac{1}{2}h\right) - a(AD - B)(\widehat{C_{\mathcal{W}}} - 1)\dot{\xi}^{\mathrm{S}} \\ & + (D - w)\delta - a\big(AD - B\big)(1+w)\theta, \end{aligned} \tag{7.17}$$

where we have defined three frequently appearing ratios,

$$D \equiv \frac{B_{\mathcal{Y}}}{A_{\mathcal{Y}}}, \quad A \equiv \frac{A_{\mathcal{V}}}{C_{\mathcal{Y}}}, \quad B \equiv \frac{B_{\mathcal{V}}}{C_{\mathcal{Y}}}, \tag{7.18}$$

and the combination,

$$\mathcal{Z} \equiv \frac{1}{3}\widehat{P} + \widehat{D_{\mathcal{W}}} + \frac{2}{3}\widehat{E_{\mathcal{W}}}. \tag{7.19}$$

The problem with these expressions is that they are not manifestly closed. We will show how this is remedied in a simplified theory, where we impose reparameterization invariance so that we can obtain equations of state, and then move back to the general theory to show the existence of such an equation of state.

7.4.1 Reparameterization Invariant Scalar Fields

If we now apply the decoupling conditions (3.62) to (7.16), (7.17), (7.18), then we find the following perturbed fluid variables for a reparameterization invariant first order scalar field theory

$$\delta = -\frac{1}{2}\left(\widehat{A_{\mathcal{V}}}\delta\phi + \frac{1}{a}\widehat{A_{\mathcal{Y}}}\dot{\delta\phi}\right), \tag{7.20a}$$

$$\theta = -\frac{1}{2a(1+w)}\widehat{C_{\mathcal{Y}}}\delta\phi, \tag{7.20b}$$

$$\delta P = \frac{1}{2a}\left((\dot{C}_{\mathcal{Y}} + 3\mathcal{H}C_{\mathcal{Y}})\delta\phi + C_{\mathcal{Y}}\dot{\delta\phi}\right), \tag{7.20c}$$

$$\Pi^{\mathrm{S}} = \theta^{\mathrm{V}} = \Pi^{\mathrm{V}} = \Pi^{\mathrm{T}} = 0, \tag{7.20d}$$

$$w\Gamma = (D - w)\delta - a(AD - B)(1+w)\theta, \tag{7.20e}$$

where

$$D = -\frac{C_{\mathcal{Y}}}{A_{\mathcal{Y}}}, \quad A = \frac{A_{\mathcal{V}}}{C_{\mathcal{Y}}}, \quad B = -\frac{\dot{C}_{\mathcal{Y}} + 3\mathcal{H}C_{\mathcal{Y}}}{aC_{\mathcal{Y}}}. \tag{7.21}$$

As a direct consequence of employing the decoupling conditions, the anisotropic stress has now vanished, and the entropy is specified in closed form. The perturbed fluid variables for theories with $\mathcal{L} = \mathcal{L}(\mathcal{X}, \phi)$ are special cases of (7.20).

There are a number of explicit scalar field theories we will explicitly study which will enable us to get a feel for the typical form of the functions which appear in the entropy. The theories we consider have Lagrangian densities given by

$$\mathcal{L} = F(\mathcal{X})V(\phi), \quad \mathcal{L} = F(\mathcal{X}) - V(\phi). \tag{7.22}$$

The first theory encompasses k-essence [6, 7], and the second theory encompasses (for example) canonical scalar field theory. It is useful to realize that the coefficients that appear in the decomposition of $\delta U^{\mu\nu}$ for a generic kinetic scalar field theory $\mathcal{L} = \mathcal{L}(\phi, \mathcal{X})$ can be written as

$$A_{\mathcal{V}} = -2\big(2\mathcal{X}\mathcal{L}_{,\mathcal{X}\phi} - \mathcal{L}_{,\phi}\big), \quad B_{\mathcal{V}} = -2\mathcal{L}_{,\phi}, \tag{7.23a}$$

$$A_{\mathcal{Y}} = -2\sqrt{2\mathcal{X}}\big(2\mathcal{X}\mathcal{L}_{,\mathcal{X}\mathcal{X}} + \mathcal{L}_{,\mathcal{X}}\big), \quad B_{\mathcal{Y}} = -C_{\mathcal{Y}} = -2\mathcal{L}_{,\mathcal{X}}\sqrt{2\mathcal{X}}. \tag{7.23b}$$

and the energy density and pressure are computed via

$$\rho = 2\mathcal{X}\mathcal{L}_{,\mathcal{X}} - \mathcal{L}, \quad P = \mathcal{L}. \tag{7.23c}$$

Our notation is $F' \equiv \mathrm{d}F/\mathrm{d}\mathcal{X}$, $V' \equiv \mathrm{d}V/\mathrm{d}\phi$.

For a theory described by $\mathcal{L} = F(\mathcal{X}) - V(\phi)$ one obtains

$$w = -\left(1 - 2\mathcal{X}\frac{F'}{F - V}\right)^{-1}, \quad D = \left(1 + 2\mathcal{X}\frac{F''}{F'}\right)^{-1}, \tag{7.24a}$$

$$AD - B = -\frac{2V'}{F'\sqrt{2\mathcal{X}}}\left(1 + 2\mathcal{X}\frac{F''}{F'}\right)^{-1}\left(1 + \mathcal{X}\frac{F''}{F'}\right). \tag{7.24b}$$

For the canonical scalar theory, $\mathcal{L} = \mathcal{X} - V(\phi)$, one obtains

$$D = 1, \quad AD - B = -\frac{2V'}{\sqrt{2\mathcal{X}}}, \tag{7.25}$$

where V' can be rewritten in terms of $\dot{\rho} = -3\mathcal{H}\rho(1 + w)$ and $\dot{w}$, to obtain the formula

$$w\Gamma = (1 - w)\Big[\delta - 3\mathcal{H}(1 + w)\theta\Big] - \dot{w}\theta. \tag{7.26}$$

This is clearly of the same form as the expression suggested by [1, 2], (7.9) when we take $c_{\mathrm{s}}^2 = 1$, $\dot{w} = 0$.

For a theory described by $\mathcal{L} = F(\mathcal{X})V(\phi)$ one obtains

$$w = -\left(1 - 2\mathcal{X}\frac{F'}{F}\right)^{-1}, \quad D = \left(1 + 2\mathcal{X}\frac{F''}{F'}\right)^{-1}, \tag{7.27a}$$

$$AD - B = \frac{2V'}{F'\sqrt{2\mathcal{X}}}\frac{F}{V}\left(1 + 2\mathcal{X}\frac{F''}{F'}\right)^{-1}\left[1 + \mathcal{X}\left(\frac{F''}{F'} - \frac{F'}{F}\right)\right]. \tag{7.27b}$$

It is interesting to note that two of the frequently appearing combinations are

$$\frac{\mathcal{X}F'}{F} = \frac{\mathrm{d}\log F}{\mathrm{d}\log\mathcal{X}}, \quad \frac{\mathcal{X}F''}{F'} = \frac{\mathrm{d}\log F'}{\mathrm{d}\log\mathcal{X}}, \tag{7.28}$$

which are the logarithmic slopes of F and F' respectively and bear a resemblance to "slow-roll" parameters. For a pure k-essence theory, $\mathcal{L} = F(\mathcal{X})$, we have that $V' = 0$ and thus from (7.27) notice that the entropy becomes $w\Gamma = (D - w)\delta$, which bears a strong resemblance to the form of the entropy in the no extra fields case (7.14), but it is not of the form suggested by [1, 2].

It is possible to construct a Lagrangian density which has a specific equation of state. If we impose $P = w\rho$ upon (7.23c) then after trivial rearrangement we obtain

$$\frac{1}{\mathcal{L}}\frac{\partial\mathcal{L}}{\partial\mathcal{X}} = \frac{1+w}{2w\mathcal{X}}, \tag{7.29}$$

which can be integrated, yielding

$$\mathcal{L}(\mathcal{X}, \phi) = \tilde{V}(\phi)e^{\frac{1}{2}\int\frac{1+w}{w}\frac{\mathrm{d}\mathcal{X}}{\mathcal{X}}}, \tag{7.30}$$

where $\tilde{V}(\phi)$ appears as a "constant of integration". We can further impose $w_{,\mathcal{X}} = 0$, perform the integral, and find

$$\mathcal{L}(\mathcal{X}, \phi) = V(\phi)\mathcal{X}^{\frac{1+w}{2w}}. \tag{7.31}$$

This theory has been constructed to have an equation of state w which is independent of the kinetic term. In this case one is able to deduce that

$$D = w, \quad AD - B = 0, \tag{7.32}$$

which means that the entropy for this particular theory vanishes.

7.4.2 Equations of State for Dark Sector Perturbations

We will show how the entropy $w\Gamma$ and anisotropic stress Π^{S} can be specified as *equations of state for dark sector perturbations*, by showing how to determine the pressure perturbation, δP, and the anisotropic stress, Π, from the perturbed fluid variables $\{\delta, \theta, \dot{\delta}, \dot{\theta}\}$. Once these relations are given, the perturbed fluid equations (7.6) becomes a closed system.

Our strategy is to write down the scalar decomposition of the perturbed conservation equation, and write down the components of the perturbed dark energy momentum tensor. We then differentiate the components of $\delta_{\mathrm{E}}U^{\mu}{}_{\nu}$ and use the conservation equation to eliminate the ensuing second time derivatives. We then use these equations to show how the perturbed pressure and stress can be obtained.

We begin by noting that the perturbed fluid variables for a general first order scalar field theory (7.16) can be written schematically as

$$\delta = A_1\xi^{\mathrm{S}} + C_1\delta\phi + D_1\dot{\delta\phi} + E_1 h, \tag{7.33a}$$
$$\theta = B_2\dot{\xi}^{\mathrm{S}} + C_2\delta\phi, \tag{7.33b}$$
$$\delta P = X_1\xi^{\mathrm{S}} + Y_1\delta\phi + Z_1\dot{\delta\phi} + E_3 h, \tag{7.33c}$$
$$\Pi^{\mathrm{S}} = X_2\xi^{\mathrm{S}} + F_4\eta. \tag{7.33d}$$

We move the perturbed metric field variables h, η over to the LHS (because they are determined from the perturbed Einstein equations, and therefore do not constitute "unknowns"). This forms the "tilded" peturbed fluid variables; for example

$$\tilde{\delta P} = X_1\xi^{\mathrm{S}} + Y_1\delta\phi + Z_1\dot{\delta\phi}, \quad \tilde{\Pi}^{\mathrm{S}} = X_2\xi^{\mathrm{S}}. \tag{7.34}$$

Explicitly, the tilded perturbed pressure and stress are

$$\tilde{\delta P} = \rho\left[\frac{1}{3}w + \widehat{D_{\mathcal{W}}} + \frac{2}{3}\widehat{E_{\mathcal{W}}}\right]\left(k^2\xi^{\mathrm{S}}\right) - \frac{1}{2}\left(B_{\mathcal{V}}\delta\phi + \frac{1}{a}B_{\mathcal{Y}}\dot{\delta\phi}\right), \tag{7.35a}$$

$$\tilde{\Pi}^{\mathrm{S}} = 2\frac{w - \widehat{E_{\mathcal{W}}}}{w}\left(\xi^{\mathrm{S}}\right). \tag{7.35b}$$

The explicit relationships between the tilde and un-tilded fluid variables are

$$\tilde{\delta} = \delta + \frac{1}{2}(1 + \widehat{B_{\mathcal{W}}})h, \quad \tilde{\theta} = \theta, \tag{7.36a}$$

$$\tilde{\delta P} = \delta P + \frac{1}{2}\rho\left[\frac{1}{3}w + \widehat{D_{\mathcal{W}}} + \frac{2}{3}\widehat{E_{\mathcal{W}}}\right]h, \quad \tilde{\Pi}^{\mathrm{S}} = \Pi^{\mathrm{S}} - \frac{w - \widehat{E_{\mathcal{W}}}}{w}\eta, \tag{7.36b}$$

$$\dot{\tilde{\delta}} = \dot{\delta} + \frac{1}{2}\left\{\dot{\widehat{B_{\mathcal{W}}}} - \frac{1}{\rho}\left[\dot{B}_{\mathcal{W}} + \mathcal{H}(3B_{\mathcal{W}} + 3D_{\mathcal{W}} + 2E_{\mathcal{W}} - 2P)\right]\right\}h$$
$$+\frac{1}{2}\left\{\frac{1}{\rho}[P - B_{\mathcal{W}}] + (1 + \widehat{B_{\mathcal{W}}})\right\}\dot{h}, \tag{7.36c}$$

$$\dot{\tilde{\theta}} = \dot{\theta} + \frac{\widehat{C_{\mathcal{W}}} - 1}{1 + w}\left\{\frac{1}{2}\frac{1}{\rho - C_{\mathcal{W}}}\left[\frac{1}{3}P + D_{\mathcal{W}} + \frac{2}{3}E_{\mathcal{W}}\right]h + \frac{2}{3}\frac{1}{\rho - C_{\mathcal{W}}}\left[P - E_{\mathcal{W}}\right]k^2\eta\right\}. \tag{7.36d}$$

If the "unknown fields" $\{\xi^{\rm S}, \dot{\xi}^{\rm S}, \delta\phi, \dot{\delta\phi}\}$ can be written in terms of the "known" fluid variables $\{\delta, \theta, \dot{\delta}, \dot{\theta}\}$, then a closed system of equations can be found, which will be entirely written in terms of the perturbed fluid variables. To do this, we obtain $\dot{\delta}, \dot{\theta}$ by directly differentiating the fluid variables, and then insert $\ddot{\xi}^{\rm S}, \ddot{\delta\phi}$ from the conservation equations to remove all second time derivatives. This procedure yields

$$\tilde{\delta} = \left(1 + \widehat{B_{\mathcal{W}}}\right)k^2\xi^{\rm S} - \frac{1}{2}\left[\widehat{A_{\mathcal{V}}}\delta\phi + \frac{1}{a}\widehat{A_{\mathcal{Y}}}\dot{\delta\phi}\right], \tag{7.37a}$$

$$\tilde{\theta} = -\frac{1}{1+w}\left[(\widehat{C_{\mathcal{W}}} - 1)\dot{\xi}^{\rm S} + \frac{1}{2a}\widehat{C_{\mathcal{Y}}}\delta\phi\right], \tag{7.37b}$$

$$\begin{aligned}\tilde{\dot{\delta}} = {}& \frac{3\mathcal{H}}{2a}\left[\widehat{B_{\mathcal{Y}}} - \widehat{A_{\mathcal{Y}}}w\right]\dot{\delta\phi} + \frac{1}{2}\left[3\mathcal{H}(\widehat{B_{\mathcal{V}}} - w\widehat{A_{\mathcal{V}}}) - \frac{1}{a}\widehat{C_{\mathcal{Y}}}k^2\right]\delta\phi \\ & + \left[3\mathcal{H}\widehat{B_{\mathcal{W}}w} - \mathcal{H}(3\widehat{D_{\mathcal{W}}} + 2\widehat{E_{\mathcal{W}}} - 2w)\right]k^2\xi^{\rm S} - \left[\widehat{C_{\mathcal{W}}} - 1\right]k^2\dot{\xi}^{\rm S},\end{aligned} \tag{7.37c}$$

$$\begin{aligned}\tilde{\dot{\theta}} = {}& \left[w - 2\widehat{E_{\mathcal{W}}} - \widehat{D_{\mathcal{W}}}\right]k^2\xi^{\rm S} + \left[\mathcal{H}(\widehat{C_{\mathcal{W}}} - 1)(1 - 3w) + \frac{1-w}{1+w}\dot{\widehat{C_{\mathcal{W}}}}\right]\dot{\xi}^{\rm S} \\ & + \frac{1}{2a}\left[\frac{w}{1+w}\dot{\widehat{C_{\mathcal{Y}}}} + a\widehat{B_{\mathcal{V}}} + \mathcal{H}\widehat{C_{\mathcal{Y}}}\left(\frac{1}{1+w} - 3w\right)\right]\delta\phi \\ & + \frac{1}{2a}\left[\widehat{B_{\mathcal{Y}}} + \frac{w}{1+w}\widehat{C_{\mathcal{Y}}}\right]\dot{\delta\phi}.\end{aligned} \tag{7.37d}$$

The system of equations (7.37) is just a set of linear equations, which can be written as $\mathbf{F} = \mathsf{N}\mathbf{U}$, where

$$\mathbf{F} \equiv \begin{pmatrix} \tilde{\delta} \\ \tilde{\theta} \\ \tilde{\dot{\delta}} \\ \tilde{\dot{\theta}} \end{pmatrix}, \quad \mathsf{N} \equiv \begin{pmatrix} A_1 & 0 & C_1 & D_1 \\ 0 & B_2 & C_2 & 0 \\ A_3 & B_3 & C_3 & D_3 \\ A_4 & B_4 & C_4 & D_4 \end{pmatrix}, \quad \mathbf{U} \equiv \begin{pmatrix} \xi^{\rm S} \\ \dot{\xi}^{\rm S} \\ \delta\phi \\ \dot{\delta\phi} \end{pmatrix}. \tag{7.38}$$

The components of N can be read off from (7.37). If $\det \mathsf{N} \neq 0$ then the matrix N is, at least in principle, invertible and the "unknown" variables can be written in terms of the "known" fluid variables:

$$\xi^{\rm S} = m_{11}\tilde{\delta} + m_{12}\tilde{\theta} + m_{13}\tilde{\dot{\delta}} + m_{14}\tilde{\dot{\theta}}, \quad \dot{\xi}^{\rm S} = m_{21}\tilde{\delta} + m_{22}\tilde{\theta} + m_{23}\tilde{\dot{\delta}} + m_{24}\tilde{\dot{\theta}}, \tag{7.39a}$$

$$\delta\phi = m_{31}\tilde{\delta} + m_{32}\tilde{\theta} + m_{33}\dot{\tilde{\delta}} + m_{34}\dot{\tilde{\theta}}, \quad \dot{\delta\phi} = m_{41}\tilde{\delta} + m_{42}\tilde{\theta} + m_{43}\dot{\tilde{\delta}} + m_{44}\dot{\tilde{\theta}}, \tag{7.39b}$$

where the m_{AB} are components of N^{-1}. These expressions can then be inserted into (7.34) to obtain

$$\tilde{\delta P} = \tilde{\delta P}(\tilde{\delta}, \tilde{\theta}, \dot{\tilde{\delta}}, \dot{\tilde{\theta}}), \qquad \tilde{\Pi}^{\mathrm{S}} = \tilde{\Pi}^{\mathrm{S}}(\tilde{\delta}, \tilde{\theta}, \dot{\tilde{\delta}}, \dot{\tilde{\theta}}), \tag{7.40}$$

which can be thought of as being equations of state which close the dynamical system of equations. The general inversion of the matrix N is rather cumbersome and difficult to extract meaningful information from. However, the point remains that in principle it is possible to obtain general equations of state of perturbations in the dark sector. We will now proceed with special cases.

7.5 Possible Phenomenological Parameterizations for Scalar Perturbations

We are now in a position to write down equations of state for dark sector perturbations. These will be phenomenological parameterizations and were motivated from the different field contents of the generalized theories we developed. The physical interpretation of using equations of state will be explained at the end of this chapter.

7.5.1 *The $(\kappa, \lambda, \varepsilon)$-Parameterization*

It is simple to obtain equations of state for dark sector perturbations in the no-extra fields case. All perturbations are parameterized by the values of three parameters, which we call $(\kappa, \lambda, \varepsilon)$. Rearranging the perturbed fluid variables (7.13) yields

$$w\Gamma = (\kappa - w)\delta, \quad w\Pi^{\mathrm{S}} = (w - \lambda)\left[\delta - 3(1+w)\varepsilon\eta\right], \tag{7.41}$$

where one can explicitly obtain

$$\kappa \equiv \frac{(D_{\mathcal{W}} + \frac{2}{3}E_{\mathcal{W}} + \frac{1}{3}P)}{\rho + B_{\mathcal{W}}}, \qquad \lambda \equiv w + \frac{2P - E_{\mathcal{W}}}{\rho + B_{\mathcal{W}}}, \qquad \varepsilon \equiv \frac{B_{\mathcal{W}} + \rho}{w(1+w)}. \tag{7.42}$$

We note that elastic dark energy has

$$\kappa = w, \quad \lambda = \frac{1}{2}(3c_{\mathrm{s}}^2 - w), \quad \varepsilon = 1, \tag{7.43}$$

so that all perturbations in elastic dark energy are parameterized by the sound speed of the elastic medium, c_s^2, with vanishing entropy.

It should also be realized that if one were to study the cosmology of massive gravity theories (see Eq. 6.51), then one could not find out the value of the individual masses, and can only probe the combinations of masses represented by the parameters $(\kappa, \lambda, \epsilon)$.

7.5.2 The (α, β)-Parameterization

Motivated by the results in the previous subsections, we write down an expression for the parameterization of $w\Gamma$, which has been derived from the Lagrangian for perturbations for the first order scalar field theory in the decoupling limit, and which we believe represents a wider range of theories than the parameterization of [1, 2]. Our parameterization for the entropy contribution is

$$w\Gamma = (\alpha - w)\left[\delta - 3\mathcal{H}\beta(1+w)\theta\right], \tag{7.44}$$

where α and β are both *a priori* time-dependant background quantities which we have explicitly determined for the no extra fields and scalar field theories, although the hope would be that they would only be slowly varying functions of time. In Table 7.1 we provide explicit formulae for these quantities for generic example theories. It should be noted that in the parameterization of [1, 2] k-essence is excluded, and the parameter α is always taken to be constant. It is rather simple to determine whether or not α can be expected to be constant or not for particular examples, as we show below.

For a general scalar field theory one can obtain

$$\alpha = -\frac{C_{\mathcal{Y}}}{A_{\mathcal{Y}}}, \tag{7.45a}$$

$$\beta = \frac{A_{\mathcal{Y}}}{3\mathcal{H}a(C_{\mathcal{Y}} + wA_{\mathcal{Y}})}\left(\frac{A_{\mathcal{V}}}{A_{\mathcal{Y}}} + \frac{\dot{C}_{\mathcal{Y}} + 3\mathcal{H}C_{\mathcal{Y}}}{aC_{\mathcal{Y}}}\right). \tag{7.45b}$$

For the theory $\mathcal{L} = \mathcal{L}(\phi, \mathcal{X})$,

$$\alpha = \left(1 + 2\mathcal{X}\frac{\mathcal{L}_{,\mathcal{X}\mathcal{X}}}{\mathcal{L}_{,\mathcal{X}}}\right)^{-1}, \tag{7.46a}$$

Table 7.1 Collection of the quantities which determine the entropy contribution; the function C is defined in (7.15); in the final line we give the parameterization of [1, 2]

Theory	α	β
$\mathcal{L} = \mathcal{L}(g_{\mu\nu})$	C	0
$\mathcal{L} = \mathcal{L}(\phi, \mathcal{X})$	D	$\frac{AD - B}{3\mathcal{H}(D - w)}a$
$\mathcal{L} = \mathcal{X} - V(\phi)$	1	1
$\mathcal{L} = V(\phi)\mathcal{X}^{\frac{1+w}{2w}}$	w	0
$\mathcal{L} = \mathcal{L}(\mathcal{X})$	D	0
[1, 2]	c_s^2	1

$$\beta = \frac{2a\mathcal{L}_{,\phi}}{3\mathcal{H}\mathcal{L}_{,\mathcal{X}}\sqrt{2\mathcal{X}}}\left[1 + \mathcal{X}\left(\frac{\mathcal{L}_{,\mathcal{X}\mathcal{X}}}{\mathcal{L}_{,\mathcal{X}}} - \frac{\mathcal{L}_{,\mathcal{X}\phi}}{\mathcal{L}_{,\phi}}\right)\right]\frac{\alpha}{\alpha - w}. \tag{7.46b}$$

Explicitly, for the theory $\mathcal{L} = F(\mathcal{X}) - V(\phi)$ one obtains

$$\alpha = \left(1 + \frac{2\mathcal{X}F''}{F'}\right)^{-1}, \tag{7.47a}$$

$$\beta = -\frac{1}{F'}\left[\frac{2aV'}{3\mathcal{H}\sqrt{2\mathcal{X}}}\right]\left(1 + \mathcal{X}\frac{F''}{F}\right)\frac{\alpha}{\alpha - w}. \tag{7.47b}$$

We have written β to isolate a parameter combination, $2aV'/(3\mathcal{H}\sqrt{2\mathcal{X}})$.

For the theory $\mathcal{L} = F(\mathcal{X})V(\phi)$, one obtains

$$\alpha = \left(1 + \frac{2\mathcal{X}F''}{F'}\right)^{-1}, \tag{7.48a}$$

$$\beta = \frac{1}{F'}\left[\frac{2aV'}{3\mathcal{H}\sqrt{2\mathcal{X}}}\right]\frac{F}{V}\left(1 + \mathcal{X}\left(\frac{F''}{F'} - \frac{F'}{F}\right)\right)\frac{\alpha}{\alpha - w}, \tag{7.48b}$$

which is the same as (7.47) but with an extra factor in the expression for β.

7.5.3 The Second Order Scalar Field Theory Parameterization

In the discussion leading up to Eq. (4.88) we found a closed expression for the perturbed pressure for a second order scalar field theory imposed with (i) second order field equations and (ii) reparameterization invariance (see Sect.4.3.4). This motivates us to parameterize the entropy perturbation as

$$w\Gamma = (\alpha - w)\left[\delta - 3\mathcal{H}(1 + w)(\beta + k^2\gamma)\theta\right] - \frac{3}{2\rho\mathcal{H}}\zeta\dot{h}, \tag{7.49}$$

where there are four free parameters, $(\alpha, \beta, \gamma, \zeta)$. This represents a significant generalization of the (α, β)-parameterization (7.44), which was itself a significant generalization of the c_s^2-parameterization of [1, 2]. With (7.49) we are now able to test, probe and compute observables for theories with high order derivatives.

7.6 Summary

In this chapter we presented three different equations of state for dark sector perturbations, motivated by different generalized theories for the dark sector. These equations of state are given in (7.41), (7.44), (7.49), and are in some sense the main results of the entire thesis. These expressions close the perturbed fluid equations and allow explicit calculation of quantities which are have been observed in our universe. We should point out that the values that the parameters $(\kappa, \lambda, \varepsilon)$, $(\alpha, \beta, \gamma, \zeta)$ can take are likely to be restricted by stability. For instance, when $\kappa = w$, $\varepsilon = 1$ the stability conditions on EDE derived by [3] impose constraints on the allowed value of λ. These equations of state are the key to detecting and discriminating between different types of dark sector theories.

Physically speaking, using equations of state to close a system of equations is rather interesting since the modified gravity theories can be thought of as some "gas" whose effects are entirely encapsulated by the specific way in which the equation of state is constructed. From our derivations it is clear that an equation of state is the "observable" part of rather complicated theories. Indeed, different theories may well produce identical values for the equations of state. The point is that an equation of state provides a macroscopic coarse grained description of potentially very complicated microphysics.

References

1. J. Weller, A. M. Lewis, Large scale cosmic microwave background anisotropies and dark energy. Mon. Not. Roy. Astron. Soc. **346**, 987–993 (2003). [astro-ph/0307104]
2. R. Bean, O. Dore, Probing dark energy perturbations: the dark energy equation of state and speed of sound as measured by WMAP. Phys. Rev. **D69**, 083503 (2004). [astro-ph/0307100]
3. R.A. Battye, A. Moss, Cosmological perturbations in elastic dark energy models. Phys. Rev. **D76**, 023005 (2007). [astro-ph/0703744]
4. C.-P. Ma, E. Bertschinger, Cosmological perturbation theory in the synchronous versus conformal Newtonian gauge. Astrophys. J. (1994). [astro-ph/9401007]
5. W. Hu, Structure formation with generalized dark matter. Astrophys. J. **506**, 485–494 (1998). [astro-ph/9801234]
6. C. Armendariz-Picon, T. Damour, V. F. Mukhanov, k-Inflation. Phys. Lett. **B458**, 209–218 (1999). [hep-th/9904075]
7. C. Armendariz-Picon, V. F. Mukhanov, P. J. Steinhardt, Essentials of k essence. Phys. Rev. **D63**, 103510 (2001). [astro-ph/0006373]

Chapter 8
Observational Signatures of Generalized Cosmological Perturbations

8.1 Introduction

Our aim is to understand how cosmological observables are modified by different values of the "free" parameters in the equations of state for dark sector perturbations we constructed in Chap. 7. If a particular parameter only modifies the large-scale behavior of a particular observable then our ability to observationally distinguish that parameter will be hampered by limitations imposed by cosmic variance, and so it would be preferable to search for parameters which have a "large" effect on observables on "small" scales. It is rather unfortunate, however, that it is most natural for dark sector theories to induce modifications on large scales, since it is only at late times when the dark sector dominates when those large scale modes cross the horizon. A similar study was performed in [1] with a "sound speed" parameterization of the dark energy.

In addition to the usual calculation of the angular power spectrum of temperature (TT) anisotropies in the cosmic microwave background (CMB) we will compute the gravitational lensing of the CMB (dd) and correlation of the lensing with temperature (dT) angular power spectra. We also compute the matter power spectrum $P_{\rm tot}(k)$. These spectra have all either been observed to exquisite precision over a range of scales, or will be for upcoming data releases from various experiments (such as *Planck, CFHTLS* and *Euclid*).

These spectra are directly observable quantities, but it is also useful to obtain an understanding of the behavior of other important quantities. The one which has received a large amount of attention in the literature is the effective gravitational coupling, $G_{\rm eff} = \mu(k, z)G_{\rm N}$, where $G_{\rm N}$ is Newton's gravitational coupling constant. We will briefly show how modified gravity and dark energy theories give rise to an "effective" gravitational coupling. In the conformal Newtonian gauge, the perturbed Einstein equations are

$$k^2\Phi = -4\pi G_{\rm N}a^2\rho\Delta, \qquad k^2(\Phi - \Psi) = 12\pi G_{\rm N}a^2(\rho + P)\sigma. \tag{8.1}$$

J. Pearson, *Generalized Perturbations in Modified Gravity and Dark Energy*,
Springer Theses, DOI: 10.1007/978-3-319-01210-0_8,

In these formulae, ρ, P are the *total* density and pressure, $\Delta \equiv \delta + 3\mathcal{H}(\rho + P)\theta$ is the *total* comoving density perturbation, δ the *total* density contrast and σ the *total* anisotropic stress. These equations hold when all gravitating species are known and are taken into account. The dark sector can be thought of as GR with a "known" matter sector (indicated below with a suffix "kn") and an effective, environmentally dependent, gravitational coupling

$$G_{\text{eff}} = \mu(k, a)G_{\text{N}} \tag{8.2}$$

and a slip parameter, $Q = Q(k, a)$ quantifying the relative difference between the Newtonian potentials (which we will not make use of here). With these modifications, and explicitly identifying the known matter sector (8.1) becomes

$$k^2\Phi = -4\pi\mu G_{\text{N}}a^2\rho_{\text{kn}}\Delta_{\text{kn}}, \qquad k^2(\Phi - Q\Psi) = 12\pi\mu G_{\text{N}}a^2(\rho_{\text{kn}} + P_{\text{kn}})\sigma_{\text{kn}}. \tag{8.3}$$

This allows one to infer the existence of a dark sector if, by comparison to data, $\mu \neq 1$ and/or $Q \neq 1$ were found. In [2–7] parameterizations of μ were given and confronted with observational data. We will not impose any parameterization upon μ. Instead we will compute its value in various models so that we can get a feel for typical values of μ that generalized models naturally produce; our study is more akin to [8], where the effective gravitational coupling was computed for Galileon theries.

The equations of state we use are

- $(\kappa, \lambda, \epsilon)$-parameterization—probes "no extra fields":

$$w\Gamma = (\kappa - w)\delta, \qquad w\Pi^{\text{S}} = (w - \lambda)\Big[\delta - 3(1 + w)\varepsilon\eta\Big]. \tag{8.4}$$

  In our investigations we will fix $\lambda = \epsilon = 1$ and vary κ. We also study interesting subclass, elastic dark energy, parameterized by the sound speed, c_{s}^2 of the medium which has $\kappa = w$, $\lambda = \frac{1}{2}(3c_{\text{s}}^2 - w)$, $\varepsilon = 1$.
- $(\alpha, \beta, \gamma, \zeta)$-parameterization – probes the restricted second order scalar field theories:

$$w\Gamma = (\alpha - w)\Big[\delta - 3\mathcal{H}(1 + w)(\beta + k^2\gamma)\theta\Big] - \frac{3}{2\rho\mathcal{H}}\zeta\dot{h}, \qquad \Pi = 0. \tag{8.5}$$

  We also study the (α, β) parameterization, obtained from (8.5) by setting $\gamma = \zeta = 0$, which is valid for reparameterization invariant first order scalar field theories.

We will take the values of all parameters to be constant, but in principle they could be time dependent. We make the important caveat that this is a preliminary investigation.

8.2 Spectra

We use a modified version of the publicly available CAMB [9] software to obtain observable spectra for three equations of state for dark sector perturbations. We will be conservative and fix the cosmology with constant $w = -0.9$. Whilst having w so close to $w = -1$ suppresses the possible observed effects, the current data suggests that w is rather close to $w = -1$. We will be computing the difference in spectra between a given model and a fiducial model,

$$\frac{\Delta X}{X} \equiv \frac{X_{\rm model} - X_{\rm fid}}{X_{\rm fid}}, \tag{8.6}$$

where the fiducial model is taken to be a model with $(\alpha, \beta, \gamma, \zeta) = (1, 1, 0, 0)$, since that corresponds to a quintessence model. Note that α, β, κ, λ and ε are dimensionless, γ has units $\mathrm{Mpc}^2 h^{-2}$ and ζ has units Mpc^{-2}.

8.2.1 *Observing Spectra*

Here we will briefly review how the cosmological spectra which we compute using CAMB are actually observed.

The matter power spectrum, $P(k)$, is calculated from surveys of large scale structure after measuring the density contrast field $\delta(\mathbf{x}, t) = (\rho(\mathbf{x}, t) - \bar{\rho}(t))/\bar{\rho}(t)$, which is inferred observationally from galaxy number counts of large scale structure surveys (e.g. [10–13]). Once the density contrast field has been observed, the Fourier transform of the density contrast is computed

$$\delta_{\mathbf{k}} = \frac{1}{V} \int \mathrm{d}^3 x\, \delta(\mathbf{x}, t) e^{\mathrm{i}\mathbf{k}\cdot\mathbf{x}}, \tag{8.7}$$

and used to construct the the power spectrum,

$$P(k, t) = |\delta_{\mathbf{k}}(t)|^2. \tag{8.8}$$

It turns out that the power spectrum is the Fourier transform of the two point correlation function,

$$\xi(r) = \frac{V}{(2\pi)^3} \sum_{\mathbf{k}} P(k) e^{-\mathrm{i}\mathbf{k}\cdot\mathbf{r}} = \frac{1}{V} \int \mathrm{d}^3 x\, \delta(\mathbf{x})\delta(\mathbf{x} + \mathbf{r}). \tag{8.9}$$

Simplistically, the information that the matter power spectrum contains is of how many structures of a given size reside in the universe. Of course, this will be directly effected by changing the gravitational physics.

Anisotropies in the temperature of the CMB can be measured directly, which provides a map of temperature fluctuations, $\Delta T(\theta,\phi)/\bar{T}$, where (θ,ϕ) are angular positions on the sky and $\bar{T}$ the background temperature of the CMB. These temperature fluctuations are then decomposed using spherical harmonics, $Y_{\ell m}$,

$$\frac{\Delta T}{\bar{T}}(\theta,\phi) = \sum_{\ell=0}^{\infty}\sum_{m=-\ell}^{\ell} a_{\ell m} Y_{\ell m}(\theta,\phi), \tag{8.10}$$

where coefficients in the expansion, $a_{\ell m}$, can be found from the map via

$$a_{\ell m} = \int \mathrm{d}\Omega\ Y^*_{\ell m}(\theta,\phi)\frac{\Delta T}{\bar{T}}(\theta,\phi). \tag{8.11}$$

The correlation between fluctuations separated by an angular size ϑ is given by

$$C^{TT}(\vartheta) = \left\langle \frac{\Delta T}{\bar{T}}(\mathbf{n})\frac{\Delta T}{\bar{T}}(\mathbf{n}')\right\rangle = \frac{1}{4\pi}\sum_{\ell}(2\ell+1)C_\ell^{TT}P_\ell(\cos\vartheta), \tag{8.12}$$

where $\langle\cdots\rangle$ denotes an average over temperature fluctuations on the sky separated by $\vartheta \equiv |\mathbf{n}-\mathbf{n}'|$, $\cos\vartheta = \mathbf{n}\cdot\mathbf{n}'$, $P_\ell(x)$ is the Legendre polynomial and

$$C_\ell^{TT} \equiv \frac{1}{2\ell+1}\sum_m |a_{\ell m}|^2. \tag{8.13}$$

The C_ℓ^{TT} describe the power of correlation between fluctuations of a given angular size ℓ.

The CMB photons released from the surface of last scattering have been gravitationally lensed by structure along the line of sight. This is weak lensing of the CMB (see e.g. [14, 15] for reviews). Since the precise form of the structure that CMB photons encounter is a direct consequence of gravitational physics, the lensing of the CMB will be different for different gravitational theories. The lensing can be thought of as introducing location-dependant alterations into the temperature field of the CMB: $\tilde{T}(\mathbf{x}) = T(\mathbf{x}+\boldsymbol{\alpha}(\mathbf{x}))$. The deflection field $\boldsymbol{\alpha}(\mathbf{x})$ is given by an integral along the line of sight,

$$\boldsymbol{\alpha}(\mathbf{x}) = -2\int_0^{\chi_*}\mathrm{d}\chi\,\frac{\chi_*-\chi}{\chi_*}\nabla_\perp\Psi(\chi\mathbf{x};\eta_0-\chi), \qquad \Psi \equiv \frac{1}{2}(\phi+\psi). \tag{8.14}$$

In this expression, χ_* is the conformal distance to the surface of last scattering, η_0 is the current conformal time, $\nabla_\perp$ the gradient transverse to the propagation direction, and Ψ interpreted as the Weyl potential. The lensing potential φ is defined such that $\boldsymbol{\alpha} = \nabla\varphi$. The angular power spectrum of the weak lensing is then given by

$$C_\ell^{\psi\psi} = 16\pi \int \frac{dk}{k}\mathcal{P}_{\mathcal{R}}(k)\left[\int_0^{\chi_*} d\chi\, T_\Psi(k;\eta_0-\chi)j_\ell(k\chi)\frac{\chi_*-\chi}{\chi_*\chi}\right]^2, \quad (8.15)$$

where $\mathcal{P}_{\mathcal{R}}(k)$ is the primordial power spectrum, T_Ψ the transfer function and $j_\ell(x)$ a spherical Bessel function. The lensing potential serves to shift the unlensed spectrum, C_ℓ^{TT} to the lensed spectrum, $C_\ell^{\tilde{T}\tilde{T}}$, where

$$C_\ell^{\tilde{T}\tilde{T}} = (1-\ell^2R^\psi)C_\ell^{TT} + \int \frac{d^2\boldsymbol{\ell}'}{(2\pi)^2}\left[\boldsymbol{\ell}'\cdot(\boldsymbol{\ell}-\boldsymbol{\ell}')\right]^2 C_{|\boldsymbol{\ell}-\boldsymbol{\ell}'|}^{\psi\psi}C_{\ell'}^{TT}, \quad (8.16)$$

where

$$R^\psi \equiv \frac{1}{4\pi}\int \frac{d\ell}{\ell}\ell^4 C_\ell^{\psi\psi}. \quad (8.17)$$

It is common to plot $C_\ell^{dd} = \left[\ell(\ell+1)\right]^2 C_\ell^{\psi\psi}/2\pi$. Since the CMB temperature field that is actually observed is already lensed, the unlensed part of the field must be extracted to enable identification of the lensing potential. This tends to be done by assuming that the unlensed CMB field is due to a set of Gaussian primordial perturbations, whereas the lensed CMB is non-Gaussian. Details of the analysis can be found in [16, 15] and detection papers [17, 18].

8.2.2 Matter Power Spectra

In Fig. 8.1 we plot the total matter power spectra for various values of the free parameters in the equations of state for dark sector perturbations: one can clearly observe that these different types of theories have very different clustering properties. For the different types of parameterizations we make the following observations:

- We observe as the value of κ becomes very small, power on small scales becomes enhanced. Varying the values of (λ, ϵ) does not have an appreciable effect.
- When the sound speed of the elastic dark energy medium is close to zero the clustering on intermediate scales becomes significantly enhanced. This is simply because a medium with very small sound speed acts like CDM rather than dark energy.
- It is clear that $P_{\rm tot}(k)$ is mildly sensitive at large scales to the values of (α, β).
- The value of ζ appears to change the overall scale of the matter power spectrum, where large values of ζ substantially suppresses the large scale (small k) power.
- The value of γ only modifies the very small scales (large k). As γ is increased the power on small scales rapidly rises. Caution should be applied here because such small scales may not be accurately modeled within linear perturbation theory.

In Fig. 8.2 we plot the fractional difference for a given model relative to a fiducial $(\alpha, \beta) = (1, 1)$ model (one should recall that this point in parameter space corre-

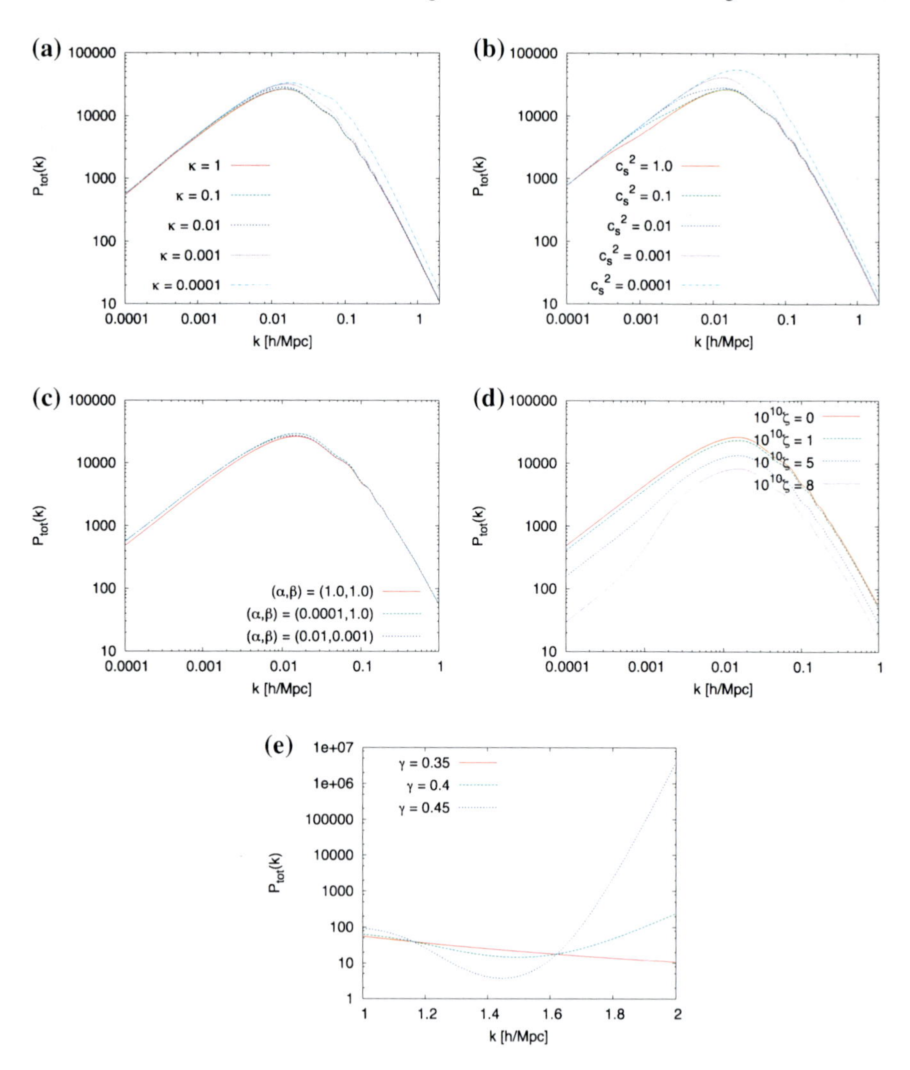

Fig. 8.1 Matter power spectra. In (**c**) we plot the spectra for values of the (α, β)-parameterizations. In (**d**) we fix $(\alpha, \beta, \gamma) = (1, 1, 0)$ and vary the value of ζ. In (**e**) We fix $(\alpha, \beta, \zeta) = (1, 1, 0)$ and vary γ (notice the difference in scale to the previous figures). One can clearly observe that the different types of theories have very different clustering properties

sponds to a quintessence model). The figures show that if the matter power spectrum can be measured to an accuracy of $\sim$15 % then models with different values of (α, β) might be observationally distinguished and/or ruled out. Models with $\beta < 0.1$ are close to being degenerate which means that it will be difficult to put a lower bound on the value of β from the matter power spectrum alone. We see that there is a dramatic effect on the matter power spectrum when κ and c_s^2 is decreased: this should allow lower bounds to be placed on the sound speed of the elastic dark energy and κ.

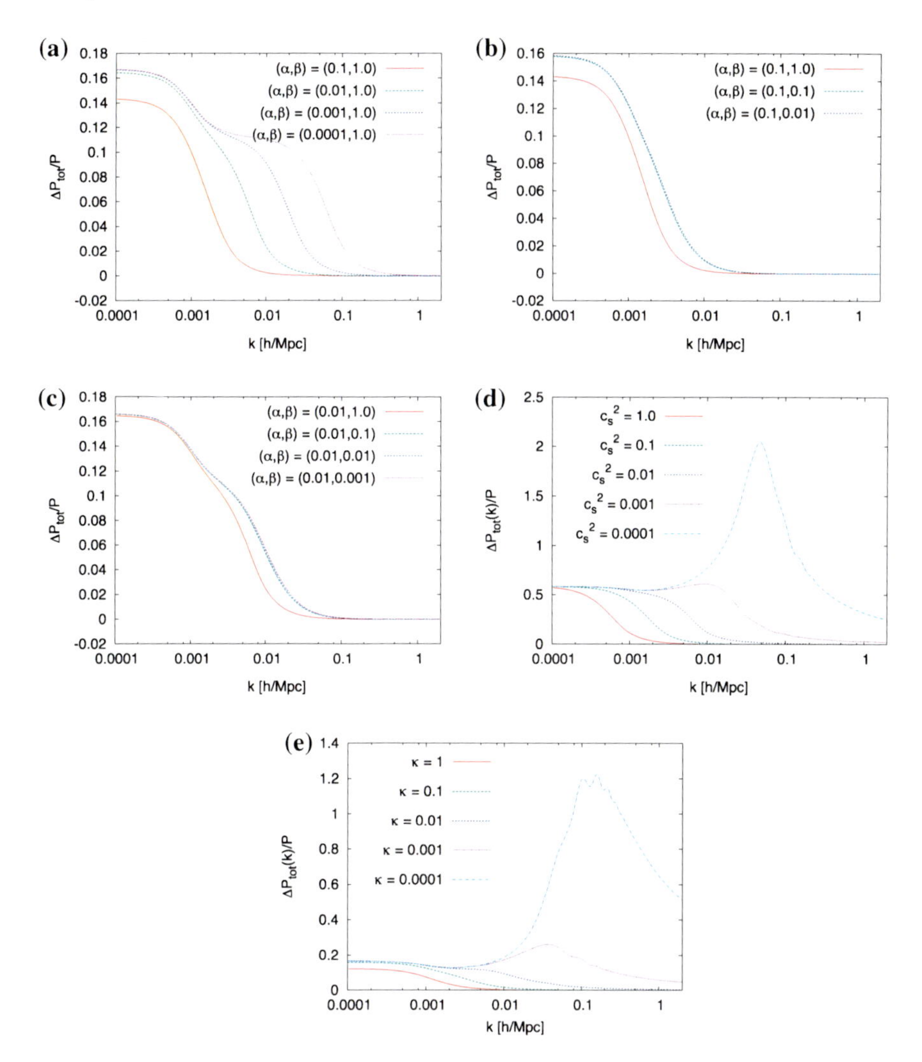

Fig. 8.2 Fractional difference in the total matter power spectra and the equivalent quintessence model, **a-c** for the (α, β)-parameterizations, **d** elastic dark energy and **e** for $(\lambda, \epsilon) = (1.1)$ and varying κ. Notice that $\beta < 0.1$ models are approximately degenerate

8.2.3 CMB Temperature Anisotropies

In Fig, 8.3 we plot the C_ℓ^{TT} spectrum of temperature anisotropies in the CMB for different values of ζ whilst fixing $(\alpha, \beta, \gamma) = (0, 0, 0)$. It is clear that power gets enhanced on large scales. In Fig. 8.4 we give the relative differences in C_ℓ^{TT} for different values of κ, c_s^2 and (α, β) relative to a fiducial $(\alpha, \beta) = (1, 1)$ model. We observe that only the large scale power is modified. Decreasing the sound speed

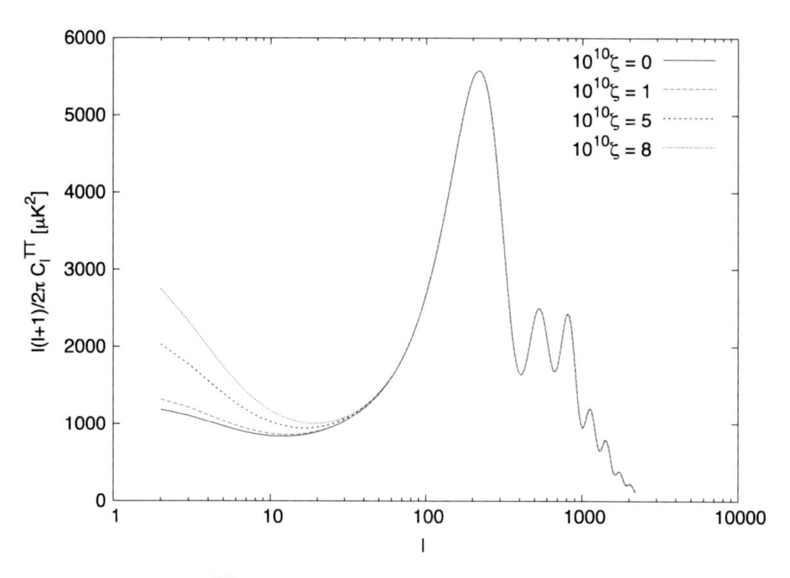

Fig. 8.3 The effect on the C_ℓ^{TT} spectrum for $(\alpha, \beta, \gamma) = (1, 1, 0)$ models. It is clear that increasing ζ enhances power on very large scales. The parameter γ does not affect this spectrum

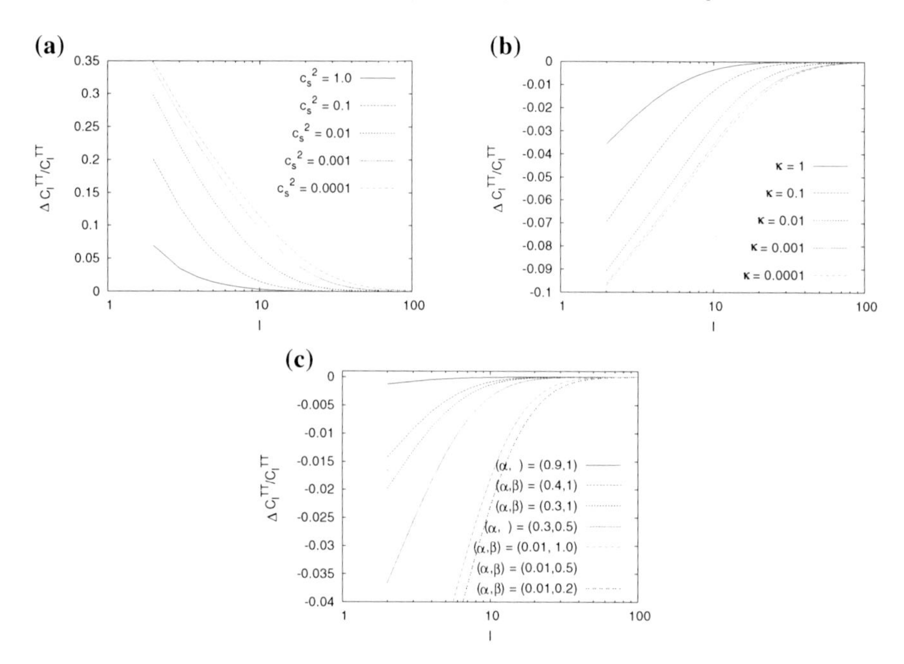

Fig. 8.4 The fractional shifts on the C_ℓ^{TT} spectrum for κ, elastic dark energy and (α, β) relative to the fiducial $(\alpha, \beta) = (1, 1)$ model. It is clear that decreasing κ suppresses power on large scales, decreasing c_s^2 enhances power on very large scales, whereas modifying (α, β) away from $(1, 1)$ suppresses power on very large scales. **a** Elastic dark energy. **b** Varying κ. **c** (α, β)

enhances power, but decreasing κ and away from $(\alpha, \beta) = (1, 1)$ both have the effect of suppressing power.

8.2.4 CMB Lensing Spectra

In Fig 8.5 we plot the lensing C_ℓ^{dd} spectra for various values of the free parameters. It is clear that varying β does not appreciably effect the spectrum, but the sound speed of elastic dark energy and ζ both modify the large scale (small ℓ) power.

In Fig. 8.6 we plot the lensing-temperature correlation C_ℓ^{dT} spectra for different values of the free parameters. We observe that α does not have an appreciable effect on the spectrum, and when it does it only affects very large scales. Varying c_s^2 has a large effect from large to mid-scales; this is also true of varying ζ.

To obtain a better understanding of the effects of varying (α, β) relative to a fiducial $(\alpha, \beta) = (1, 1)$ model we plot the relative differences in Fig. 8.7, whilst in Fig. 8.8 we plot the relative differences for various values of κ and the sound speeds of the elastic dark energy models. We observe that $\beta < 0.1$ models are approximately degenerate in the lensing C_ℓ^{dd} spectra, but the degeneracy is broken in the lensing-temperature correlation C_ℓ^{dT} spectra. Observing C_ℓ^{dT} to an accuracy of better than

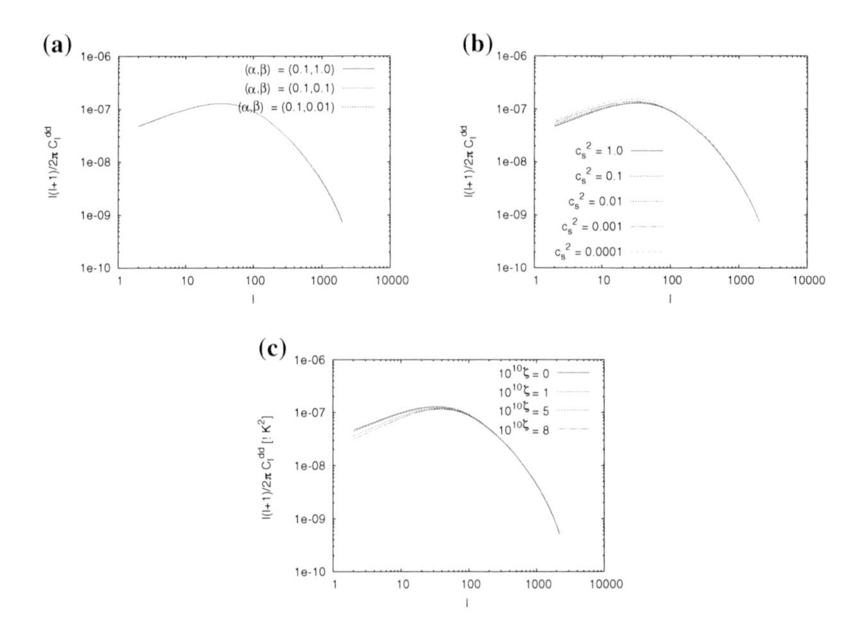

Fig. 8.5 Gravitational lensing of the CMB, C_ℓ^{dd}. In (**a**) we fix $\alpha = 0.1$ and vary the value of β, to enable an understanding of the overall shape of the spectrum to be obtained; in Fig. 8.7b we show what happens when we vary α. Varying γ does not appear to have an effect on C_ℓ^{dd} which is why we have not plotted its spectrum

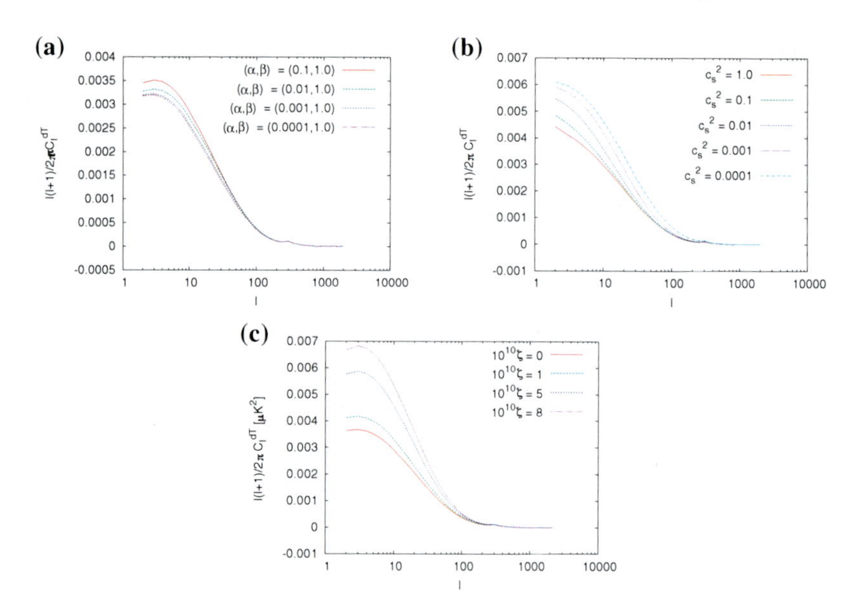

Fig. 8.6 Spectrum of correlation between temperature and gravitational lensing in the CMB, C_ℓ^{dT} for various models. We observe that it is mainly large scales which are affected. **a** Vary α.**b** C_ℓ^{dT}, elastic dark energy. **c** Vary C_ℓ^{dT}, elastic dark energy. **c** Vary

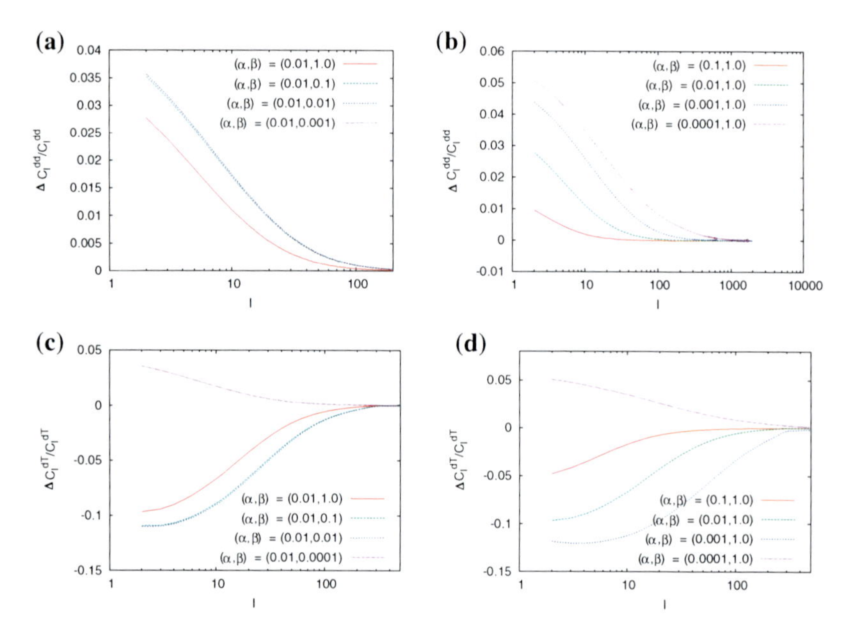

Fig. 8.7 Fractional shifts in lensing and temperature-lensing correlation spectra, from a fiducial $(\alpha, \beta) = (1, 1)$ model. Notice that $\beta < 0.1$ models are approximately degenerate in the C_ℓ^{dd} spectra, but the degneracy is softly broken by the C_ℓ^{dT} correlation spectra. **a** ΔC_ℓ^{dd}, $\alpha = 0.01$. **b** ΔC_ℓ^{dd}, $\beta = 1.0$. **c** ΔC_ℓ^{dT}, $\alpha = 0.01$. **d** ΔC_ℓ^{dT}, $\beta = 1.0$

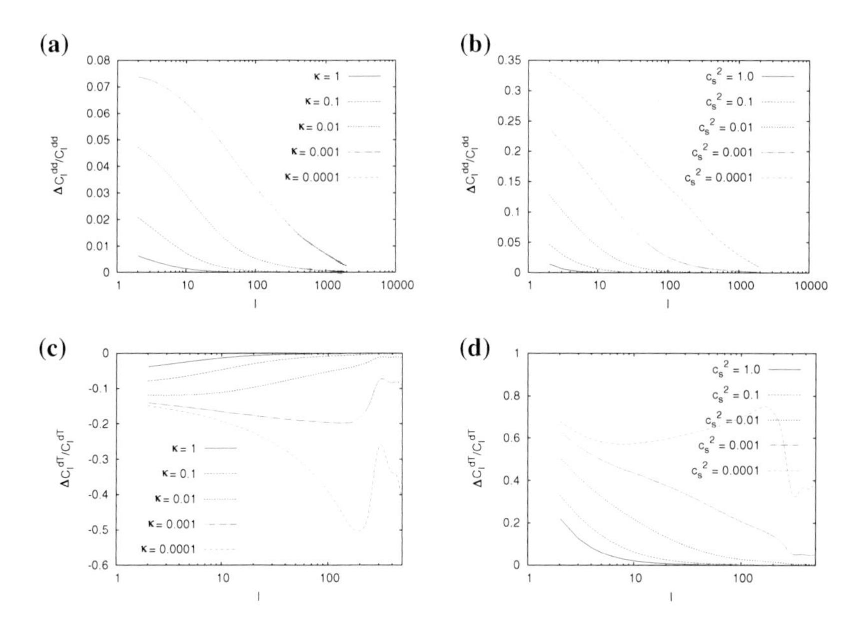

Fig. 8.8 Fractional shifts in lensing spectra for varying κ (fixing $(\lambda, \epsilon) = (1, 1)$) and elastic dark energy models. Notice that the spectra are modified on a wide range of angular scales. Also note that C_ℓ^{dT} is suppressed when varying κ but is enhanced when varying the elastic dark energy sound speed. **a** ΔC_ℓ^{dd}, vary κ. **b** ΔC_ℓ^{dd}, EDE. **c** ΔC_ℓ^{dT}, vary κ. **d** ΔC_ℓ^{dT}, EDE

15 % would be sufficient to be able to discriminate between the (α, β)-models. All lensing spectra are sensitive to κ and the elastic dark energy sound speed, where there are deviations of up to 80 % between theories.

8.3 Effective Gravitational Coupling

We now turn to the calculation of the effective gravitational coupling parameter, $\mu(k, z)$ for various values of the parameters in the generalized dark theories. To begin, in Fig. 8.9 we provide $\mu(k, z)$ for two different (α, β)-models. We observe that the effective coupling is mostly enhanced at large scales and small redshifts. Smaller scales become enhanced as α is decreased. The enhancement has a maximum of the order 5 % from its value in a GR + standard matter scenario. In Fig. 8.10 we show $\mu(k, z)$ for two values of (γ, ζ) (we fix $(\alpha, \beta = (1, 1)$ for ease). We see that ζ has the effect of suppressing the effective gravitational coupling on small scales, with a suppression of the order 30 %, whereas γ enhances on large scales, to the order 5 %.

In Figs 8.11 and 8.12 we plot the effective gravitational coupling μ for slices through k, z for various (α, β)-parameterizations. As $\alpha \to 0$ the effective gravitational coupling increases from unity. We note that for all theories, the enhancement of the effective gravitational coupling from its value in GR is only of the order 5 %. In

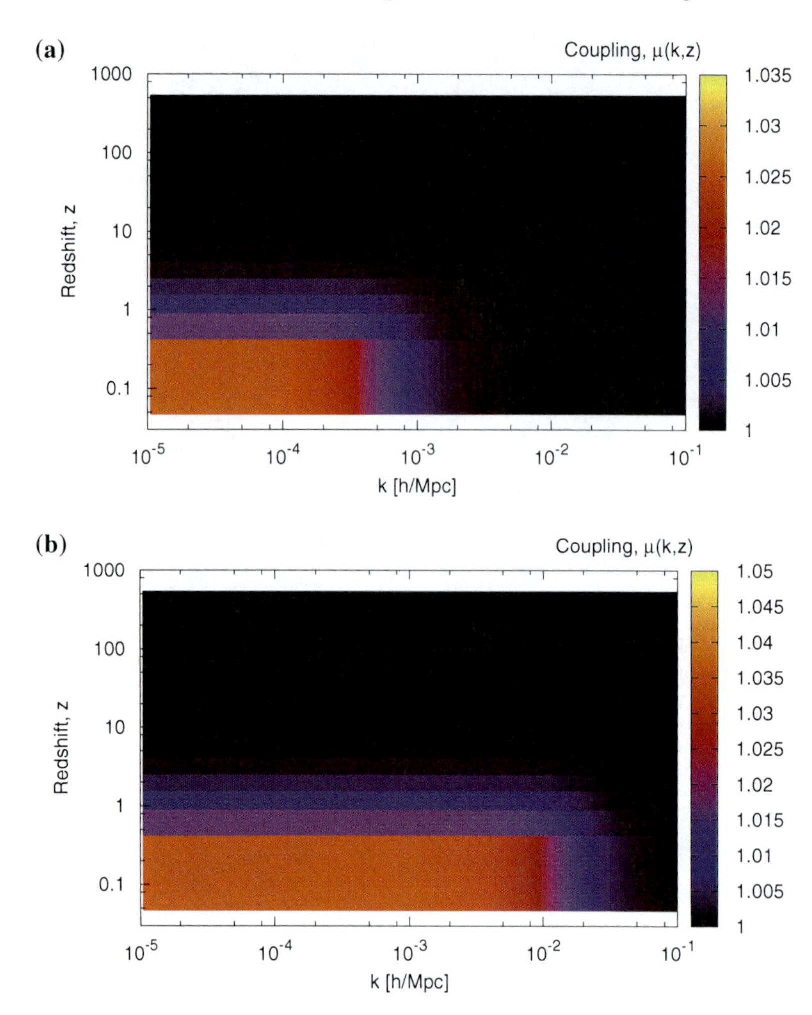

Fig. 8.9 The effective gravitational coupling parameter as a function of scale and redshift. We show two (α, β)-models. **a** $(\alpha, \beta) = (1.0, 1.0)$. **b** $(\alpha, \beta) = (0.001, 1.0)$

Fig. 8.13 we present similar plots but for various elastic dark energy models. Theories with $c_s^2 \to 0$ produce substantial evolution of the effective gravitational coupling and have a non-trivial k-dependance. These effects should be taken in stark contrast to those of the (α, β)-parameterization in Fig. 8.11.

(a)

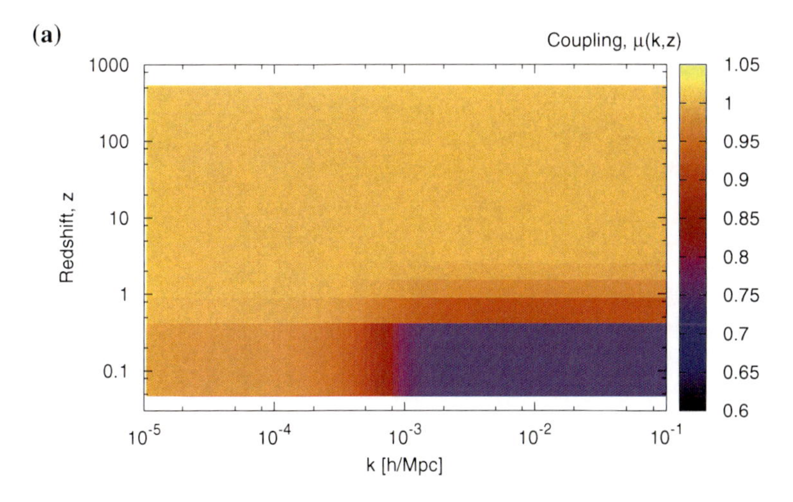

(b)

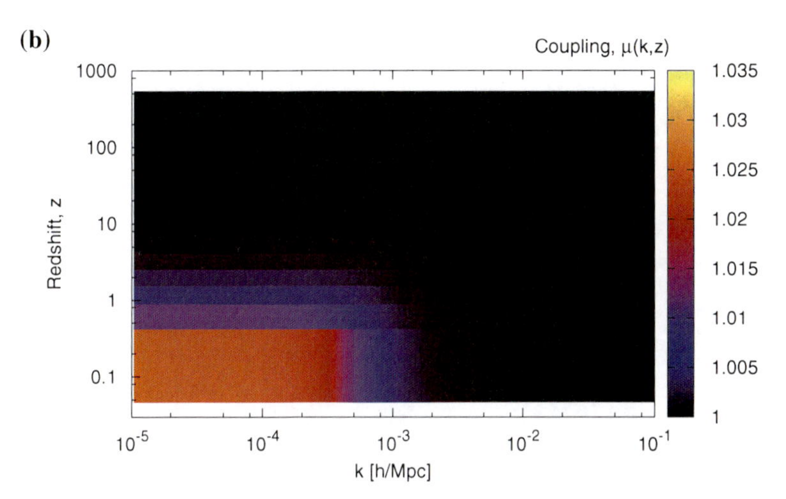

Fig. 8.10 The effective gravitational coupling parameter as a function of scale and redshift.Here we show two $(\alpha, \beta, \gamma, \zeta)$ models where we fixed $(\alpha, \beta) = (1, 1)$. **a** $(\gamma, \zeta) = (0, 8 \times 10^{10})$. **b** $(\gamma, \zeta) = (0, 0.5)$

(a)

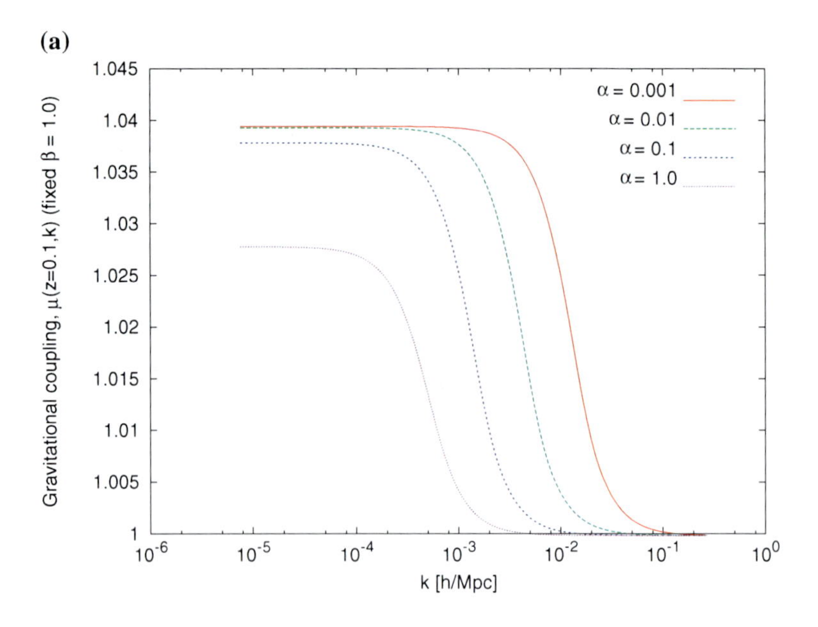

(b)

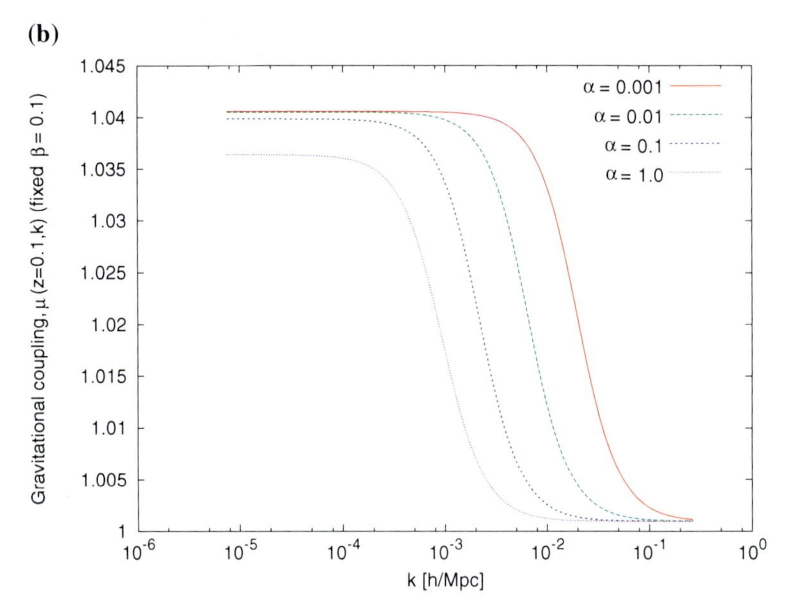

Fig. 8.11 The effective gravitational coupling parameter as a function of scale, for a range of models. These figures are at fixed redshift, $z = 0.1$. **a** $\beta = 1.0$ ($z = 0.1$). **b** $\beta = 0.1$ ($z = 0.1$)

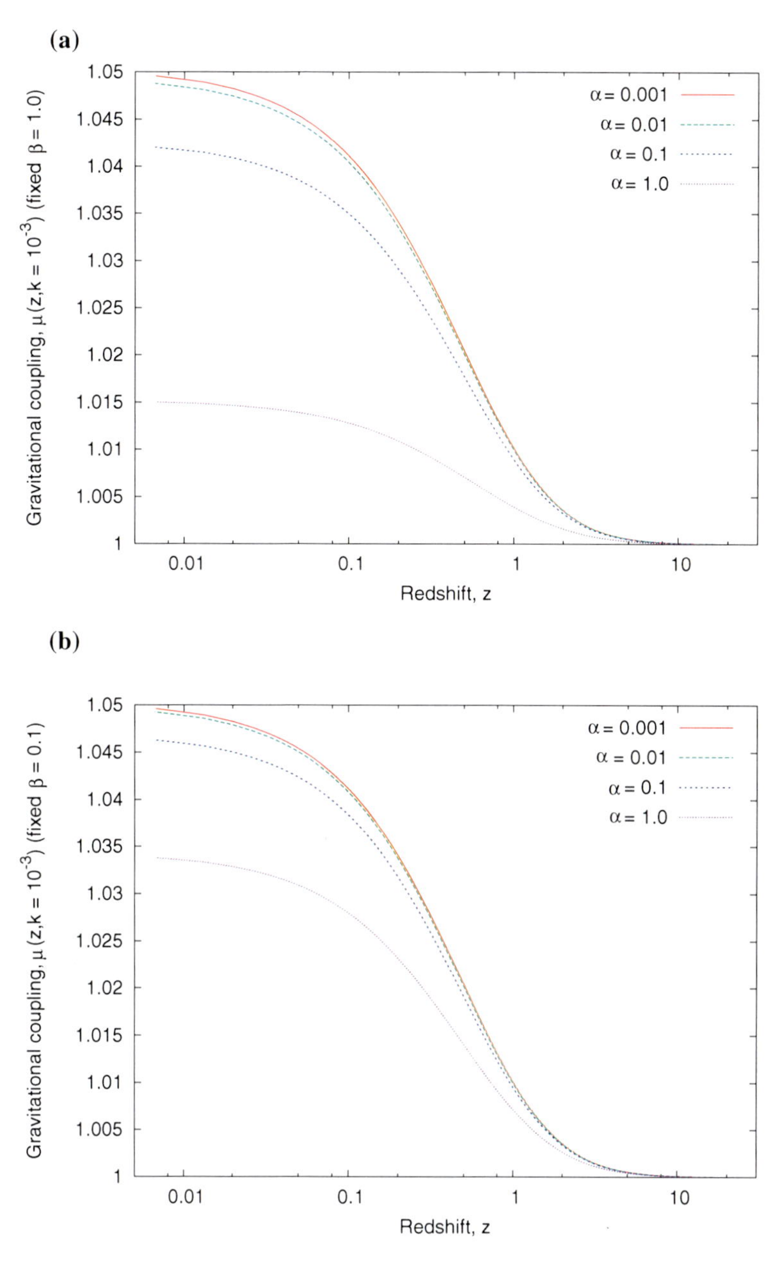

Fig. 8.12 The effective gravitational coupling parameter as a function of redshift, for a range of models. These figures are at a fixed scale, $k = 10^{-3}$ h/Mpc. **a** $\beta = 1.0\,(k = 10^{-3}$h/Mpc). **b** $\beta = 0.1$ $(k = 10^{-3}$h/Mpc)

(a)

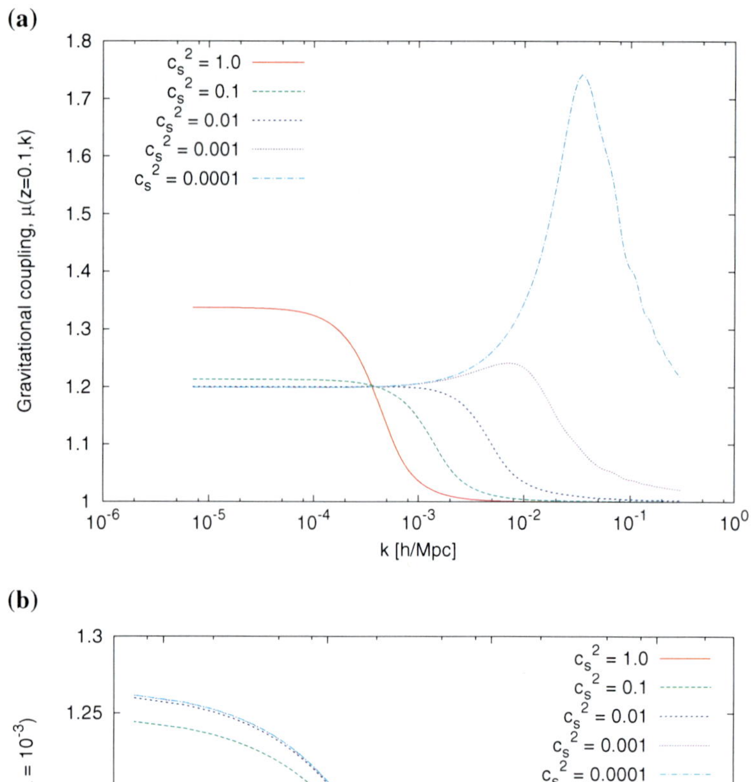

(b)

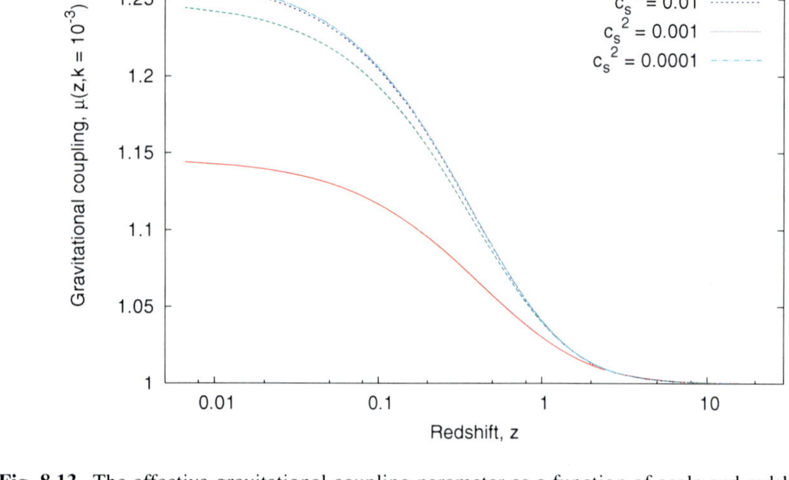

Fig. 8.13 The effective gravitational coupling parameter as a function of scale and redshift for a range of elastic dark energy models. **a** $z = 0.1$. **b** $k = 10^{-3}$h/Mpc

8.4 Summary

In this chapter we discussed how the parameters of our generalized equations of state for perturbations alter observable spectra. It should be made perfectly clear that we have performed neither extensive or exhaustive parameter space searches, but we have found some interesting regions of parameter space. This has enabled us to obtain an understanding of what quality of data is required to discriminate between models, and which spectra are best suited to do so. We also found that certain spectra are insensitive to changes in certain parameters.

In Table 8.1 we give a very schematic qualitative summary of our findings. This table makes one of the main issues apparent: most observations are affected on large scales by dark sector theories, but it is those scales where the quality of observations is limited by cosmic variance. The key is to find theories which alter the spectra on intermediate and small scales. The lensing spectra are most affected on large scales, where we *currently* do not have good data, but this will be remedied by upcoming releases of high quality lensing data from *CFHTLS* and *Planck*.

This chapter serves as a good illustration of the starting point for a much larger program of parameter space searches and subsequent confrontation of the parameters of a generalized dark theory to data. This is a work in progress [19].

Table 8.1 A very schematic and qualitative summary showing which spectra are sensitive to which parameters on which scales.

	$P(k)$			C_ℓ^{TT}		C_ℓ^{dd}		C_ℓ^{dT}	
Parameter	l	m	s	l	s	l	s	l	s
c_s^2	✗	✓	✗	✓	✗	✓	✗	✓	∼
κ	✓	✓	✗	✓	✗	∼	✗	✓	✓
α	✓	✗	✗	∼	✗	✓	✗	∼	✗
β	∼	✗	✗	✗	✗	✗	✗	∼	✗
γ	✗	✗	✓	✗	✗	✗	✗	✗	✗
ζ	✓	✓	✓	✓	✗	✓	✗	✓	∼

The $c_s^2, \ldots, \zeta$ are the free parameters in the equations of state for dark sector perturbations (8.4, 8.5), $P(k)$ is the matter power spectrum, C_ℓ^{TT} is the angular power spectrum of temperature anisotropies in the CMB, C_ℓ^{dd} is the angular power spectrum of gravitational lensing of the CMB and C_ℓ^{dT} the angular power spectrum of the correlation between the temperature anisotropies and gravitational lensing of the CMB. We use s for small, m for medium, l for large scales. The symbol "∼" is used to denote when an observable changes, but not by an appreciable amount, and so not being a particularly good probe

References

1. G. Ballesteros, J. Lesgourgues, Dark energy with non-adiabatic sound speed: initial conditions and detectability. JCAP **1010**, 014 (2010). [arXiv:1004.5509]
2. L. Amendola, M. Kunz, D. Sapone, Measuring the dark side (with weak lensing). JCAP **0804**, 013 (2008). [arXiv:0704.2421]
3. L. Pogosian, A. Silvestri, K. Koyama, G.-B. Zhao, How to optimally parametrize deviations from General Relativity in the evolution of cosmological perturbations?. Phys. Rev. **D81**, 104023 (2010). [arXiv:1002.2382]
4. A. Hojjati, L. Pogosian, G.-B. Zhao, Testing gravity with CAMB and CosmoMC. JCAP **1108**, 005 (2011). [arXiv:1106.4543]
5. E. Bertschinger, P. Zukin, Distinguishing modified gravity from dark energy. Phys. Rev. **D78**, 024015 (2008). [arXiv:0801.2431]
6. G.-B. Zhao, L. Pogosian, A. Silvestri, J. Zylberberg, Searching for modified growth patterns with tomographic surveys. Phys. Rev. **D79**, 083513 (2009). [arXiv:0809.3791]
7. E. Di Valentino, A. Melchiorri, V. Salvatelli, A. Silvestri, Parametrised modified gravity and the CMB Bispectrum. arXiv:1204.5352
8. S.A. Appleby, E.V. Linder, Galileons on, Trial. arXiv:1204.4314
9. A. Lewis, A. Challinor, A. Lasenby, Efficient Computation of CMB anisotropies in closed FRW models. Astrophys. J. **538**, 473–476 (2000). [astro-ph/9911177]
10. R.A. Croft, D.H. Weinberg, M. Bolte, S. Burles, L. Hernquist, et al. Towards a precise measurement of matter clustering: Lyman alpha forest data at redshifts 2–4. Astrophys. J. **581**, 20–52 (2002), [astro-ph/0012324]
11. SDSS Collaboration Collaboration, M. Tegmark et al. Cosmological constraints from the SDSS luminous red galaxies. Phys. Rev. **D74**, 123507 (2006). [astro-ph/0608632]
12. B.A. Reid, W.J. Percival, D.J. Eisenstein, L. Verde, D.N. Spergel, et al. Cosmological constraints from the clustering of the sloan digital sky survey DR7 luminous red galaxies. Mon. Not. Roy. Astron. Soc. **404**, 60–85 (2010). [arXiv:0907.1659]
13. S. Nuza, A. Sanchez, F. Prada, A. Klypin, D. Schlegel, et al. The clustering of galaxies at z 0.5 in the SDSS-III data release 9 BOSS-CMASS sample: a test for the LCDM cosmology. arXiv:1202.6057
14. A. Lewis, A. Challinor, Weak gravitational lensing of the CMB. Phys. Rept. **429**, 1–65 (2006). [astro-ph/0601594]
15. D. Hanson, A. Challinor, A. Lewis, Weak lensing of the CMB, Gen. Rel. Grav. **42**, 2197–2218 (2010). [arXiv:0911.0612]
16. W. Hu, T. Okamoto, Mass reconstruction with cmb polarization. Astrophys. J. **574**, 566–574 (2002). [astro-ph/0111606]
17. A. van Engelen, R. Keisler, O. Zahn, K. Aird, B. Benson, et al. A measurement of gravitational lensing of the microwave background using south pole telescope data. Astrophys. J. **756**, 142 (2012). [arXiv:1202.0546]
18. S. Das, B.D. Sherwin, P. Aguirre, J.W. Appel, J.R. Bond, et al. Detection of the power spectrum of cosmic microwave background lensing by the atacama cosmology telescope. Phys. Rev. Lett. **107**, 021301 (2011). [arXiv:1103.2124]
19. R.A. Battye, J.A. Pearson, A. Moss, (Constraints and observational signatures of generalized cosmological perturbations. in prep, 2012)

Chapter 9
Discussion and Final Remarks

9.1 Summary

In this thesis we outlined an approach for computing consistent perturbations to the generalized gravitational field equations for physically meaningful theories. The approach requires a field content, but does not require a particular Lagrangian to be presented for calculations to be performed. Once the field content has been specified we have shown how to write down an effective action for linearized perturbations, and from that how to compute the generalized perturbed gravitational field equations.

In Chap. 2 we reviewed the current methods for parameterizing perturbations in the dark sector, and their shortcomings. We also provided an overview of our formalism, describing the technology we used throughout the thesis. We provided an argument for why we use the Eulerian coordinate system to construct the perturbations which correspond to physically relevant quantities. The reason is that quantities in cosmology are perturbed about some known background, which is the same as the statement of using the Eulerian perturbation scheme.

We have given a number of examples in varying detail, each illustrating how to use our formalism. In Chap. 3 we presented the two simplest theories: first where the field content of the dark sector is entirely composed of the metric, and secondly where the dark sector contains a scalar field, its derivative and the metric. *A priori* we did not constrain these fields to combine into scalar quantities (such as a kinetic term). We gave formulae for the Lagrangian for perturbations in these examples, which enabled us to write down the perturbed dark energy momentum tensor, $\delta_{\rm L} U^{\mu}{}_{\nu}$. These expressions involved a number of coupling coefficients which could be split with an isotropic $(3+1)$ decomposition; the number of coefficients in this decomposition determines an absolute upper bound on the number of functions which must be provided for perturbations on an isotropic background. The possible functions in the decomposition is further constrained by applying the perturbed conservation equation, $\delta(\nabla_{\mu} U^{\mu\nu}) = 0$ and by imposing, for example, theoretical restrictions or decoupling between the relevant field and the vector field ξ^{μ} so that the theory is reparameterization invariant.

J. Pearson, *Generalized Perturbations in Modified Gravity and Dark Energy*,
Springer Theses, DOI: 10.1007/978-3-319-01210-0_9,

In Chap. 4 we further developed the formalism for rather complicated but theoretically very interesting cases, where the theory contains different types of fields (scalars, vectors and tensors), different ways in which high numbers of derivatives enter the theory (partial or covariant derivatives, or via Christoffel symbols or curvature tensors) and for reparameterization invariant theories. These general theories were incredibly complicated with a huge amount of freedom, but it was shown that the freedom dramatically reduces to a manageable number of functions when theoretical restrictions were imposed. Due to our generality, the discussion was able to highlight very interesting structures within theory space by providing expressions for the field equations for entirely general classes of theories and obtaining covariant expressions for reparameterization invariant theories.

For the examples we studied it was clear that the number of parameters that were required to be specified depended upon what theoretical restrictions we imposed. For example: (a) for the theory $\mathcal{L} = \mathcal{L}(g_{\mu\nu})$ at the level of the effective action there were 5 functions. Once we imposed the decoupling conditions the 5 functions reduced to just 2. (b) For the theory $\mathcal{L} = \mathcal{L}(g_{\mu\nu}, \phi, \nabla_\mu\phi)$ at the level of the effective action there were 14 functions. When we applied the linking conditions, the 14 functions reduced to 11. When we imposed the decoupling conditions the 14 functions reduced to just 3. It is worth noting that all scalar field theories of the form $\mathcal{L} = \mathcal{L}(\phi, \mathcal{X})$ satisfy the decoupling condition.

In Chap. 5 we provided explicit examples: kinetic scalar field theories with $\mathcal{L} = \mathcal{L}(\phi, \mathcal{X})$, example second order scalar field theories and $F(R)$ gravities. Using these explicit calculations we have been able to justify our expansions of $\delta U^{\mu\nu}$ and provide an insight into how the dependancies of the background Lagrangian filter into the Lagrangian for perturbations and observable quantities.

In Chap. 6 we pointed out how our formalism is useful for studying more "theoretical" problems, such as those which arise in massive gravity. In the chapter we used the $(3+1)$ decomposition to show that the ghost is the time-like part $\chi \equiv u^\mu \xi_\mu$ and that was removed for the entirety of our studies.

Chapter 7 is where we drew our phenomenologically useful findings together, enabling us to obtain expressions for computing distinguishing signatures which can be found in cosmologically observable quantities for very general classes of theories. We did this by performing a study of the structure of the generalized perturbed fluid equations and obtaining equations of state for dark sector perturbations for theories with a given field content and various theoretical restrictions. These equations of state close the perturbed fluid equations and are provided in the form of the entropy, $w\Gamma$, and anisotropic stress source, Π. Our main results are the parameterizations of the equations of state: (7.41), (7.44) and (7.49). We used these equations of state to compute cosmological spectra in Chap. 8.

The formalism we have developed can be used as a stepping-stone to create a tool which can be used to discriminate different gravity and dark energy theories using experiment. All perturbations in our formalism have a obvious, consistent and concrete origin from an effective action.

9.2 The Big Picture: Theory Versus Observation

The last decade in cosmology has seen an explosion in the detection and high precision measurement of a large number of observables which are sensitive to the details of the dark sector and its perturbations. The spectrum of anisotropies in the cosmic microwave background (CMB) has been observed to fantastically high precision using ground [1, 2], balloon [3] and space [4] based telescopes. Gravitational lensing of the CMB due to structure along the line of sight to the surface of last scattering has very recently been detected by the South Pole Telescope [5] and the Atacama Cosmology Telescope [6]. Weak gravitational lensing (see e.g., [7–9]) and its statistical properties are sensitive to the details of the dark sector because perturbations in the dark sector will alter the growth of structure on large scales. Weak lensing was first observed by [10, 11] and further refined using the Canada-France-Hawaii Telescope Legacy Survey (*CFHTLS*) [12, 13]. There has also been progress in measurements of the power spectrum of matter in the universe [14–18] and of the baryon acoustic peak in the matter power spectrum [19]. Upcoming data releases from *CFHTLS* and *Planck* at the beginning of 2013 should increase the accuracy of all of these measurements, significantly reducing error bars; *Euclid* [20, 21] (which is due to fly in 2019) will provide much higher quality measurements of all of these spectra. These experiments have all come to the conclusion that the dark sector exists, where some exotic substance must be added into the gravitating content of the universe.

To properly use the data to tell us about which of the plethora of theories which have been dreamt up are actually realized by nature, some model must be used for the gravitational dynamics of the dark sector. Both the behavior of the equation of state w and perturbations in the dark sector must be modeled. These models are necessary ingredients and particular choices will limit the regions of theory space which are probed with a given analysis. For instance, if the data analysis is performed whilst imposing $\dot{w} = 0$, or a particular choice for the parameterization of the perturbations (we reviewed popular choices in Sect. 2.2), then only the theories which fall into those categories will be probed. The important point is that the choice of parameterization will strongly dictate the regions of theory space which are tested against observations (indeed, "arbitrary" parameterizations may not correspond to any theory at all, and may contain pathological inconsistencies by construction). As for the background evolution of the dark sector, recent results from a principle component analysis study [22] provide a model independent and parameterization independent detection and constraint on the current value and evolution of w.

What remains, therefore, is a way to parameterize the perturbations in the dark sector in the most (feasibly) model independent way, whilst retaining a good link to theory space. In this thesis we have formulated and developed exactly this technology. We have isolated the precise way in which wide classes of theories alter cosmological observables via the form of the equation of state for dark sector perturbations. This culminated in Sect. 7.42 where we were able to confidently write down three different equations of state for dark sector perturbations for theories with a very small amount of theoretical priors imposed. These equations of state are

- No extra fields:

$$w\Gamma = (\kappa - w)\delta, \qquad w\Pi^{S} = (w - \lambda)\left[\delta - 3(1+w)\varepsilon\eta\right]. \tag{9.1a}$$

- Reparameterization invariant scalar fields with one derivative:

$$w\Gamma = (\alpha - w)\left[\delta - 3\mathcal{H}\beta(1+w)\theta\right], \qquad \Pi = 0. \tag{9.1b}$$

- Reparameterization invariant scalar fields with two derivatives and second order field equations:

$$w\Gamma = (\alpha - w)\left[\delta - 3\mathcal{H}(1+w)(\beta + k^2\gamma)\theta\right] - \frac{3}{2\rho\mathcal{H}}\zeta\dot{h}, \qquad \Pi = 0. \tag{9.1c}$$

Importantly, each of these parameter sets $(\kappa, \lambda, \epsilon)$, (α, β) and $(\alpha, \beta, \gamma, \zeta)$ probe dark sectors with fundamentally different field contents. Confronting upcoming cosmological data sets with the parameters and determining the allowed values that these parameters can take would represent a significant and important step to determining the nature of the dark sector. We believe that the equations of state (9.1) represent the best phenomenological connection between theory and observation to date.

References

1. H. Chiang, P. Ade, D. Barkats, J. Battle, E. Bierman, et al. Measurement of CMB polarization power spectra from two Years of BICEP data. Astrophys. J. **711**, 1123–1140 (2010). [arXiv:0906.1181]
2. QUaD collaboration Collaboration, P. Castro et al. Cosmological parameters from the QUaD CMB polarization experiment. Astrophys. J. **701**, 857–864 (2009). [arXiv:0901.0810]
3. MaxiBoom Collaboration Collaboration, J.R. Bond, et al. CMB analysis of boomerang and maxima and the cosmic parameters Omega(tot), Omega(b) h**2, Omega(cdm) h**2, Omega(Lambda), n(s). astro-ph/0011378
4. N. Jarosik, C. Bennett, J. Dunkley, B. Gold, M. Greason, et al. Seven-Year Wilkinson microwave anisotropy probe (WMAP) observations: Sky Maps, systematic errors, and basic results. Astrophys. J.Suppl. **192**, 14 (2011). [arXiv:1001.4744]
5. A. van Engelen, R. Keisler, O. Zahn, K. Aird, B. Benson, et al. A measurement of gravitational lensing of the microwave background using south pole telescope data. Astrophys. J. **756**, 142 (2012). [arXiv:1202.0546]
6. S. Das, B.D. Sherwin, P. Aguirre, J.W. Appel, J.R. Bond, et al. Detection of the power spectrum of cosmic microwave background lensing by the atacama cosmology telescope. Phys. Rev. Lett. **107**, 021301 (2011). [arXiv:1103.2124]
7. D. Hanson, A. Challinor, A. Lewis, Weak lensing of the CMB. Gen. Rel. Grav. **42**, 2197–2218 (2010). [arXiv:0911.0612]
8. D. Huterer, Weak lensing, dark matter and dark energy. Gen. Rel. Grav. **42**, 2177–2195 (2010). [arXiv:1001.1758]

9. J.-P. Uzan, Tests of general relativity on astrophysical scales. Gen. Rel. Grav. **42**, 2219–2246 (2010). [arXiv:0908.2243]
10. L. van Waerbeke, Y. Mellier, T. Erben, J. Cuillandre, F. Bernardeau, et al. Detection of correlated galaxy ellipticities on CFHT data: first evidence for gravitational lensing by large scale structures. Astron. Astrophys. **358**, 30–44 (2000). [astro-ph/0002500]
11. D.J. Bacon, A.R. Refregier, R.S. Ellis, Detection of weak gravitational lensing by large-scale structure. Mon. Not. Roy. Astron. Soc. **318**, 625 (2000). [astro-ph/0003008]
12. H. Hoekstra, Y. Mellier, L. Van Waerbeke, E. Semboloni, L. Fu, et al. First cosmic shear results from the canada-france-hawaii telescope wide synoptic legacy survey. Astrophys. J. **647**, 116–127 (2006). [astro-ph/0511089]
13. J. Benjamin, C. Heymans, E. Semboloni, L. Van Waerbeke, H. Hoekstra, et al. Cosmological constraints from the 100 square degree weak lensing survey. Mon. Not. Roy. Astron. Soc. **381**, 702–712 (2007). [astro-ph/0703570]
14. R.A. Croft, D.H. Weinberg, M. Bolte, S. Burles, L. Hernquist, et al. Towards a precise measurement of matter clustering: Lyman alpha forest data at redshifts 2–4. Astrophys. J. **581**, 20–52 (2002). [astro-ph/0012324]
15. SDSS Collaboration Collaboration, M. Tegmark et al. Cosmological constraints from the SDSS luminous red galaxies. Phys. Rev. **D74**, 123507 (2006). [astro-ph/0608632]
16. B.A. Reid, W.J. Percival, D.J. Eisenstein, L. Verde, D.N. Spergel, et al. Cosmological constraints from the clustering of the sloan digital sky survey DR7 luminous red galaxies. Mon. Not. Roy. Astron. Soc. **404**, 60–85 (2010). [arXiv:0907.1659]
17. R. Hlozek, J. Dunkley, G. Addison, J.W. Appel, J.R. Bond, et al. The atacama cosmology telescope: a measurement of the primordial power spectrum. Astrophys. J. **749**, 90 (2012). [arXiv:1105.4887]
18. S. Nuza, A. Sanchez, F. Prada, A. Klypin, D. Schlegel, et al. The clustering of galaxies at z 0.5 in the SDSS-III data release 9 BOSS-CMASS sample: a test for the LCDM cosmology. arXiv:1202.6057
19. SDSS Collaboration Collaboration, D.J. Eisenstein et al. Detection of the baryon acoustic peak in the large-scale correlation function of SDSS luminous red galaxies. Astrophys. J. **633**, 560–574 (2005). [astro-ph/0501171]
20. R. Laureijs, J. Amiaux, S. Arduini, J.-L. Augueres, J. Brinchmann, et al. Euclid definition study report. arXiv:1110.3193
21. L. Amendola, Cosmology and fundamental physics with the Euclid satellite. arXiv:1206.1225
22. G.-B. Zhao, R.G. Crittenden, L. Pogosian, X. Zhang, Examining the evidence for dynamical dark energy. Phys. Rev. Lett. **109**, 171301 (2012). [arXiv:1207.3804]

About the Author

The author obtained a first class Masters degree in Physics with Theoretical Physics at the University of Manchester in 2009. He remained at the University of Manchester, beginning his PhD under the supervision of Prof. Richard Battye in September 2009. His research initially focused on the dynamics of domain wall networks and vortons, the result of which culminated in the publishing of three papers [1–3], one conference proceeding [4] and a number of papers which are currently being prepared for publication [5, 6]. The work presented in this thesis is the culmination of a project which was undertaken by the author during his PhD.

Publications

The work in this thesis is partly based upon the following publications:

- *Massive gravity, the elasticity of space-time and perturbations in the dark sector* [7], with Richard Battye.
- *Effective action approach to cosmological perturbations in dark energy and modified gravity* [8], published in JCAP with Richard Battye.
- *Effective field theory for perturbations in dark energy and modified gravity* [9], conference proceedings.
- *Equations of state for dark sector perturbations for general high derivative scalar-tensor theories*, in preparation with Richard Battye.
- *Parameterizing dark sector perturbations via equations of state*, in preparation with Richard Battye.
- *Observational signatures and constraints on generalized cosmological perturbations*, in preparation with Richard Battye and Adam Moss.

J. Pearson, *Generalized Perturbations in Modified Gravity and Dark Energy*,
Springer Theses, DOI: 10.1007/978-3-319-01210-0,

References

1. R.A. Battye, J.A. Pearson, S. Pike, P.M. Sutcliffe, Formation and evolution of kinky vortons. JCAP 0909, 039 (2009). [arXiv:0908.1865].
2. R.A. Battye, J.A. Pearson, Charge, junctions and the scaling dynamics of domain wall networks. Phys. Rev. D 82, 125001 (2010). [arXiv:1010.2328].
3. R.A. Battye, J.A. Pearson, A. Moss, X-type and Y-type junction stability in domain wall networks. Phys. Rev. D 84, 125032 (2011). [arXiv:1107.1325].
4. J.A. Pearson, Charge, domain walls and dark energy. [arXiv:1010.3195].
5. R.A. Battye, J.A. Pearson, P.M. Sutcliffe, Stationary gauged vortons (2012). (in prep).
6. R.A. Battye, J.A. Pearson, Fate of random domain wall, networks (2012). (in prep).
7. R.A. Battye, J.A. Pearson, Massive gravity, the elasticity of space-time and perturbations in the dark sector. [arXiv:1301.5042]. <!- Missing/Wrong Year ->. <!- Missing/Wrong Year ->. <!- Missing/Wrong Year ->. <!- Missing/Wrong Year ->. <!– Missing/Wrong Year –>
8. R.A. Battye, J.A. Pearson, Effective action approach to cosmological perturbations in dark energy and modified gravity. JCAP 1207, 019 (2012). [arXiv:1203.0398].
9. J.A. Pearson, Effective field theory for perturbations in dark energy and modified gravity. [arXiv:1205.3611]. <!- Missing/Wrong Year ->. <!- Missing/Wrong Year ->. <!- Missing/Wrong Year ->. <!- Missing/Wrong Year ->. <!– Missing/Wrong Year –>